Biophysical Chemistry of Dioxygen Reactions in Respiration and Photosynthesis

Chemica Scripta: Volume 28A, 1988

Biophysical Chemistry of Dioxygen Reactions in Respiration and Photosynthesis

Proceedings of the Nobel Conference held at Fiskebäckskil, Sweden, 1–4 July 1987

Editor: Tore Vänngård

Published on behalf of The Royal Swedish Academy of Sciences
by Cambridge University Press

The right of the
University of Cambridge
to print and sell
all manner of books
was granted by
Henry VIII in 1534.
The University has printed
and published continuously
since 1584.

Cambridge University Press

Cambridge   New York   New Rochelle   Melbourne   Sydney

CAMBRIDGE UNIVERSITY PRESS
Cambridge, New York, Melbourne, Madrid, Cape Town,
Singapore, São Paulo, Delhi, Tokyo, Mexico City

Cambridge University Press
The Edinburgh Building, Cambridge CB2 8RU, UK

Published in the United States of America by Cambridge University Press, New York

www.cambridge.org
Information on this title: www.cambridge.org/9780521107051

First published 1988
First paperback edition 2011

A catalogue record for this publication is available from the British Library

ISBN 978-0-521-36604-5 Hardback
ISBN 978-0-521-10705-1 Paperback

Contents

Preface

During the last few years two electron-transfer processes of fundamental biological importance have received intense scientific attention, namely, photosynthetic oxygen evolution as it occurs in Photosystem II in plants and algae and, what might be considered as the reverse process, the reduction of dioxygen to water as catalyzed by cytochrome c oxidase in mitochondrial and some microbial respiration. With the current state of knowledge about the molecular and mechanistic details of these reactions it was of considerable interest to try to bring together some of the leading researchers in both fields for an exchange of ideas and to discuss to what extent the two processes might be governed by common principles.

The Nobel Conference on 'The Biophysical Chemistry of Dioxygen Reactions in Respiration and Photosynthesis' was held at the Hotel Gullmarsstrand, Fiskebäckskil, a resort on the Swedish west coast, 1–4 July, 1987. In addition to the twenty-one invited speakers, some twenty observers attended the Conference. The Conference was made possible by a major grant from the Nobel Institute of Chemistry and additional financial support from The International Union of Pure and Applied Biophysics, The Royal Society of Arts and Science in Göteborg and from AB Hässle, Göteborg.

Göteborg, September 1987

Lars-Erik Andréasson　　*Tore Vänngård*

List of participants

Roland Aasa, Göteborg, Sweden
Bertil Andersson, Stockholm, Sweden
Lars-Erik Andréasson, Göteborg, Sweden
Gerald T. Babcock, East Lansing, USA
Helmut Beinert, Milwaukee, USA
Gary W. Brudvig, New Haven, USA
Maurizio Brunori, Rome, Italy
Sunney I. Chan, Pasadena, USA
Charles Dismukes, Princeton, USA
Anders Ehrenberg, Stockholm, Sweden
James Fee, Los Alamos, USA
Sture Forsén, Lund, Sweden
Harry B. Gray, Pasadena, USA
Colin Greenwood, Norwich, England
Örjan Hansson, Göteborg, Sweden
Wolfgang J. Junge, Osnabruck, West Germany
Bruno Källebring, Göteborg, Sweden
Göran Lagenfelt, Göteborg, Sweden
Sven Larsson, Göteborg, Sweden
Märtha Larsson-Raźnikiewicz, Uppsala, Sweden
Ylva Lindqvist, Uppsala, Sweden

Sven Lindskog, Umeå, Sweden
Brigitta Maison-Peteri, Göteborg, Sweden
Bo G. Malmström, Göteborg, Sweden
Thomas Nilsson, Göteborg, Sweden
Yutaka Orii, Kyoto, Japan
Graham Palmer, Houston, USA
Karl-Gustav Paul, Umeå, Sweden
Bengt Reinhammar, Göteborg, Sweden
Gernot Renger, Berlin, West Germany
A. William Rutherford, Gif-sur-Yvette, Cedex, France
Kenneth Sauer, Berkeley, USA
Stenbjörn Styring, Göteborg, Sweden
Per-Eric Thörnström, Göteborg, Sweden
Tore Vänngård, Göteborg, Sweden
Hans J. van Gorkom, Leiden, The Netherlands
Ron Wever, Amsterdam, The Netherlands
Mårten Wikström, Helsinki, Finland
R. J. P. Williams, Oxford, England
Michael T. Wilson, Colchester, England
Tom Wydrzynski, Göteborg, Sweden
Hans-Erik Åkerlund, Lund, Sweden

Dedication

Metal ions have been good to Bo Malmström. And he has been good to them: even in his earliest work, as a graduate student in histochemistry with David Glick, he managed to brighten the spectroscopically dull zinc ion by attaching it to a very colorful ligand.

After finishing his Ph.D. at Minnesota, Bo began a research program in Uppsala aimed at understanding the activation of the enzyme enolase. This work, which led to a real, honest-to-God doctorate (Uppsala, 1956), exploited metal ions whose spectroscopic signatures are much richer than zinc's (manganese, for example). The collection of original papers that became his dissertation established that it is the enolase enzyme itself, not the substrate, that is activated by magnesium.

Bo's pioneering research on metal-ion activation of enzymes led quite naturally to an interest in the role of zinc in carbonic anhydrase, and in the early 1960s he and Sven Lindskog collaborated on studies of the problem. The tremendous contribution that Swedish science has made to the spectacular development of our understanding of the structure and mechanism of carbonic anhydrase is due in large measure to the foundation laid by the Malmström–Lindskog work.

Although manganese and zinc had come through for Bo, he adopted copper as his main metal for the 1960s. Indeed, the modern era of investigation of copper proteins began in 1960 when Bo and Tore Vänngård demonstrated that the blue copper in fungal laccase is a highly covalent cupric species. Bo stepped up the pace of his work on copper proteins after he moved to Göteborg in 1963, and in the late 1960s (aided by a visit to Eraldo Antonini's laboratory in Rome) he identified the other two electron-acceptor sites in laccase. One of these sites, designated type 2 copper, is a mononuclear (tetragonal) species; the other, not surprisingly called type 3 copper, turned out to be a binuclear active center. Everyone who works seriously on copper proteins uses the Malmström notation: most learned it first by reading the 1970 Malkin–Malmström classic in *Advances in Enzymology*.

It was during one of his visits to Rome that Bo started his love affair with cytochrome oxidase. And what a love affair it has been! (Not to worry, Betty, the enzyme only makes water, and it can't play the harpsicord.) The importance of the binuclear (iron–copper) site in dioxygen reduction, the pathways of electron flow through the enzyme, and the nature of the catalytic intermediates, are matters that have been elucidated by Bo and his co-workers in a series of original and highly penetrating investigations.

Armed with a good model for the four-electron reduction of dioxygen by the enzyme, Bo has turned his attention in recent years to the mechanism of proton pumping. Characteristically, he has thrown himself in the midst of the fray by introducing a new idea, in this case a conformationally gated long-range electron-transfer step as the driver of the pump. Like all his previous work, this hypothesis is stimulating many new experiments in the field.

Bo's enthusiasm for his work is truly contagious. Sunney Chan and I will never forget his visit to Pasadena during 1980–81. Bo took charge on his first day in the Noyes Laboratory, and soon our students and postdocs were all marching to the beat of the Malmström drum. During this period, we learned that no one can match Bo's intensity in give-and-take discussions; but we kept trying, and we had a lot of fun doing science together. I'm sure that others who have worked with Bo over the years – Bob Bray, Jim Fee, Maurizio Brunori, Colin Greenwood, Emil Smith, Mike Wilson, J. F. G. Vliegenthart, to name just a few of his collaborators outside of Sweden – have had similar experiences.

It would not be unreasonable to conclude that Bo's devotion to his work and those who work with him is total. But, don't go quite that far. It is no secret that Betty occupies the top spot in his book, and honorable mention goes to his music (including an incredible recorder collection) and to any tournament tennis match in which at least one Swede is playing.

This volume contains reports of work on the roles of metal ions in biological oxidation-reduction processes involving dioxygen and water as substrates. It features three metals – manganese, iron, and copper – that have been Bo's partners for thirty-five years. The science in the volume is dedicated to Bo in the hope that his sixty-first year and the ones to follow will be just as rich and productive as those we have experienced with him.

HARRY GRAY

Pasadena, California, 31 August 1987

Chemica Scripta 1988, **28A**, 5–13

The Where, the When, the How and the Why of Biological Oxygen Reactions

R. J. P. Williams

Inorganic Chemistry Laboratory, Oxford University, South Parks Road, Oxford OX1 3QR, England

Paper presented at the Nobel Conference 'Biophysical Chemistry of Dioxygen Reactions in Respiration and Photosynthesis', Fiskebäckskil, Sweden, 1–4 July, 1987

Abstract

Oxygen is in the atmosphere and relative to the accumulation of reducing equivalents in biology it is clearly a huge energy store and available metabolite. It enters biology directly or through carriers. Taking carrier systems first copper systems operate outside and iron inside cells. Activation of O_2 outside cells or in internal vesicles is by copper enzymes largely. Inside cells iron is the chief catalyst. However the major energy generating systems are in organelle not cytoplasmic membranes. The distribution in space of oxygen reactions is intriguing. However some systems do not operate except under stimulus – the protective flavin oxygen reactions of leucocytes. Both time and place are important. The activation does not yield a single product since O_2^-, H_2O_2 and H_2O are produced. Where, when and how is control exerted? Producing O_2^- and H_2O_2 needs clearance devices. These are Cu in the cytoplasm and Fe and Mn (see above) in the organelles. Why? Oxygen is also used as an hydroxylating agent and once again different enzymes are differently placed, Cu outside the cytoplasm, Fe inside. Note also the distribution of Mo and flavin,. Oxygen reactions are further coupled to other devices and we must look at this machinery, especially in organelles. The full appreciation of the reactions of oxygen in biology needs inspection of the four critical question given in the title.

Introduction

This article concerns the general biochemistry of dioxygen. I have just written a paper 'From the Fire Air of Scheele to the Dioxygen of Today' [1] to celebrate the 200th anniversary of the pioneer Swedish chemist, Scheele, and I should like the present paper to be seen as a celebration of the 60th birthday of a distinguished Swedish biochemist and immediate colleague, Professor Bo Malmström, who has worked for much of his life on the biochemistry of dioxygen. I have enjoyed many an exchange with him. My theme will be a general review of the enzymes associated with dioxygen since the rest of this volume will provide a more detailed examination of particular cases.

The conventional approach to such biochemical themes is to separate molecular species large or small and then, having analysed them by physico-chemical methods, to describe their biological function. There is a danger here that the effect of biological organization or even of external fields, gravitational for example, will be overlooked. Believing this to be a very real danger I shall concentrate firstly upon the distribution of the enzyme reaction sites for O_2, O_2^-, H_2O_2, i.e. *where* they are in space. Again all are not functional at the same time so we must see *when* they are employed and what signals are used in order to switch them on. Looking at the picture of the space/time relationships of these activities we shall find that the patterns are associated with a very fixed distribution of specific catalyst centres, mainly different metal ions, and we are then faced with the query *why* copper here and *why* iron there, etc? The subtle nature of biology is more

extensive for we find that even when we have understood these features the actual reaction using a given metal ion and oxygen goes in one particular way $O_2 \rightarrow O_2^-$, or $O_2 \rightarrow H_2O_2$, or $O_2 \rightarrow H_2O$ so that particular products are generated in special places. We now touch upon the hardest problem – *how a catalyst works in the place and at the time observed* so that the correct product is generated without contamination from other species. Of course this means that each metal ion or metal cluster is associated with a special protein.

The fact that reaction and product generation is local and not general means that the local concentration of reactants and products is controlled and there will be gradients of all kinds of molecules including O_2^-, H^+, H_2O_2 and even O_2 in a biological system. These gradients, biological fields, are of the utmost importance in bio-energetics and in morphogenesis. The small molecules diffuse relative to their sources and sinks making all kinds of current-dependent devices [2].

So much for the general theme. The next step is to establish that there are such generalities and that we are not failing to see more or less random or chaotic biology while concentrating at one moment on one species and the next moment on another. Figure 1 and Table I make some general assertions. A few comparative examples now make the case. Copper not iron oxidases are outside cells and copper oxygen-carriers are also extracellular while the comparable iron carriers are inside cells. The copper superoxide dismutase

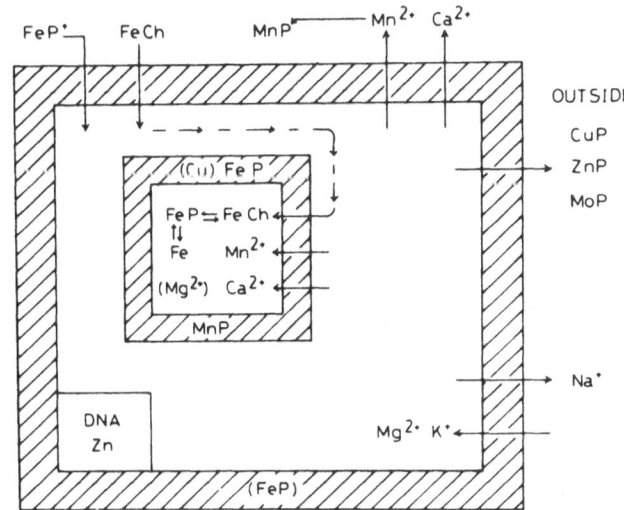

Fig. 1. The distribution of metal ions in cells showing the aqueous and membrane phases and the transport mechanisms which control distribution. P, Protein; Ch, Chelate. Note that Cu and Mo are not associated with the cytoplasm of cells. The use of metals in organelles is of special concern here.

6   *R. J. P. Williams*

Table I. *Locations of enzyme sites not in membranes*

Enzyme	Co-factor	Cellular site
Oxidases	Cu	Extracellular (eukaryotes or in special vesicles)
Oxidases	Fe	Intracellular (eukaryotes and prokaryotes) Extracellular (some prokaryotes)
Oxidases	Flavin	Vesicular (eukaryotes)
Oxidases	Mo	Extracellular (eukaryotes) Periplasm (prokaryotes)
Superoxide dismutases	Cu	Cytoplasmic (eukaryotes)
Superoxide dismutases	Fe, Mn	Cytoplasmic (prokaryotes)
Peroxidases	Fe	Vesicular or extracellular (eukaryotes)
Catalases	Fe	Vesicular (eukaryotes)
Electron transfer	Cu	Exterior face of membranes or in vesicles
Electron transfer	Fe	Inside cells or in membranes as well as in periplasm

is in the cytoplasm of all eukaryotic cells while prokaryotes and organelles such as mitochondria and chloroplasts contain either manganese or iron superoxide dismutases. The oxygen producing reactions of photosynthesis are based in membranes of special organelles of prokaryotes and the iron, copper and manganese in these membranes are differently placed. Flavin oxidases where flavin reacts directly with oxygen, i.e. not dehydrogenases, are to be found in vesicles or outside cells and they generate much superoxide in special regions. I give more details of several of these cases below.

The best example of a timed response is that of the neutrophil which after phagocytosis of a target bacteria attacks the target in a vesicle using superoxide from a flavin oxidase. The enzymic system is latent. However there may be other oxidases under the control of timed DNA expression or metabolic events, e.g. calcium pulses. We know little about the timed control of oxidases but we do know that some dehydrogenase are so controlled.

I turn next to some detailed examples of functional separation in space and time of some metal-containing proteins related to the use of dioxygen in biology. Of course the positioning of the metal ions and clusters is related to the nature of the proteins to which they are bound and I shall make comments upon the selection of metal and protein not only opposite catalytic function but also opposite position in a cell.

(A) The oxygen carriers: inside and outside cells

The easiest group of metalloproteins to look at in a comparative way are the oxygen-carriers. Two features stand out. All iron oxygen carriers are inside cells and we take it that this means that in order to maintain the Fe(II) state required for O_2 binding there are associated reductases; and/or, Fe(II) of proteins cannot be sufficiently strongly held except by membrane barriers. The inside of a cell is a reducing medium. We suspect that Fe(III) proteins based on binding through protein side-chains only and which undergo easy reduction to Fe(II) can not be made such that they retain Fe(II) very effectively. Hemerythrin is always in danger of losing Fe(II) and is therefore in a special cellular container which, maybe, has a high free Fe(II) concentration. One way

of retaining Fe(II) and which also has the advantage of generating the high-spin $\rightleftharpoons$ low-spin switch which allows for a large allosteric features in the oxygen carriers is to provide the special ligand protoporphyrin for the iron. Unfortunately this ligand itself is open to oxidative attack especially in the Fe(III) form and the haemoglobin O_2-carriers are therefore protected in cells by reductases.

In the case of copper complexes the situation is very different in that strong binding to both Cu(II) and Cu(I), stronger to the latter, is easily generated by neutral, e.g. imidazole, side-chain donors of proteins. The maintenance of the Cu(I) state with little danger of oxidation or dissociation from a protein is then manageable. The copper oxygen carrier proteins can be and are kept in the bloodstream and not in cells. Now we see why two types of oxygen carrier can co-exist and in principle could serve slightly different distributive functions in one organism since a cellular O_2-carrier has a different type of interaction with its biological containing organism from that of an extracellular protein. Certain crustacea provide interesting examples.

Although the three well-known O_2-carriers are very different in the above respects they have one feature in common – they are all helical proteins. I shall maintain that this type of construction readily generates a set of interacting sites due to the possibility of helix/helix rotational/translational motion. This gives allosteric properties allowing strain at the metal, on O_2-binding for example, to be transmitted [3, 4]. Such rearrangements must be at the expense of site selectivity and the O_2-carriers are not very discriminating except in terms of the size of molecule bound. They bind NO, CO, CN^-, RNC, imidazole, N_3^-, SCN^- and so on. Selectivity of binding is more a property of the tightly controlled metal proteins described below and which are β-sheet containing enzymes (Table II). We next extend these ideas to enzymes.

(B) Oxidative enzymes: inside and outside the cytoplasm: iron and copper oxidases

The most obvious feature of the metal of the oxidative enzymes is that copper enzymes are outside the eukaryotic cytoplasm. They may be placed in organelles (i.e. comparable with prokaryotes) or in vesicles or completely outside the cell (Table I). This means that where the last step in the oxidation of an organic molecule is required outside the cytoplasm in

Table II. *β-sheet enzymes*

Apart from the general point that enzymes are generally based on β-sheets Brandén has shown that the active site is frequently close to emergent loops, more than one, of the sheet and controlled by the sheet (personal communication). In those cases where enzymes are made from a mixed set of sheets and helices it is the sheets which are the basis of the active site while the helices provide any communicating (allosteric) network.

Enzyme	Metal	Function
Superoxide dismutase	Zn, Cu	O_2^- disproportionation
Plastocyanin	Cu	Electron transfer
Carbonic anhydrase	Zn	Hydration of CO_2
Many peptidases	Zn	Peptide hydrolysis
Alcohol dehydrogenase	Zn	Alcohol oxidation
Rubredoxin	Fe	Electron transfer
Ferredoxins Fe_2S_2, Fe_4S_4	Fe	Electron transfer

Chemica Scripta 28A

order to form a functional molecule it is done by copper not iron enzymes. Examples are adrenaline, tyrosine, proline and lysine (of collagen), and ascorbic acid oxidases. We turn to the special relationships between copper enzymes and hormones and copper enzymes and extracellular filamentous structures later. Copper oxidases can also be used to clear unwanted chemicals from the extracellular fluids, e.g. amine oxidases.

The iron oxidative enzymes are often in the cytoplasm and are used in primary attack on very resistant organic molecules such as terminal methyl groups, phenylalanine and steroids. They are either haem or non-haem proteins. The reaction requires the special oxene, MO, oxidation state, see below. Here as in many of the copper oxidases the final product from the dioxygen is water or an hydroxyl group incorporated in an organic molecule but it is never H_2O_2 or O_2^-. We return to the ways in which the iron is retained in these enzymes since it is not obvious that the non-heme iron should not be lost during cycling to Fe(II).

Molybdenum containing oxidases are outside cells too and often are associated with flavin. They frequently produce H_2O_2 and O_2^-. We shall have to enquire why it is that both copper and molybdenum are used but are used so differently. The molybdenum enzymes resemble the flavin oxidases of neutrophils which are held in readiness to attack bacterial invasions using oxidative degradation and the flavin oxidases of the peroxyzome which give H_2O_2.

It is interesting next to see how the above oxidases are used in the different parts of space to control a whole range of biological functions.

Hormones, transmitters and oxidation

It is an intriguing feature of the production of some hormones that they require the activity of oxidases either for their formation or their destruction or for both (Table III). Other hormones based on peptide or ester bonds require, dehydration for synthesis, usually driven by ATP, and hydrolysis for degradation. Now it is well known that amongst hydrolysable hormones most are packaged in vesicles and that their enzymes for degradation are in closely related parts of space to the points of release. One example is the acetylcholine vesicles for release and the acetylcholine esterases for destruction but a more general case is the location of the storage vesicles and release systems for hormonal peptides and their hydrolysis after use. As we have stressed elsewhere [5] the use of the *metallo*peptidases is associated with the aggressive, i.e. fast, catalytic activity of relatively low selectivity which is so valuable in enzymes employed both in the generation of the peptide hormones, e.g. carboxypeptidases to give endorphins, and in clearance of a range of these molecules, e.g. peptidases 24.11. Both sets of enzymes are localized by containment in vesicles or positioning on surfaces of special cells.

Table III. *Hormones and oxidative metabolism*

Hormones	Synthesis	Destruction
Hydroxy-sterols	(Cytochrome P-450)	Cytochrome P-450
Dopamine	Dopa hydroxylase	Amine oxidases
5-OH tryptamin	Iron enzymes	Amine oxidases
Indole acetic acid	Iron enzymes	Peroxidases
Prostaglandins	Non-haem Fe enzymes	?

Turning to oxidation reactions we should look carefully at the disposition in space of the oxidases and peroxidases used to generate and to remove hormones. Hormones acting quickly must be distinguished from those of long life. We must expect a transmitter to be oxidized very close to its point of release. An example must be dopamine. It is prepared by a series of oxidases, some iron enzymes in the cytoplasm and for the last step of reaction some copper enzymes (β-hydroxylase) in the storage vesicle. (Note that I take a vesicle volume to be outside the cell in the sense that it is removed from the cytoplasm.) The destructive amine oxidases should be located very close to the terminals for release, e.g. in the brain. They too are copper enzymes and also outside cells. I do not know where they are positioned.

While passing we notice that the connection of oxidation of a metabolite with specifically located oxidases is a way of generating apparent pumping systems. If a chemical cannot pass through a membrane its synthesis in a given location, a vesicle, is a way of generating a steep gradient in the chemical without a pump through the membrane. The location of the catalyst is then the secret of the hormone gradient. Is this true for adrenaline?

A totally different class of hormone is the sterols. They are maintained in organisms at an almost constant level with fluctuations only over long periods of time, i.e. even days. Their synthesis and destruction relies upon the microsomal oxidases often of the P-450 class of which there are many. These are membrane bound enzymes which are intracellular and we notice that they are iron heme oxidases. We return to these enzymes again later.

The mechanisms of the iron oxidases are quite different from the copper oxidases in that the iron oxidase (P-450-like) reduce dioxygen to the oxene–iron complex before inserting an oxygen atom in the substrate. The copper oxidases, e.g. laccase, reduce the oxygen to water and the substrate is oxidized by one-electron oxidations. There is a double risk in the latter in that the partially reduced organic molecules could produce O_2^- or H_2O_2 and the substrate is usually allowed to escape as a free radical. Can copper not easily form a safe CuO unit? Other similar copper oxidases, e.g. ascorbate and phenolate oxidases deliberately produce ascorbate free radicals (possibly as protective scavenging radicals) or phenolate and indole free radicals in the production of protective melanin and other polymers especially in plants but all of these reactions are outside cells. Notice that the very dangerous O_2^- and H_2O_2 are not produced however so that copper enzymes may well deliberately generate only *protective* radicals! [6].

Soft connective tissue and oxidases

Let us suppose that the primitive cell organization was mediated by inorganic cross-linking of biopolymers on cell surfaces, e.g. polysaccharides supporting SiO_2 (sponges) or $CaCO_3$ (corals). There is considerable evidence for this. Now this is not a flexible system. If Nature wished to dispense with the inorganic rigid component it had to be possible to cross-link the extracellular biopolymers to make cell–cell networks of sufficient organizational strength while keeping flexibility. Cross-linking of the linear polymers of biology requires oxidation and was hardly possible before the advent of a high O_2 content in the air. This oxidative cross-linking reaction outside cells requires enzymes and given the above criteria for metallo-enzyme stability the obvious choice was copper. Most, perhaps all, soft tissue outside cells is made stable through the cross-linking reactions catalysed by copper enzymes, e.g. laccase, melanin-producing enzymes, collagen cross-linking enzymes. However *initial* hydroxylation, e.g. of phenylalanine to tyrosine or of proline and lysine to hydroxy amino-acids is done by iron enzymes inside the cell. Here again the appropriate enzyme must be in the proper place. There is the strong suggestion here that perhaps copper enzymes evolved only after the appearance of oxygen in the atmosphere while iron enzymes developed from older systems.

Organelles

The very strong functional separation of copper and iron is not so strongly maintained in prokaryotes or in the organelles of eukaryotes. Cytochrome oxidase is a Cu/Fe protein and several blue copper electron-transfer (ET) proteins are found in chloroplasts. Such organisms do not have to guard against mutation (due to O_2^- and H_2O_2) very closely. The most interesting feature of these proteins is that they are used to generate proton gradients across membranes [7, 8]. This means that the oxidase and ET proteins are deliberately placed in membranes with an orientation such that fields of charge and of protein concentration result. The oxidations produce water as the ultimate product but the intermediates are H^+ and OH^- separated locally in space. This is the link to ATP production (Fig. 2). The chloroplast organelles are also responsible for O_2 generation from H_2O and we return to this reaction later.

The major point here is the distribution of metal ions and now especially clusters along and across a membrane while remaining in it, as opposed to the previously observed distributions described above which are in different aqueous zones isolated by membranes (see Figs. 1–3).

The flavin oxidases

The use of flavin as a catalyst for O_2 reactions is obviously quite different from the uses of copper and iron. Flavin oxidases are again associated with special locations and we notice immediately where they are to be found, Fig. 3.

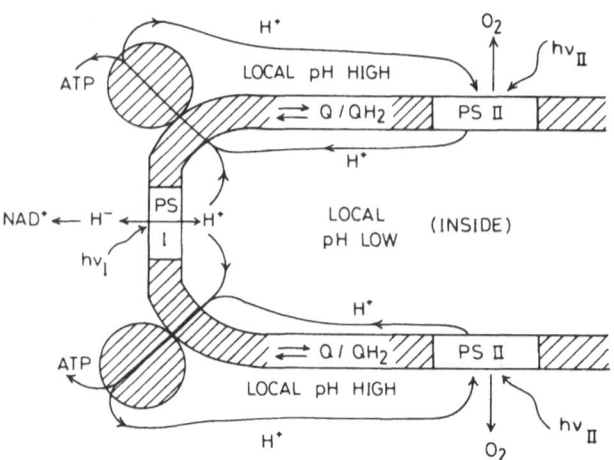

Fig. 2. The edge of a lamellar structure of the chloroplast thylakoid membrane. Mn is associated with photosystem PS II; Fe with both photosystems; Cu with the inside aqueous phase; Q, Quinones. The lateral distribution of metals used in redox reactions is intriguing. Note that the organelles pump in iron and manganese.

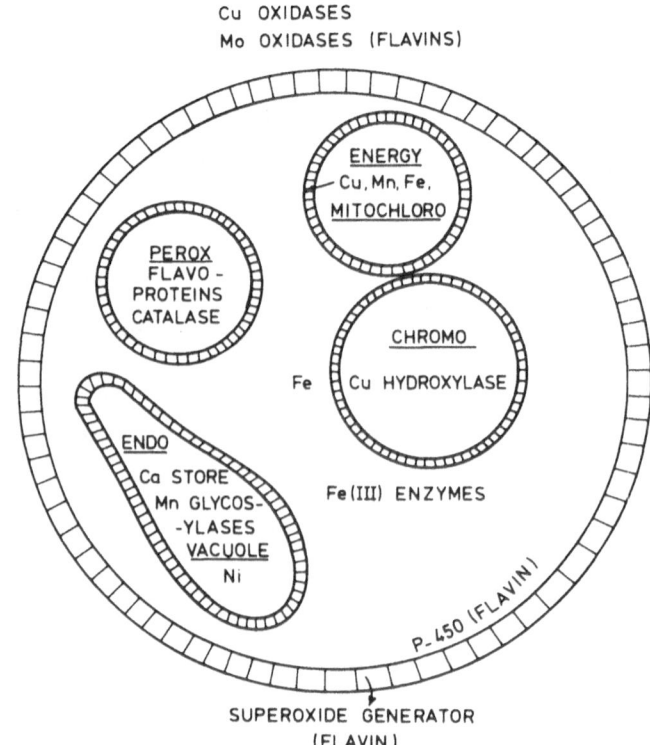

Fig. 3. An amplification of features of Fig. 1 to illustrate the distribution of different metal ions in different organelles. Chrome is the chromaffin granule, Mitochloro is either a chloroplast or a mitochondria, Perox is a peroxyzome, and Endo is the complex reticula of all cells.

(1) In peroxyzomes where H_2O_2 is generated as a product,
(2) In phagocytosed vesicles where O_2^- is generated as a product,
(3) Outside cells in a few special enzymes.

The most obvious new feature of all the reactions apart from their spatial localisation is the partial use of the oxidising equivalents of O_2. There are two possible reasons. The first is that H_2O_2 or O_2^- can be the most important product (not the oxidized form of some substrate) as it is in certain protective systems which oxidize NADH simultaneously. The second is that a mild oxidation by O-atom insertion may be required and that this can not be secured through the FeO systems which are extremely aggressive. Notice the differences between the two oxidases. In the case of FeO oxidase (P-450) O_2 is reduced first to $H_2O + FeO$ while the flavin oxidases the first atom of O_2 to be used is involved in oxidation. Flavin oxidases as well as haem oxidases can insert dioxygen into organic molecules. In organelles the flavins work differently and usually in dehydrogenases.

Molybdenum oxidases

The use of molybdenum in oxidases illustrates the same theme in a different way. The molybdenum does not undergo reaction with oxygen and has little capacity for binding it in any oxidation state. Molybdenum containing oxidases are flavoproteins which can oxidize (or reduce) Mo centres and which use flavin to react with O_2. The high oxidation state Mo(VI) can then insert O atoms giving Mo(IV) as follows

$$Mo(VI)O + HR \rightarrow Mo(IV) OH^- + R^+ \rightarrow Mo(IV) + ROH$$

or

$$Mo(VI)O + R \rightarrow Mo(IV) + RO$$

The molybdenum reactions do not use O_2. They do use the easy M–O bond break in two-electron reactions which are common in higher oxidation states of second row but not first row transition metals. Note that the mechanism is not free radical as for cytochrome P-450. Finally the reaction occurs at low potential (-250 mV). These molybdenum enzymes are extracellular and may have appeared after oxygen came into the air. Notice that molybdenum handles molecules generated by oxidation, N_2, NO_3^-, SO_4^{2-}, and which were not present in the original environment before the advent of O_2, compare the copper enzymes which are used to handle O_2, NO_2^- and so on.

Mechanistic questions

We must next wonder *how* it is that the non-haem iron proteins can be used in the cytoplasm of cells when there is the apparent risk of loss of Fe(II) during a redox cycle. The risk is that unrestricted Fe(II) sites can always react with O_2 to give radicals which damage DNA or that free Fe(II) will be released from the enzymes [9]. Even in the case of O_2-carrier haem proteins there is still risk and this appears to be guarded against in the peculiar feature of some red-cells that the DNA has been rejected. There are certain other ways in which risk can be diminished.

(1) Use of a cycle in which Fe(II) does not occur. This is the case for many soil bacteria oxygenases where the enzyme activates substrate (usually a phenol) so as to allow attack directly by O_2 to give a dioxygenated product. The attack is not possible on inert substrates. But notice that this is an extra cellular reaction and may well be risky if it were introduced into cells.

(2) Controlled order of reactions such that Fe(III) is not reduced to Fe(II) until the substrate which is to be attacked is bound, see below. The substrate subsequently protects from the dangers of Fe(II) release. Here very inert substrates can be attacked.

(3) Formation of multi-nuclear complexes, e.g. Fe–O–Fe, Fe–O–Cu, Mn–O–Mn, see organelles.

(4) Rapid removal of a dangerous product. Notice that peroxyzomes which generate H_2O_2 are full of catalase for its destruction and that all cells and organelles destroy superoxide. The placing of generating and destroying systems close together is again obvious but here too the reactions are not in the cytoplasm. Here organelles (Fe, Mn) are quite different from the eukaryotic cytoplasm (Cu) in their superoxide dismutases.

The most interesting example immediately is that of controlled order (2) above, since may be it is the only mechanism which allows the extensive use of iron oxidases in the cytoplasm of eukaryotes.

Controlled order

As we have stressed a major problem with the use by biology of iron or manganese enzymes is the loss of metal ions (Table IV) [9]. This is overcome in complex iron enzymes such as heme enzymes, in FeOFe proteins and in many $Fe_n S_n$ proteins since mostly they do not exchange metal ions. For most other simple iron or manganese enzymes the danger is the loss of the weakly bound divalent cation which is not well retained by protein side-chain ligands unlike copper, zinc and nickel (see the Irving–Williams stability constant order). This

Table IV. *Copper, iron and manganese proteins*

Protein	Location	Ligands
Haemerythrin (Fe)	Special cells	$3N, 2CO_2^{-a}$
Superoxide dismutase (Fe)	Mitochondria	$3N(His), 1CO_2^-$
Iron uptake protein (Fe)	*E. coli* cytoplasm	$2N(His)$
Superoxide dismutase (Cu)	Cytoplasm	
Haemocyanin (Cu)	Extracellular	$3N(His)$
Superoxide dismutase (Mn)	Chloroplasts	$3N(His), 1CO_2^-$
O_2-forming protein (Mn)	Chloroplasts	$3N(His), 1CO_2^-$ (?)
Concanavalin A	Plant lectin	$N(His), 2CO_2^-$

^{a} The two iron atoms are different.

means that the cytosol of eukaryotic cells which is in a part of biological space which includes the nucleus – cannot have simple iron or manganese enzymes which freely cycle through the Fe(II) or Mn(II) states since these ions could cause oxidative damage to DNA. In fact this restriction applies to iron enzymes outside the cell since here loss of iron means the destruction of the enzyme [9]. The problem has been overcome by controlled orders of reaction of the metallo-enzymes. One scheme for iron oxidases is

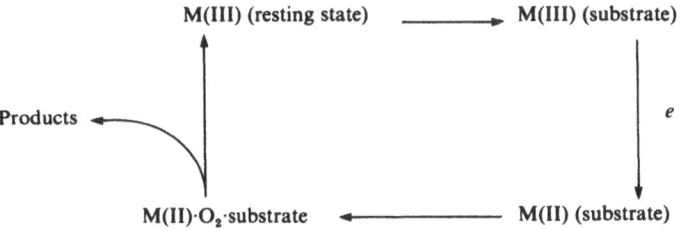

Here Fe(II) cannot dissociate sine Fe(II) is not formed excepted on substrate binding and dissociation is then blocked by substrate. It is also the case that the very reactive intermediate, the oxene FeO, is never able to attack the protein but can only attack substrate. The path is familiar in cytochrome P-450 but I believe it must be common to many hydroxylases such as phenylalanine hydroxylase, proline hydroxylase, the oxidative enzymes of penicillin synthesis, sterol hydroxylases, and prostaglandin synthesis, many of which are not haem proteins.

Outside the cell and especially in soil bacteria attack can take place in a different way since the substrates to be attacked are more reactive, e.g. phenolic materials of lignins. Here oxidative attack can use metal-ion activation of the ligand as mentioned above.

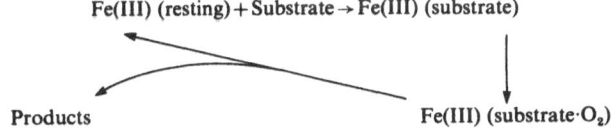

Reactive substrate radicals or dangerous species such as reduced oxygen compounds are not menacing outside the cell. [It is even possible to use Mn(III) or Fe(III) enzymes in acid media since their stability is high. We see this in the acid phosphatase of the lysozomes while we note that alkaline phosphatases in the cytosol are zinc enzymes.]

The conclusion is that the free iron and manganese in the cytosol of eukaryotic cells is very low in order to protect DNA. This has meant that only very special iron (and manganese) enzymes are present in this compartment. In the organelles of the same cells and in prokaryotes we envisage that the standing levels of Fe(II) and Mn(II) are higher (mutation is not a real problem), that these ions are regulatory both at the level of direct effects in exchange from enzyme sites and in their communication DNA through special proteins. I am told that future publications of Professor J. B. Nielands will demonstrate this [9, 10].

In order to bring about this type of controlled redox state chemistry specially evolved ligand systems are required.

The protein part of O_2-utilizing enzymes

Before leaving the oxidases we should look at their proteins. We have commented that O_2-carriers are all helical proteins. It is also the case that may control proteins such as calmodulins and histones and here we should probably include the iron uptake regulatory protein, FUR, of Nielands [10] and the storage protein ferritin are helical. The proteins which are enzymes are usually based on β-sheets though if they are regulated there are helical sections too. The overall impression is that the stability of a β-sheet gives better definition to function in an enzyme, not only in so far as specificity of substrate recognition is concerned but also in the special site geometries which can be forced on metal ions, entatic state. There are however two or three *enzymes* which have a largely helical structure. These enzymes are those which require steric adjustment along the reaction path to generate controlled order. Cytochrome P-450 is one such enzyme, Table V. Is it generally true that such adjustments need helical structures?

An interesting point too is that it would appear as if it is quite difficult to build an enzyme site actually in a membrane bilayer. The oxidation sites are often on one side or the other. The reason for this may be that the part of a protein which is in the membrane is helical and hence is unsuitable for enzyme activity but very suitable for mechanical transmission. The parts of the enzymes on the surface of the membrane are then β-sheet at least in part and give rise to localized enzyme sites on one side of the membrane or another. This automatically and invariably leads to strong gradients of chemicals across the membrane, of protons, for example, but also of all products and substrates. Secondly, the helices of the membranous part of the protein provide both mechanical connections, helix/helix movement, and controlled channels allowing movement, including pumping, of ions and molecules with or against the gradients of concentration and charge. Thus ET sites on helices can be in the membrane. These notions are interesting in the description of the functions of many other classes of membrane enzymes as well as oxidases (Table V).

Summary of oxidases

The oxidases are now seen to be a very diverse set of enzymes in the light of metal ions used, substrates attacked, and

Table V. *Largely helical enzymes*

Protein (enzyme)	Special feature
Citrate synthetase	Compulsory order of substrate addition
Cytochrome P-450	Compulsory order of substrate addition
Cytochrome c	Immobilized by cross-linking
Cytochrome b and a	Compulsory connection with proton or electron pumps with a compulsory order

products produced. Their protein structures are adapted to enable controlled order of reaction. However, the stress in this article is on the location of these enzymes inside the cytoplasm, in membranes, outside cells or in vesicles. Now this location should be seen not just as in a volume but actually on a three-dimensional grid. The enzymes can be so located due to the restrictions of the curvature of membranes and the network of filaments inside and outside cells in which the enzymes are held [2]. These constructions associated with locations for enzymes mean that local gradients of all kinds of chemicals occur and these gradients provide driving forces for many biochemical and biophysical transformations. Finally, the organization can be changed with time as the organism develops or as it meets new circumstances so that the gradients of chemicals can be totally altered by switching mechanisms.

(C) The superoxide dismutases

This group of three enzymes again allows an easy approach to the wider problems of the use of copper, iron and manganese in oxygen reactions. The three dismutases are completely different in different species and in location within a species (Table I) [11]. Virtually no prokaryote, no mitochondrion and few chloroplasts have a copper dismutase. In all probability and within evolutionary time the copper protein is a newcomer. Moreover Fe and Mn dismutases are absent from the cytoplasm of all higher cells but are always present in their organelles. The striking contrast in species and spatial distribution of the three superoxide dismutases has parallels with the distribution of elements central to the above oxygen reactions. First we observe that in the enzyme reactions all the metals cycle through oxidation states. I take the reasons for the distribution of the dismutases to be

(1) Cu superoxide dismutases do not equilibrate with a concentration of free copper ions. Mn and Fe superoxide dismutases lose divalent metal ions relatively easily and require a standing concentration of these ions for their stability. There is no menace from copper since there are virtually no free copper ions.

(2) All three metals as free ions are a source of free radical damage to DNA by mutation, probably via hydroxyl radicals from oxygen.

(3) The DNA of prokaryotes does not require too much protection from mutagens. Protection is essential for eukaryotes and the copper enzyme is therefore the least harmful in their cytoplasm.

(4) The standing concentration of free Mn and Fe is quite high in mitochondria, chloroplasts and prokaryotes but is lower in the cytoplasm due to pumping and 'deliberate' association with control mechanisms.

It follows that eukaryotes, especially differentiated multicellular systems, could develop securely from prokaryotes only if the protection from superoxide was by a better reagent than the Mn or Fe dismutases. To prevent the exchange of copper we note especially the enormously strong binding of Cu dismutase in a β-barrel fold cross-linked by zinc compared with the relatively loosely folded more helical Mn and Fe dismutases. (Note, however, that all three enzyme active sites are associated with β-sheet structures – a general but not quite universal rule for enzymes, see above).

Now the coordination chemistry of these three dismutases

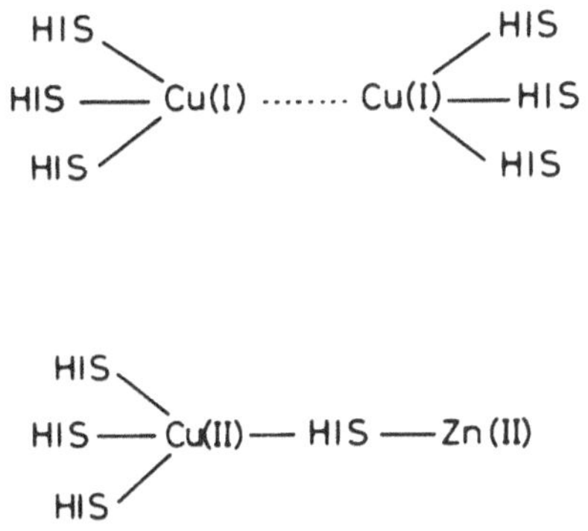

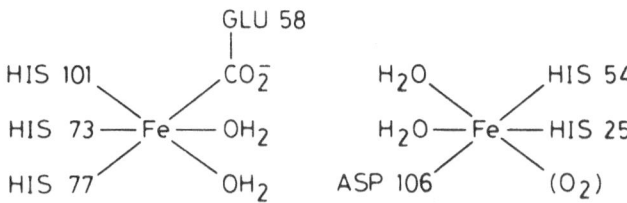

Fig. 4. The active site of Cu superoxide dismutase (below) compared with that of haemocyanin (above).

is of great interest since the redox potential of the metal couples should be halfway between the E° of two reactions $O_2^- \rightarrow O_2$ and $O_2^- \rightarrow H_2O_2$ so that the centre can manage both reduction and oxidation steps. The actual value for all three is at it should be, about $+0.4$ V – a high potential for metalloproteins. The first point of interest then is that the ligands that had to evolve must not stabilize too strongly the higher oxidation state. For this to be the case and for all these metals the site must be made largely from neutral (imidazole) ligands. Now for the copper enzyme the observed choice is four imidazoles (Fig 4) [12], which actually leads to a stabilization of the lower oxidation state, Cu(I). We note too the entactic geometry of the distorted (from planar), copper site which also helps to create the high redox potential through strain, giving a reduced binding stability for Cu(II). The geometry also provides an open site for reaction. The Fe and Mn sites are also open-sided (Fig. 5), but have three imidazoles plus one carboxylate in 5- or 6-coordinate geometry [13]. This combination can not be as powerful a binding site, either in a thermodynamic or in a kinetic release sense, for divalent metal ions as the copper site of copper superoxide dismutase, the zinc site of carbonic anhydrase, or the second metal tetrahedral site of copper dismutase occupied by zinc, since, as observed, zinc is not a good competitor for the dismutase sites of the iron and manganese dismutases given the free ion concentrations of all the metal ions in the systems. The inevitable consequence is a low-binding constant for Fe(II) and Mn(II) and therefore a relatively high free concentration of Fe(II) and Mn(II). This high concentration is maintained by pumping the elements into the organelles and prokaryotes. (If it is asked why

Fig. 6. A comparison can be made between the Cu and Fe oxygen-carrier sites of haemocyanin (Fig. 4), and that of haemerythrin here. Beneath is shown the potentially dangerous monomeric Fe(II) structures which if formed could lead to the loss of iron.

copper was not used in the prokaryotes the answer possibly lies in the lack of availability of copper in early evolution since it was precipitated as CuS but also in the useful control connection of enzyme synthesis in prokaryotes to the levels of free metal ions, Fe and Mn) [9].

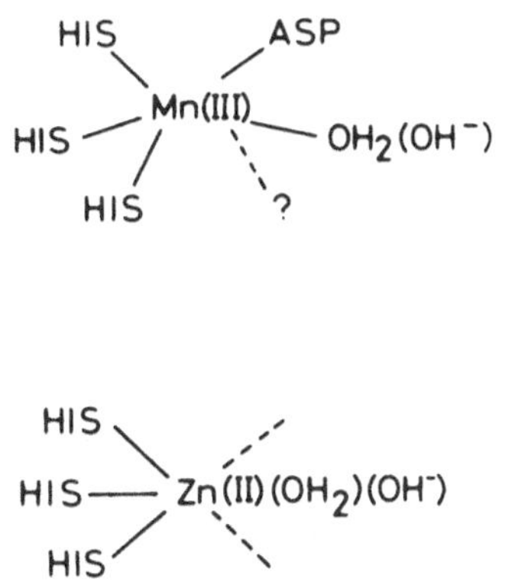

Now the mechanistic significance of these sites is also apparent. First they are open sided to accept substrates. Second they do not bind O_2 in the reduced state although they form strong $Cu^{II} \cdot O_2^-$ and $M^{III} \cdot O_2^-$ complexes. This arises from the instability of the lower oxidation state $M.O_2$ complexes in comparison with dimeric centres. Interestingly, there are relatively similar (in coordination geometry) metal centres of both iron and copper in haemerythrin and haemocyanin, respectively (see Fig. 6), which do carry oxygen but these centres are dimeric in metal [14]. We turn to the deliberate production of superoxide in a later section.

The compartmental trapping of metal ions in different regions of biological systems just discussed removes some of the mystery of why several metals are found with very similar coordinating groups but are in different proteins (Fig. 4–6). The metals are in part kinetically controlled in their combinations and do not equilibrate freely with all possible ligands. Of course thermodynamic factors are a major part of the control over binding but the pumping of many metal ions and of proteins directly, or through the use of receptors and movement across membranes, gives preferred zones of reaction.

(D) Hydrogen peroxide reactions

Hydrogen peroxide, produced in the manner described later, is eliminated from cells by catalases usually iron but sometimes manganese enzymes. These enzymes do not cycle through the M(II) state and they do not bind dioxygen. The coordination site is described below. The location of the enzymes in higher cells is in peroxyzomes so that there is rapid removal of H_2O_2 as it is formed from the flavoproteins of that organelle.

Higher alkyl peroxides are cleared in the cytoplasm by glutathione peroxidase which is a selenium enzyme and as such does not go through the dangerous one electron steps of a catalase. This is a further protection from mutation.

Fig. 5. The active site of Fe or Mn superoxide dismutase compared with that of carbonic anhydrase.

Chemica Scripta 28A

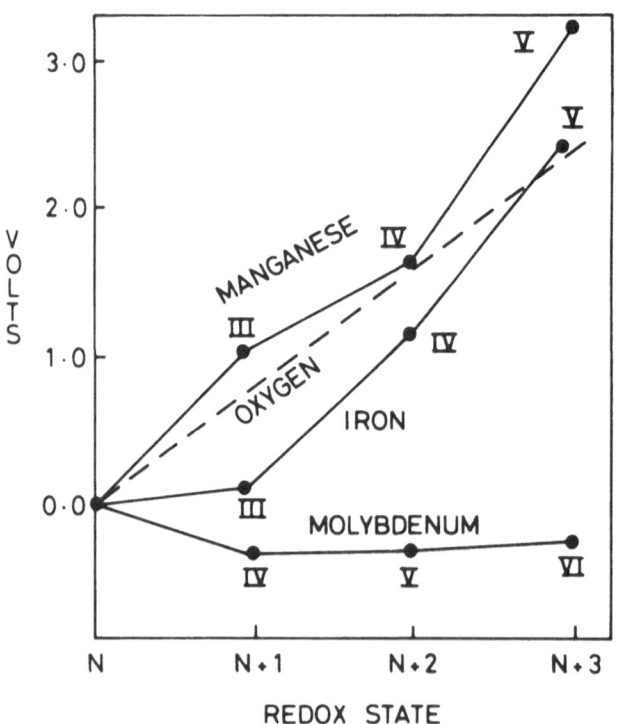

Fig. 7. A comparative plot of the redox states of iron, molybdenum and manganese showing their specific features. How are these potentials altered especially in clusters?

(E) Use of manganese and O_2-production

In a carrier of dioxygen it is important that the uptake of dioxygen does *not* lead to the reaction

$$2M + O_2 \rightarrow 2MO$$

and this is prevented if the intermediate states e.g. $2M \cdot O_2^{2-}$ are very stable. In the case of copper and iron cluster complexes this is true, e.g. $2Cu(\text{II}) \cdot O_2^{2-}$ and $2Fe(\text{III}) \cdot O_2^{2-}$ as seen in hemocyanin and haemerythrin, but the unstable $2Mn(\text{III}) \cdot O_2^{2-}$ cannot be formed since it would go on to two $Mn(\text{IV})O$ (Fig. 7). Single atoms of iron, but not so easily copper, go to their MO states after O_2 uptake at $Fe(\text{II})$, [or $Cu(\text{I})$], but only on half reduction of O_2 to H_2O. This reaction is only just possible for $Mn(\text{II})$ since it has a very poor affinity for O_2 and in any event MnO would be but a weak oxidizing agent. Here again it is the $Mn(\text{III})$ state which is not readily available in the reaction, common for iron,

$$M^{IV}O + HS \rightarrow M^{III}OH + S \cdot \rightarrow M^{II} + SOH$$

The reverse reaction

$$2MO \rightarrow 2M(\text{II}) + O_2$$

will be considered next.

Oxidation of water

The final reactant of interest to us is water which the photosystems of biology convert to dioxygen. Simultaneously a proton gradient is formed which is then used to produce ATP. Now there are two photosystems one of which, PS II, generates O_2 and the other, PS I, which generates reducing equivalents. Both make proton gradients. This apparatus is all in the thylakoid membrane but it is an organized array

such that PS II with its manganese centre is in the oppressed regions (Fig. 2), while the Fe/S reductase is in PS I. The location of the cytochromes and the Rieske centre is not known with precision. Again depending upon the light intensity light energy is shared differentially between the systems. In the dark the organisation disperses to some extent. Communication between PS I and PS II uses plasto-cyanin, a blue Cu-protein. The several metal centres are allocated different compartments and so localise different events. Notice the changes in space and with time of the system. (The copper protein in a thylakoid is in a vesicle not in the eukaryotic cytoplasm as stressed earlier.)

We must also turn to the use of manganese not iron in oxygen evolution. There are now great advantages in Mn chemistry

(1) MnO is not a powerful oxidizing agent and therefore will not destroy its protein ligand, cf. FeO.

(2) $Mn(\text{II})$ does not retain O_2.

(3) The state between $M(\text{II})$ and MO is the unstable $Mn(\text{III})$ which will therefore give $MnO + Mn(\text{II})$ in a multi-metal unit when radical reactions are not a danger, but clusters are essential.

This allows the building of the O_2 molecule in convenient one electron steps without risks. Biology using redox state chemistry has found the best solution for the problems of several redox reactions by using different metal ions in different reactions and places but the ET and connections to product, H^+, gradients are very similar.

Before proceeding to the final section I draw attention to the redox state diagrams for Fe and Mo (Fig. 7). Such a diagram shows the great differences between the functional potential of these elements in oxidation reactions which are realised in biology. Molybdenum is suitable for low redox potential O-transfer, e.g. to NO_2^-, SO_3^{2-} and $-CHO$; iron is useful for radical reactions, since the redox potentials increase uniformly, at high potential; manganese while doing some one-electron reactions at high potential, $Mn(\text{II})/Mn(\text{III})$, can also give two-electron, O-transfer, reactions since $Mn(\text{III})$ is unstable. This means that manganese undergoes hysteresis in its cycling of redox states due to the energy input from light. The O-transfer here leads to dioxygen. The redox potential diagram for copper is rather like that for iron.

(F) Control over products

Control over reactions can be exerted by different means as we have seen. The most obvious are:

(i) Retention or release of an intermediate products.

(ii) Preferred order of combination of reactants with a catalyst.

(iii) Manipulation of barriers to reaction by control of for example redox potential.

(iv) Construction of site to exclude certain combinations.

(v) Burying of the site to control ET.

Some simple examples follow

(*a*) Cytochrome P-450, cytochrome oxidase and laccase retain all intermediates between $O_2 \rightarrow H_2O$.

(*b*) Cytochrome P-450 is held resting in the $Fe(\text{III})$ state so that reduction to $Fe(\text{II})$ and O_2-binding follows after substrate binding.

(*c*) The redox potential of catalase is too low for it to be reduced to $Fe(\text{II})$ whence catalase cannot bind O_2.

(*d*) O_2 and H_2O_2 do not bind to superoxide dismutase but are released from it while O_2^- binds. Carbon monoxide does not inhibit this enzyme.

(*e*) Most ET centres of purely ET proteins such as cytochrome *c* are buried. Cytochrome *c* oxidase and its copper appear to react very selectively with cytochrome *c*. How are all these features controlled?

It is not difficult to see that control over redox potential is exerted in heme proteins by ligating anionic groups of the protein, e.g. catalase and cytochrome P-450 have low potentials, or by unsaturated ligands such as thioethers and the special porphyrin of haem a (e.g. cytochrome *c* has a high potential and so do cytochromes *a*), or by exposure to solvent or by strain in bond length. (Copper potentials are more readily manipulated by stereochemistry.) The exact intermediates which these centres retain or release is controlled partly by accessibility of redox states of single or *pairs* of metal ion centres.

Now if the required catalysts are put into definite compartments there is local control over substrate used and product released throughout the space of the cell. This is important in that for example offensive use of generally destructive free radical chemistry is kept well away from the nucleus of a eukaryotic cell. The partitioning of activity can be through the use of isolated aqueous regions such as outside the cell, the cytoplasm, the organelle inner solution, any vesicle inner solution, and so on, or the positioning of any enzyme in a containing membrane with an orientation, Table VI. However there are still further ways of dividing space in that the vesicles can be put in particular regions of a cell or that the membrane can be laterally differentiated. Finally due to the network of fibres in the cytoplasm there is the possibility of organizing enzymes locally even within the aqueous phase along these fibres [2].

Table VI. *Biological compartmentation of some metal ions*

Metal	Compartment
Copper	Usually extracellular or in vesicles. Very rare in cytoplasm of eurkaryotic cells
Calcium	Usually extracellular or in vesicles. Low in cytoplasm
Magnesium	Universally around 10^{-3} M except in special vesicles
Iron	Often in membraneous systems and organelles. Low in cytoplasm except ferritin vesicles
Zinc	Universally distributed but high in certain vesicles
Nickell	Not in eukaryotic cells except some vesicles
Manganese	Mainly confined to organelles and vesicles when in eukaryotic cells. (In the Golgi?)

Note. Every vesicle has selective uptake devices for one or two but not all metal ions.

Organization and flow (Currents)

The positioning of catalysts in space leads to a diffusion field where substrate is denuded in one region and product is high in this same region. The substrate and product can be any kind of molecule or molecular ion. Since simple inorganic ions themselves cannot be metabolized they are removed only by membrane pumps which transfer them from one part of space to another or they are admitted into a new space through a channel in a membrane. The circuits of metabolites and ions are linked to oxygen since its reactions are such as to set up in many ways diffusion gradients of organic metabolites and especially note the proton. Patterns of concentration of necessity set up patterns of current.

Patterns of oxidation/reduction

Location of a reciprocal pair of enzymes in space where reciprocal means that the product from one is the reactant for the next generates gradients. The simplest example is in bioenergetics where the photosystems or the oxidative enzymes produce H^+ and the ATP-synthetase utilizes this ion to make ATP. However the principle is general so that all biological space with its fixed enzymes and its metabolic paths has an underlying substrate and product diffusion gradient. The charge on molecules may change and again molecules or ions can be pumped across membranes. The gradients set up mean that biological space/time is a network of chemical potential fields. These fields act like gravity so that biological space is always inhomogeneous and local events dominate. Only on the basis of such a theory can we explain the complexities of differential growth, morphogenesis. I believe that oxidation and reactions of oxygen will be found to be one of these field-generating systems.

Acknowledgements

This article is the product of forty years of study in the topic of bioinorganic chemistry and has behind it not only my own work and that of my pupils but of a large and constantly increasing international body of chemists and biochemists. During this period I have had many interesting exchanges with Professor Bo Malmström who started his work at almost the same time as I did. I take this opportunity to wish him every success in the future.

References

1. Williams, R. J. P., *Chem. Scr.* **26**, 513–523 (1986).
2. Williams, R. J. P., *J. Theoret. Biol.* **121**, 1–22 (1986).
3. Perutz, M. F., *Nature* **228**, 726–739 (1970).
4. Banerjee, R., Alpert, Y., Leterrier, F. and Williams, R. J. P., *Biochemistry* **8**, 2862–2869 (1969).
5. Williams, R. J. P., *Carlsberg Res.Comm.* **52**, 1–30 (1987).
6. Williams, R. J. P., *Phil. Trans. R. Soc. Lond.* **B211**, 593–603 (1985).
7. Williams, R. J. P., *J. Theoret. Biol.* **1**, 1–13 (1961).
8. Mitchell, P., *Nature* **191**, 144–147 (1961).
9. Williams, R. J. P., *FEBS Lett.* **140**, 1–10 (1982).
10. Bagg, A. and Nielands, J. B. (accepted for publication).
11. Developments in Biochemistry, in *The Biology and Chemistry of Active Oxygen*, vol. 26 (ed. J. V. Bannister and W. H. Bannister). Elsevier, Amsterdam (1984).
12. Richardson, J. S., Thomas, K. A., Rubin, D. H. and Richardson, D. C., *Proc. Natl. Acad. Sci. USA* **72**, 1349–1353 (1975).
13. Stallings, W. C., Pattridge, K. A., Strong, R. K. and Ludwig, M. L., *J. Biol. Chem.* **260**, 16424–16432 (1985).
14. Stenkamp, R. E., Sieker, L. C. and Jensen, L. H., *Nature* **291**, 263–264 (1981).

Chemica Scripta 1988, **28A**, 15–20

Distance Dependence in Biological Electron Transfer. Theoretical Aspects

Sven Larsson

Chalmers University of Technology, Department of Physical Chemistry, S-412 96, Göteborg, Sweden

Paper presented at the Nobel Conference 'Biophysical Chemistry of Dioxygen Reactions in Respiration and Photosynthesis', Fiskebäckskil, Sweden, 1–4 July, 1987

Abstract

The distance dependence in rate constants for electron transfer (ET) between metal centres in proteins is discussed on the basis of the Marcus model. ET may be regarded as tunnelling through the protein medium but tunnelling is here treated using a molecular orbital model where the orbitals of the bridge are included. The theory is illustrated by simple examples where electrons are transferred between aromatic systems through saturated, rigid spacers. In a more flexible saturated system the ET capability depends in an intricate way on the dihedral angles. Calculations are carried out on a system consisting of two metal ions separated by a glycine oligomer with up to ten peptide units. Approximately exponential decrease of the electronic factor with distance between the redox centre is found. Possible enhancement due to for example aromatic groups, is discussed.

1. Introduction

The problem of long-distance electron transfer (LDET) in biology was discussed in 1941 by Albert Szent-Györgyi [1]. In a lecture entitled 'Towards a New Biochemistry' he called attention to the problem of interaction of oxidation enzymes, the problem of light-harvesting in photosynthesis, and some other cases where interaction over large distances must occur. Szent-Györgyi wanted to view LDET as due to electrons moving in common energy levels approximately in the same way as in semi-conductors. Attempts to measure an appreciable electric conductivity in proteins failed, however [2]. Calculations of the band gap gave values of 3–5 eV, which is much too large to be of any importance in biology [2]. The fact that proteins are bad conductors while electron transfer (ET) reactions may be very fast suggests that biological ET steps involve very specific interactions where molecular and electronic structure play a decisive role.

As late as 1979 it seems that there was no widely accepted view of ET in biological systems [3]. After this date the Marcus model [4–6], which was commonly used as a model for inorganic ET, gained in importance. The model was mostly applied to cases with a direct contact between the electron exchanging metal complexes. The same applied to other possible models, based on unimolecular decay reactions [7], which were also used to treat inorganic and biochemical ET reactions [8]. The long distances for ET in biological systems was therefore a problem until it was shown both experimentally [9, 10] and theoretically [11] that LDET is possible in organic bridges, saturated as well as unsaturated.

Two factors are of great importance in the Marcus model: reorganization during ET expressed by the parameter λ and electron interaction over large distances as expressed by the parameter Δ. In this paper we will be particularly interested in the latter quantity which plays a major role in determining the conditions for LDET in proteins.

Δ may be calculated using quantum chemical methods. So called *ab initio* methods are conceptually simple but can be carried out with accurate results only on small molecules, preferably with no atoms larger than chlorine. Their use in the case of ET systems is mainly for the important purpose of testing semi-empirical methods on small systems as has been done in our group. A problem in all quantum chemical calculations is the complexity of the results. To avoid such complexity as much as possible we are going to carry out the discussion in terms of a 'one-electron-orbital' approach where, as a further simplification, each molecular orbital (MO) is expanded as a linear combination of atomic orbitals. One must keep in mind that such a description is not entirely correct. In most cases, however, a many-electron description does not change any general conclusions made on the basis of the one-electron model.

Much of the physics of LDET is demonstrated by the simple tunnelling model. In Fig. 1 the electron is oscillating between two wells with the frequency Δ/h. Δ decreases exponentially with distance

$$\Delta = \exp(-\alpha R) \qquad (1)$$

In a more reasonable model the energy levels are different corresponding to different redox potentials. The energy levels are also varying due to bond vibrations and oscillations in the

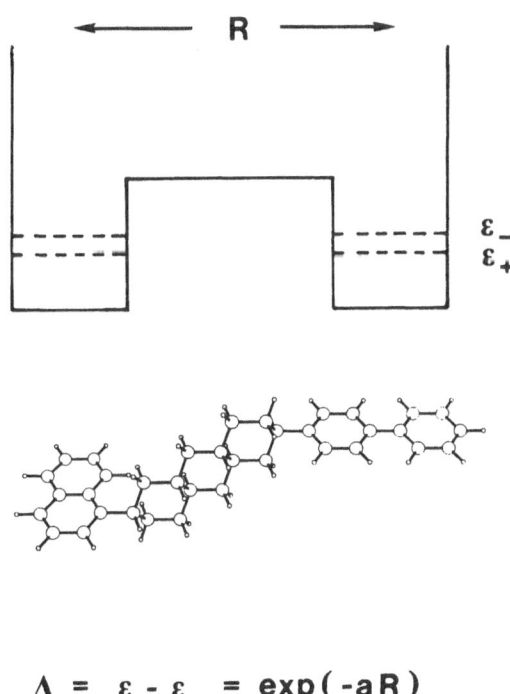

$$\Delta = \varepsilon_- - \varepsilon_+ = \exp(-aR)$$

Fig. 1. Tunnelling between two wells.

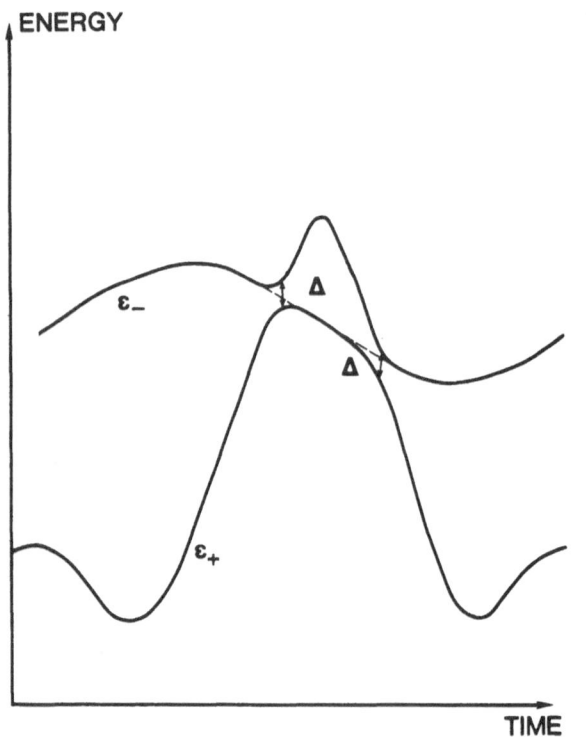

Fig. 2. Oscillation and crossing of orbital energies due to nuclear motion.

surrounding protein structure or solvent (Fig. 2). Electron transfer takes place when the energy levels pass through an avoided crossing region. The probability for ET is proportional to Δ^2 if Δ is small. In this case the electron easily jumps the small gap from the lower to the upper orbital. If Δ is large the lowest energy level will always be occupied which is equivalent to adiabatic ET.

A problem with the tunnelling model is that there is no accurate way of estimating the potential barrier between the wells (see, however, a recent paper by Kuki and Wolynes [12]). Instead we will here, as in earlier work [11, 13], use MO's and include also the orbitals of the 'bridge', i.e. the forbidden region between the wells. After a short summary of the theory in sections 2 and 3, calculations of the electronic factor will be described in sections 4 and 5. We will find that $|\Delta|$ in most cases decreases exponentially with the length of the bridge, however, with a surprisingly small α and a large pre-exponential factor.

2. Reorganization energy

The rate of constant in Marcus theory [4–6] is given by

$$k = \nu_n \kappa_e \exp\left(-\Delta G^\ddagger / RT\right) \qquad (2)$$

where ν_n is a vibrational frequency associated with the activating motion and κ_e the electronic factor which is a function of Δ and will be discussed in detail below. $\Delta G^\ddagger$ is the free energy of activation which is calculated from the reorganization free energy λ and the standard free energy of reaction as described in Fig. 3. λ consists of two parts

$$\lambda = \lambda_i + \lambda_0 \qquad (3)$$

λ_i is due to changes of the bond lengths, δq, between reactants and products during ET

$$\lambda_i = \Sigma \tfrac{1}{2} f_\mu (\delta q_\mu)^2 \qquad (4)$$

where f_μ is an effective force constant, and λ_0 due to changes in the solvent polarization [4, 5]:

$$\lambda_0 = \left(\frac{1}{2a_1} + \frac{1}{2a_2} - 1/R\right)(1/n^2 - 1/D) \qquad (5)$$

n is the refractive index and D the dielectric constant of the solvent. a_1 and a_2 are the radii of the solvent spheres around the donor and acceptor, respectively, and R is the distance between them.

If the standard free energy of reaction is ΔG (the work of bringing the reactants together is ignored) we obtain the free energy of activation (Fig. 3)

$$\Delta G^\ddagger = \lambda/4 (1 + \Delta G/\lambda)^2 \qquad (6)$$

The barrier is zero for $-\Delta G = \lambda$. For large values of $-\Delta G$, it appears again (the inverted region). In the charge separation process in photosynthesis there is a possibility that the back-reaction to the ground state is prevented due to inverted behaviour.

A distance dependence is introduced in the rate constant through equation (5). With $a_1 = a_2 = 4\,\text{Å}$ and $(1/n^2 - 1/D) = 0.55$ (water), $\lambda_0 = 7.9(0.25 - R^{-1})$ eV, where R is expressed in Å. It is likely that evolution has selected systems where the redox centres are buried in the protein (as for example in plastocyanin) in order to avoid strong caging by water. Of course if the redox centre is too far from the surface the electronic factor will be small. Thus in a biological redox system the two important factors of water exclusion and orbital overlap have to be in some balance to guarantee rapid ET. It is also important that the protein structure does not contain highly polarizable groups close to a redox centre.

A π-system as opposed to a σ-system does not change its bond lengths very much at donation or acceptance of electrons or at excitation. If one adds an electron to a π-orbital in a molecule with a single π-bond there will be a bond-length increase of about 0.1 Å. If the force constant for bond stretch is assumed to be 7.5×10^5 dyn/cm [14] this is equivalent to an energy increase of 5.4 kcal/mol. If we are dealing with ET between two π-bonds of this kind the contribution to the thermal barrier will be [equation (4)] $2 \times 5.4/4 = 2.7$ kcal/mol. If the π-systems consist of N equivalent π-bonds each bond-length change is only 0.1/N Å since their bond order is distributed equally over the bonds. The energy change in each bond will be $5.4/N^2$ kcal/mol. The barrier arising from all the bond-length changes will be only $2.7/N$ kcal/mol. Thus for large π-systems it is reasonable to neglect the bond distortion contribution to the activation energy.

Metal complexes show a great variation in bond length changes at ET and the rate of electron self-exchange reactions consequently change over several orders of magnitude [15]. In the case of metallo-enzymes one suspects that evolution has selected metal complexes as redox centres which do not have large thermal barriers for ET. Iron porphyrins [16] and the $Cu^{+/2+}$ centres in for instance plastocyanin [17] seem to have this property.

3. Electronic factor

An electronic frequency ν_e may be defined as [6]

$$\nu_e = g\Delta^2/[4(\lambda_i + \lambda_0)^{\frac{1}{2}}] \qquad (7)$$

$g = 1.5 \times 10^{14}$ kcal$^{-\frac{3}{2}}$ s^{-1} at 25 °C. Δ is the energy difference between two energy surfaces at a point of avoided crossing. The energy surfaces correspond to localization at different redox centres. Any nuclear motion which takes the system across the avoided crossing seam between these energy surfaces gives ET if the system remains on the lower surface. If ν_e of equation (7) is much smaller than the activating vibrational frequency ν_n we may write [6]

$$\nu_n \kappa_e = \nu_e \tag{8}$$

The rate constant [equation (2)] for that case is thus simply

$$k = \nu_e \exp(-\Delta G^{\ddagger}/RT) \tag{9}$$

If ν_e is of the same magnitude as ν_n we have to use [6, 15]:

$$\kappa_e = 2[1 - \exp(-\nu_e/2\nu_n)]/[2 - \exp(-\nu_e/2\nu_n] \tag{10}$$

which tends to unity as ν_e becomes large compared to ν_n. In the latter case ET occurs at every crossing of the seam between the two energy surfaces (adiabatic transfer).

The change which takes place in the wave function in the avoided crossing region usually occurs in a single MO. This orbital thus undergoes a change of character from being localized at the donating part of the system to being localized at the acceptor. An orthogonal, unoccupied orbital at the same time changes from A character to D character. At the point of avoided crossing there is 50% of each character in the two orbitals. In the calculations one has to identify the two interacting orbitals in the printout of the LCAO coefficients. The nuclear coordinates are changed in order to increase the mixing of A and D character in the orbitals. One makes use of bonding-antibonding characteristics of the orbitals or point charges and new calculations are carried out until a 50%–50% situation occurs. Δ is then the energy difference between the two equally mixed orbitals. Of course, the smaller the value of Δ, the faster is the character change in the relevant MO as a function of the nuclear coordinates. If, on the other hand, Δ is large it may not even be possible to localize the electron by changing the nuclear coordinates. In experiments this shows up as an insensitivity of the intervalence transition energies to solvents, as in the well-known case of the Creutz–Taube ion $[(NH_3)_5Ru\text{-}pyrazin\text{-}Ru(NH_3)_5]^{5+}$ [18] or, under different circumstances to high conductivity, with superconductivity as a limiting behaviour.

We have now described a way of calculating the electronic factor for electron transfer which may be applied to the case of proteins. One question to ask is whether one may use MO theory to derive an exponential decrease with distance as suggested in the tunnelling model. This is the case only for certain types of system. If one single molecule connects two metal ions one may derive the following equation for Δ [11]:

$$\Delta = 2\eta_1\eta_2\theta \tag{11}$$

where η_1 and η_2 are interaction matrix elements for the interaction between the metal orbital and its adjacent bridge atomic orbital and

$$\theta = \Sigma C_{1\nu} C_{n\nu}/(a - b_\nu) \tag{12}$$

a is the orbital energy of the electron to be transferred, b_ν the orbital energy of a bridge orbital, $C_{1\nu}$ and $C_{n\nu}$ coefficients of the bridge orbital at the point of contact with the metal ion. θ may be called the ET capability of the bridge. Due to the form of the denominator the energy gap between the HOMO

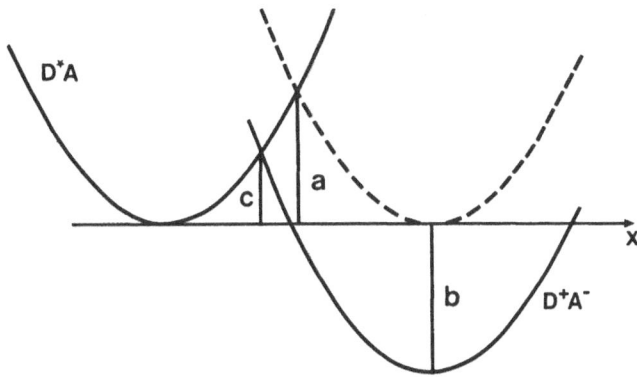

Fig. 3. Standard model (ref. [4–6]) for the interdependence of activation energy $c = \Delta G^{\ddagger}$; exergonicity, $b = \Delta G$ and activation energy for $\Delta G = 0$, $a = \Delta G^{\ddagger}(0) = \lambda/4$. If the same curvature is assumed for the parabolas we obtain $c = a(1 - b/4a)^2$.

and LUMO orbitals of the bridge should be small and the orbitals should be localized near to the contact points to the metal ion, in order to have a large value of Δ. π-systems have good ET properties if placed symmetrically between metal ions. In proteins such an orientation may or may not occur. It is incorrect to ascribe any special role in general for π-systems regarding the ET properties of a protein but it is not unreasonable to find cases where a particular π-system plays a role in the enhancement of $|\Delta|$.

In LDET in proteins, the distance is bridged by several molecular groups which are at best at van der Waals' contact. A generalization to equation (12) is possible where Δ is a product of interaction matrices and ET capability matrices for the bridging molecular groups [19]

$$\Delta = 2\eta_1 \ldots \eta_{N+1}\theta_1 \ldots \theta_N \tag{13}$$

In simple cases the η's and θ's may be represented by numbers. From equation (13) it follows that in a solvent, with many intervening identical solvent molecules in Van der Waals contact, one must see some sort of exponential decrease of Δ with distance [20].

In the final generalization of equation (11) one has to describe the redox centre too with the help of MO's [19, 21]. The electron is initially in an MO distributed over the whole centre. The MO coefficient of this MO for the atomic orbital in contact with the bridge is one of the factors which contributes to Δ. Thus the distance between two electron interchanging porphyrin rings may be counted from the edges provided one multiplies with the relevant MO coefficient in the expression for Δ.

In all discussions of the electronic factor it is important to consider which orbital is singly occupied at the redox centre, since different orbitals may have very different overlap with the bridge orbitals [21]. This may be called the way-in-way-out problem. Expressed in more general terms it is necessary for a large Δ that there is a spin density at the atoms which connect to the bridge [22]. Change of ligands or side-groups may change the energy order of the orbitals and then also their occupancy and the spin density of the redox centre. A remarkable case, demonstrated by Isied *et al.* [23], is under investigation in our group.

4. Aromatic groups connected by a saturated spacer

Although λ_0 shows a dependence of distance [equation (5)], most of the long-range behaviour of the rate k [equation (2)]

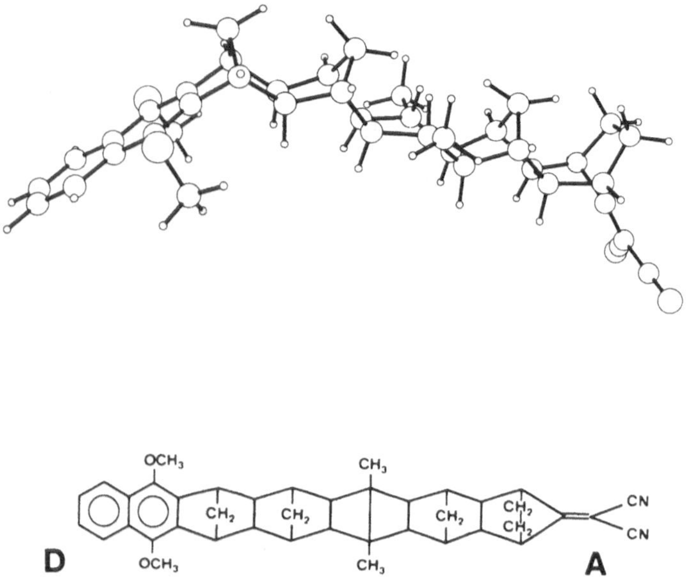

Fig. 4. Model system for photoinduced ET used in ref. [24].

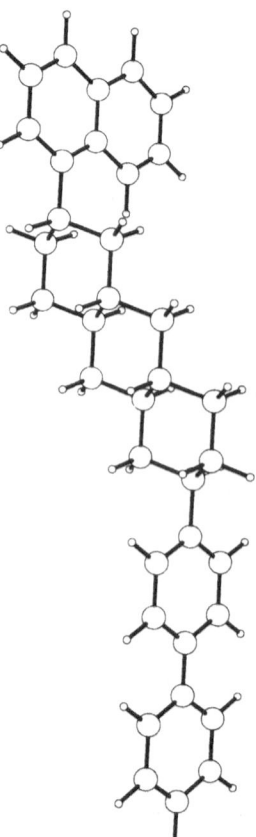

Fig. 5. Model system for ET through a saturated spacer used in ref. [26].

is in the electronic factor κ_e which is proportional to Δ^2 in the weak interaction limit [equations (7) and (8)]. As we have seen, if the bridge consists of weakly interacting molecules, an approximate exponential decrease of Δ with distance should follow. In a bridge where the atomic orbitals interact strongly, factorization as in equation (13) is not possible. In direct calculations an exponential decrease with length is still found. In one such series of bridges, which has been studied experimentally by Hush *et al.* [24], an excited electron transfers from a donor chromophore to an acceptor chromophore (Fig. 4). Calculations of Δ have been performed using a semi-empirical method (CNDO/S) which includes configuration interaction. The latter extension is necessary for a correct description of the excited states of π-systems. The calculated decrease in Δ^2 is a factor of 43 in 4 Å, which is rather close to the experimental result [25]. This corresponds to a factor of $2\alpha = 0.94$ Å^{-1} [cf. equation (1)]. An often-used reference value is a factor of 100 in 4 Å which corresponds to $2\alpha = 1.15$ Å^{-1}.

A similar type of system has been studied experimentally by Closs *et al.* [26] (Fig. 5). They find a decrease of Δ close to a factor of 100 in 4 Å. A very similar value is found in calculations using the Extended Hückel method in the same manner as described above and we may therefore have some trust in this procedure.

If the bridge is flexible but still saturated the complexity is usually too great to make possible any relevant comparisons between theoretical and experimental numbers. Calculations have been carried out with *ab initio* as well as semi-empirical methods on a $NH_2-(CH_2)n-NH_2$ in different conformations [27]. These two methods show a reasonable agreement. What is important in these calculations is that a change of conformation of the –CH$_2$-chain drastically changes the value of Δ. The largest value of Δ is obtained for the *trans* conformation. In the case of $n = 2, 4, 6$ or 8 carbon atoms it turns out to be possible to reduce $|\Delta|$ to zero by turning one or more dihedral angles from 180° (*trans*) to an angle between 60° and 0°. It may be noted that the distance between the nitrogen atoms is the largest in the *trans* form and yet this conformation has the largest value of $|\Delta|$. It should also be noticed that since Δ can have positive as well as negative

values there may be a conformation for which $|\Delta|$ is accidentally small. One may test unexpectedly small values of Δ by performing small variations of the nuclear coordinates of a magnitude which corresponds to variations in the lowest vibrational states.

5. ET between metal ions connected by oligo-peptides

Experimental results have been obtained recently by Gray *et al.* [28, 29], Isied *et al.* [30] and Hoffman *et al.* [31] for ET between metal ions in well-defined positions in proteins. Decrease factors of $\beta = 2\alpha = 0.9$ Å^{-1} were obtained for Δ^2 by Cowan and Gray [29]. Due to the complexity of the systems calculations of the kind mentioned above are difficult and are meaningful only if they can be analyzed in some simple model. Such calculations are, however, under way along with models for interpretation [19].

Here results will be presented for a simpler system consisting of two metal ions connected by a bridge of oligo-glycin [32] with the number of peptide bonds ranging from 2–10 (Fig. 6). The metal ions are placed 1.954 Å from the carbonyl oxygen atoms on the same cylindrical radii of the helics as the oxygen atoms. The Extended Hückel method was used in the procedure described above. A charge at the end of the helix is varied to obtain the 50%–50% mixing mentioned above. The parameters of Summerville and Hoffmann were used [33]. The 'way-out-orbital' was the $3d\sigma$-orbital between the metal ion and the oxygen atom.

The results are shown in Fig. 7 and 8. In Fig. 7 we see that the decrease with distance is approximately exponential, except for point 8. In Fig. 8 $\log|\Delta|$ is plotted as a function of the number of peptide bonds. After the first few points there

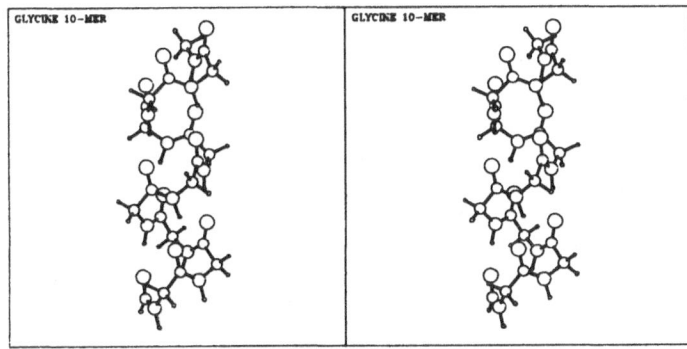

Fig. 6. α-helix of oligo-glycin (ref. [32]).

is an exponential decrease again with the exception of point 8. The latter value may be an accidental zero as described in the previous section. A variation of some arbitrary coordinates of about 0.01 Å was performed twice, however without any significant change of $\log|\Delta|$. Unexpectedly small values of $|\Delta|$ may thus occur without any easily found explanation.

6. Conclusion

A theoretical model [14] has been developed to the extent that ET in proteins can be meaningfully studied. Preliminary results suggest that the electronic factor for an α-helix spacer usually decreases a factor of 10 or slightly less in 2 Å of separation distance. ($|\Delta|$ decreases a factor of 10 in 4 Å). However, there is no guarantee that this will always happen. Occasionally very small values may be found. Conversely one may also conceive systems where occasionally large values are obtained. It is theoretically possible to place an aromatic group in a protein with a large value of the product of its transfer capability θ and the interaction matrix element η to

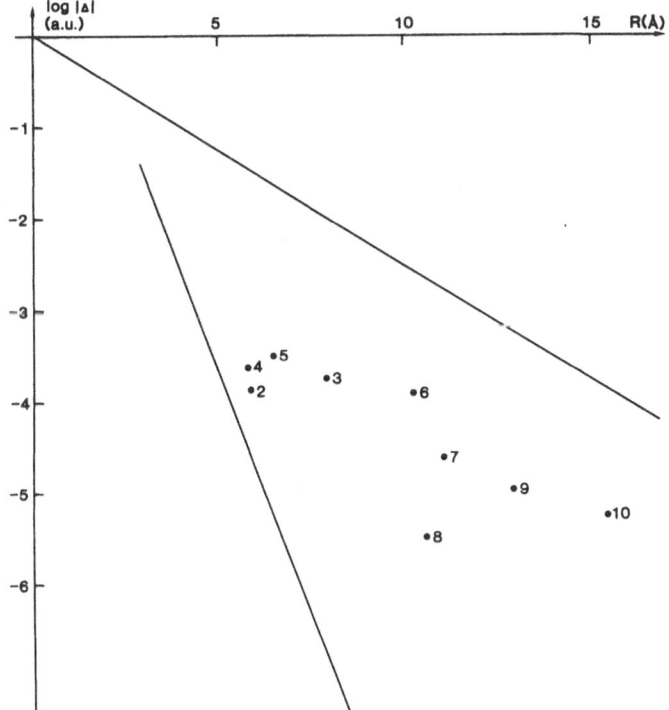

Fig. 7. Calculated values of $\log|\Delta|$ for α-helices as bridges between two metal ions. The upper line corresponds to $\beta-2\alpha = 1.15$ Å^{-1} for decrease of Δ^2 with R. The lower line represents values of $\log|\Delta|$ for direct transfer between the metal ions in the absence of oligo-glycin spacers.

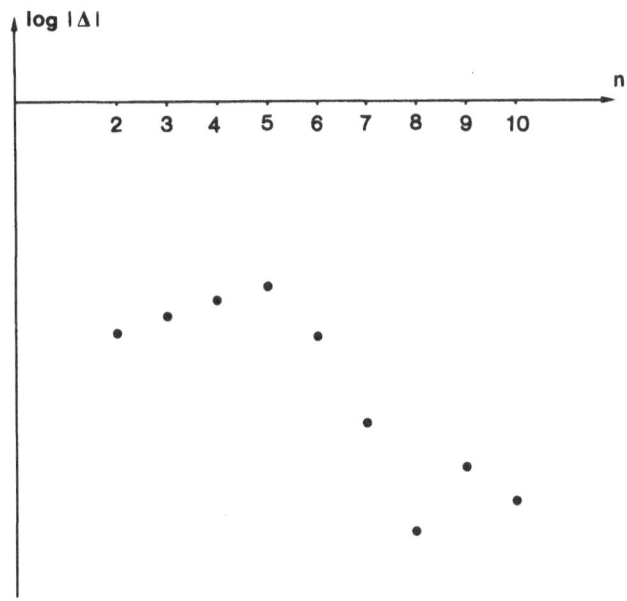

Fig. 8. The same numbers as in Fig. 7 plotted as a function the number of peptide groups in the bridge.

the next group (say equal to unity). If this group covers let us say 8 Å of the distance between the metal ions there will be an enhancement of Δ^2 of 10^4, since normally Δ^2 would decrease that factor over that distance.

Under study in our group are various model systems for ET between metal ions [19], some of the systems studied experimentally in refs. [28–31] and a photosynthetic reaction centre [34]. The ultimate goal of these studies is a detailed comparison to synthetic ET systems, which in turn reveal which factors are important in the fast biological ET reactions and to what extent evolution has been active in selecting these factors.

Acknowledgement

I am grateful for support from NFR, the Swedish Natural Science Research Council, for encouragement in various ways from Professor B. G. Malmström and Professor T. Vänngård and help from Dr A. Volosov and my students A. Broo and B. Källebring.

7. References

1. Szent-Györgyi, A., *Nature* **148**, 157 (1941); *Science* **93**, 609 (1941).
2. Cardew, M. H. and Eley, D. D., *Disc. Faraday Soc.* **27**, 115 (1959).
3. Many interesting papers may be found in *Tunneling in Biological Systems* (Ed. B. Chance, D. C. DeVault, H. Frauenfelder, R. A. Marcus, J. R. Schrieffer & N. Sutin), Academic Press (1979). An interesting review of possible models is given in Moore, G. R. and Williams, R. J. P., *Coord. Chem. Rev.* **18**, 125 (1976).
4. Marcus, R. A., *J. Chem. Phys.* **24**, 966 (1956); **26**, 867, 872 (1957).
5. Hush, N. S., *Trans. Faraday Soc.* **57**, 155, 557 (1961).
6. Sutin, N., *Progress in Inorg. Chem.* **30**, 441 (1983); Marcus, R. A. and Sutin, N., *Biochim. et Biophys. Acta* **811**, 265–322 (1985), and further references therein.
7. Levich, V. G. and Dogonadz, R. R., *Dokl. Acad. Nauk SSSR* **124**, 123 (1959) (*Proc. Acad. Sci. Phys. Chem. Sect.* **124**, 9); Dogonadze, R. R., Kuznetsov, A. M. and Levich, V. G., *Electrochim. Acta* **13**, 1025 (1968); Kestner, N. R., Logan, J. and Jortner, J., *J. Phys. Chem.* **78**, 2148 (1974); Hopfield, J. J., *Proc. Natl. Acad. Sci. USA* **7**, 3640 (1974); Ulstrup, J. and Jortner, J. *J. Chem. Phys.* **63**, 4358 (1975); Kuznetsov, A. M. and Ulstrup, J., *J. Chem. Phys.* **75**, 2047 (1980).
8. Jortner, J., *Biochim. et Biophys. Acta* **594**, 193 (1980).
9. Tabue, H. 173 of ref. [3]; Miller, J. R, *J. Phys. Chem.* **82**, 767 (1978); Miller, J. R., Beitz, J. V. and Huddleston, R. K., *J. Am. Chem. Soc.* **106**, 5057 (1984).

10. Calcaterra, L. T., Closs, G. L. and Miller, J. R., *J. Am. Chem. Soc.* **105**, 670 (1983); Miller, J. R., Calcaterra, L. T. and Closs, G. L., *J. Am. Chem. Soc.* **106**, 3074 (1984).
11. Larsson, S., *J. Am. Chem. Soc.* **103**, 4034 (1981); Beratan, D. N. and Hopfield, J. J., *J. Am. Chem. Soc.* **106**, 1584 (1984). See also Halpern, J. and Orgel, L. E., *Discuss. Faraday Soc.* **29**, 32 (1960).
12. Kuki, A. and Wolynes, P. G., *Science* **236**, 1647 (1987). See also Beratan, D. N., Onuchic, J. N. and Hopfield, J. J., *J. Chem. Phys.* **83**, 5325 (1985).
13. Larsson, S., *J. Chem. Soc., Faraday Trans. 2,* **79**, 1375 (1983); *Int. J. Quant. Chem.* **QBS9**, 385 (1982).
14. Siebrand, W., *J. Chem. Phys.* **41**, 3574 (1964).
15. Brunschwig, B. S., Logan, J., Newton, M. D. and Sutin, N., *J. Am. Chem. Soc.* **102**, 5798 (1980); Brunschwig, B. S., Creutz, C., Macartney, D. H., Sham, T.-K. and Sutin, N., *Faraday Discuss. Chem. Soc.* **74**, 113 (1982).
16. Churg, A. K., Weiss, R. M., Warshel, A. and Takano, T., *J. Phys. Chem.* **87**, 1683 (1983).
17. Gray, H. B. and Malmström, B. G., *Comments Inorg. Chem.* **2**, 203 (1983).
18. Creutz, C., *Progr. in Inorg. Chem.* **30**, 1 (1983).
19. Broo, A. and Larsson, S. (in preparation).
20. Larsson, S., *J. Phys. Chem.* **88**, 1321 (1984).
21. Larsson, S. and Matos, M., *J. Mol. Structure* **120**, 35 (1985).
22. Hoffman, M. Z. and Simic, M., *J. Am. Chem. Soc.* **94**, 1757 (1972); Neta, P., Simic, M. G. and Hoffman, M. Z., *J. Phys. Chem.* **80**, 2018 (1976); Whitburn, K. D., Hoffman, M. Z., Simic, M. G. and Brezniak, N. V., *Inorg. Chem.* **19**, 3180 (1980).
23. Bechtold, R., Kuehn, C., Lepre, C. and Isied, S. S. *Nature* **322**, 286 (1986).
24. Hush, N. S., Paddon-Row, M. N., Cotsaris, E., Oevering, H., Verhoeven, J. W. and Heppener, M., *Chem. Phys. Lett.* **117**, 8 (1985).
25. Larsson, S. and Volosov, A., *J. Chem. Phys.* **85**, 2548 (1986). Erratum: *J. Chem. Phys.* **86**, 5223 (1987); *J. Chem. Phys.* (in the press).
26. Closs, G. L., Calcaterra, L. T., Green, N. J., Penfield, K. W. and Miller, J. R., *J. Phys. Chem.* **90**, 3673 (1986).
27. Larsson, S. (to be submitted for publication).
28. Winkler, J. R., Nocera, D. G., Yocom, K. M., Bordignon, E. and Gray, H. B., *J. Am. Chem. Soc.* **104**, 5798 (1982); Lieber, C. M., Karas, J. L. and Gray, H. B., *J. Am. Chem. Soc.* **109**, 3778 (1987); Mayo, S. L., Ellis, W. R., Jr., Crutchley, R. J. and Gray, H. B., *Science* **233**, 948 (1986); Axup, A. W., Albin, M., Mayo, S. L., Crutchley, R. J. and Gray, H. B., *J. Am. Chem. Soc.* (in the press).
29. Cowan, J. A. and Gray, H. B., this volume.
30. Isied, S. S., Worosila, G. and Atherton, S. J., *J. Am. Chem. Soc.* **104**, 7659 (1982); Becthold, R., Gardiner, M. B., Kazmi, A., van Hemelryck, B. and Isied, S. S. *J. Phys. Chem.* **90**, 3800 (1986).
31. McGourty, J. L., Blough, N. V. and Hoffman, B. M., *J. Am. Chem. Soc.* **105**, 4470 (1982); Liang, N., Kang, C. H., Ho, P. S., Margoliash, E. and Hoffman, B. M., *J. Am. Chem. Soc.* **108**, 4665 (1986); Liang, N., Pielak, G. L., Mauk, A. G., Smith, M. and Hoffman, B. M., *Proc. Natl. Acad. Sci. USA* **83**, 1249 (1986).
32. Fraser, R. D. B., Harrap, B. S., McRae, T. P., Stewart, F. H. C. and Suzuki, E., *J. Mol. Biol.* **14**, 423 (1965).
33. Summerville, R. H. and Hoffman, R. *J. Am. Chem. Soc.* **98**, 7240 (1976).
34. Källebring, B. and Larsson, S., *Chem. Phys. Lett.* **138**, 76 (1987).

Chemica Scripta 1988, **28A**, 21–26

Long-Range Electron Transfer in Metal-Substituted Myoglobins

J. A. Cowan and H. B. Gray

Arthur Amos Noyes Laboratory, California Institute of Technology, Pasadena, California 91125, USA

Paper presented by Harry B. Gray at the Nobel Conference 'Biophysical Chemistry of Dioxygen Reactions in Respiration and Photosynthesis', Fiskebäckskil, Sweden, 1–4 July, 1987

Abstract

The electron-transfer (ET) properties of ruthenated magnesium mesoporphyrin IX myoglobins have been investigated. The closest distances (d) from the porphyrin-edge to the edge of the ruthenated histidines are in the 13–22 Å range, and the driving force for the reaction $^3MgP^* \rightarrow Ru^{3+}$ is similar to that of previously studied $^3ZnP^* \rightarrow M^{3+}$ (M = Fe, Ru) electron transfers in modified proteins and protein complexes. The ET rates for the Ru(His-48)Mb derivatives, reconstituted with Mg meso IX diacid and diester, are $\sim 57\,000$ s^{-1} and 32 000 s^{-1}, respectively, whereas values of ~ 75 s^{-1} and 45 s^{-1} have been determined for the 19–22 Å transfers (His-12, 81, 116). The rate decreases according to $[\exp(-0.9\,d)]$ for both the Mg meso IX diacid and diester Mb's. The activation parameters for ET were determined from the variation of rate with temperature; the values accord with the results previously obtained for the ruthenated derivatives of Mb(ZnP) [$\Delta H^* \sim 6$ kcal mol^{-1} for ET to Ru^{3+} (His-12, 81, 116) and ~ 2 kcal mol^{-1} for ET to Ru^{3+} (His-48)]. A reorganization energy $\lambda \sim 2.0$–2.5 eV was estimated from the Marcus equation relating ΔG° to ΔG^*. Experiments with myoglobins containing a variety of metal-substituted mesoporphyrin IX, protoporphyrin IX, and deuteroporphyrin IX derivatives suggest that the origin of the split Q-bands in the absorption spectra of Mb (Mg or Zn meso IX) derives from the two possible diastereomeric orientations of the porphyrin in the myoglobin pocket.

Introduction

The central role of ET reactions in chemistry and biology has been widely recognized [1]. Although many of the factors governing short-range ET processes in metal complexes have been resolved [2, 3], the field of long-range ET, of particular importance in the control of biochemical pathways, has, until recently, been less well explored.

The intensity of recent theoretical work [3–5] has increased the need for more definitive experimental investigations of long-range ET processes. In this connection, it is encouraging that progress has been made in analyzing the kinetics of ET reactions in both heme and blue-copper proteins [6–10] as well as complex model systems [11], and it is particularly significant that for certain proteins it is now possible to address the effects of driving force, distance, and medium on the rate [8, 12, 13]. An efficient means of both studying these effects and deriving comparisons would be through experiments in which each variable could be studied systematically, using a readily available protein as a reference point. Recently, we have shown that sperm whale myoglobin may constitute such a reference for studies of intramolecular ET [12, 13]. This protein contains four surface-accessible histidine residues that can be readily modified by reaction with aquopenta-ammineruthenium(II) [$a_5RuOH_2^{2+}$, a = NH$_3$]. The resulting $a_5Ru(His)Mb$ derivatives (Table I) contain redox centers at a series of fixed sites, thereby allowing both distance and medium effects to be examined; in addition, the driving force for intramolecular ET can be varied either by substitution of the natural heme with a different metalloporphyrin or through variation of the ruthenium complex attached to the protein surface.

We now report the results of ET studies of $a_5Ru(His)Mb$ derivatives that were reconstituted with magnesium mesoporphyrin IX diacid and dimethyl ester (denoted MgP in each case). The reorganization energy λ has been calculated from the pre-exponential factor in the experimentally derived rate equation, and variable temperature experiments over the range 3–38 °C have allowed the determination of the activation enthalpies and entropies for ET in each of the derivatives.

Synthesis and purification

The basic strategy involved in the synthesis of a_5-Ru(His)Mb(MgP) is summarized in Fig. 1. Native myoglobin, containing four surface-accessible histidine residues, was reacted with $a_5RuOH_2^{2+}$, under conditions that have been described previously [10, 12], to yield a mixture of a native mono-, di-, tri-, and tetraruthenated myoglobins. After an initial separation of the mixture of singly modified proteins by IEF, each of the four singly labeled species was isolated following column chromatography on carbomethoxy cellulose (CM-52, $\mu = 0.1$ M Tris). The characterization of each of these $a_5Ru(His)Mb$ derivatives has been previously described [10].

Following removal of the labile heme from the ruthenated protein by the normal acid/butan-2-one methodology [14], and consecutive dialysis against 10 mM-NaHCO$_3$ and 0.1 M ionic strength phosphate buffer, magnesium mesoporphyrin IX [15, 16] in aqueous phosphate was added to the buffered apoprotein and left stirring for 8 h at 4 °C. A little DMSO was added to aid the solubilization of the porphyrin, more being required for the dimethyl ester derivative. The

Table I. Electron-transfer distances for ruthenated myoglobins[a]

Derivative	d range (Å)	d_m range (Å)
$a_5Ru(His-48)Mb$	11.8–16.6 (12.7)	16.6–23.9 (17.1)
$a_5Ru(His-81)Mb$	18.8–19.3 (19.3)	24.1–26.6 (25.1)
$a_5Ru(His-116)Mb$	19.8–20.4 (20.1)	26.8–27.8 (27.7)
$a_5Ru(His-12)Mb$	21.5–22.3 (22.0)	27.8–30.5 (29.3)

[a] The donor-to-acceptor distances are from a calculation of the conformation of each ruthenated histidine; lower and upper values refer to 6.5 kcal above the potential energy minimum and the value at the minimum is in brackets; d is the edge–edge distance; d_m is the metal–metal distance. See ref. [12].

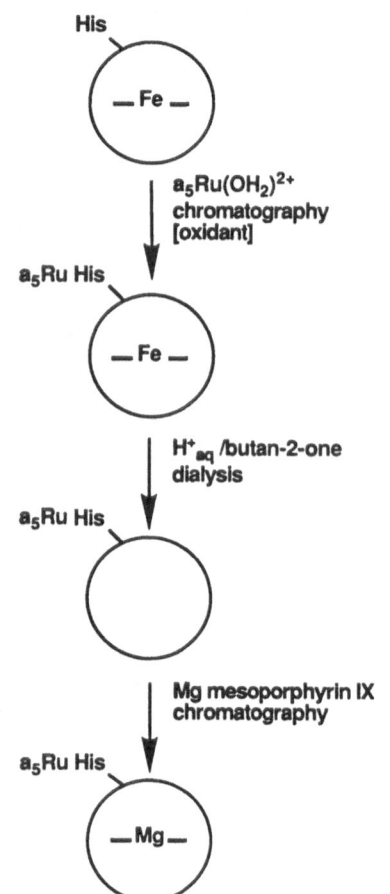

Fig. 1. Synthesis and purification of ruthenated myoglobins (at His-48, 81, 116 and 12), reconstituted with magnesium mesoporphyrin IX.

reconstituted protein was then purified by IEF and column chromatography (CM-52, $\mu = 0.1$ M Tris for the diacid porphyrin derivatives and 0.15 M Tris for the diester).

After insertion into the myoglobin pocket, the electronic absorption spectrum of the reconstituted protein displayed perturbed Q-bands and a red-shifted Soret band, indicative of coordination to the proximal histidine [17]. Further evidence for the incorporation of the magnesium porphyrin into the myoglobin pocket was provided by emission spectroscopy, using the fluorescent probe 8-anilino-1-naphthalene sulfonic acid (ANSA). Insertion into the low-polarity pocket of apomyoglobin led to a six-fold increase in intensity and 35 nm blue-shift of the emission band of ANSA relative to that in aqueous phosphate buffer. Addition of ANSA to Mb(MgP) produced no such change. An emission signal analogous to that for free ANSA in aqueous buffer was observed, thereby indicating that the magnesium porphyrin was indeed in the myoglobin pocket.

From studies performed on other metalloporphyrin systems [18], it is possible to formulate some general rules governing the likelihood of reconstitution of apomyoglobin with foreign porphyrins. A hydrophobic porphyrin can be locked into the myoglobin pocket by coordination of the metal center to the proximal histidine or through an electrostatic interaction between the carboxylate side-chains and two positively charged residues (His-97 and Arg-45) on the protein surface. If the metalloporphyrin can meet one or the other of these conditions, i.e. if it possesses a coordinatively unsaturated metal ion or carboxylate side-chains, it will be readily incorporated into the Mb pocket. If the porphyrin

contains a coordinatively saturated metal ion, *and* has ester side-chains, it will not.

Background theory

Many theories of ET have appeared in the literature [3–5, 19, 20]; the one that we employ is due to Marcus [19]. The Marcus equation for the intramolecular ET rate k_{ET} can be written as follows:

$$k_{ET} = \nu\Gamma \exp\left(-\frac{\Delta G^*}{RT}\right) \tag{1}$$

where ΔG^* represents the energy required to reach the transition state T (Fig. 2) from the ground state of the reactant potential well, Γ describes the probability of ET at the transition point T, and ν is the nuclear frequency (10^{12}–10^{13} s^{-1}) at the point where T occurs on the reactant potential well.

In non-adiabatic ET, where electronic coupling between the redox centers is weak, either through orientation or distance effects, Γ is less than unity. The distance dependence is then given by

$$\Gamma = \exp\left[-\beta(d-3)\right] \tag{2}$$

where β measures how rapidly the ET rate falls off with distance, and is therefore related to the intersite medium. The intersite distance $(d\text{-}3)$ Å includes a correction of 3 Å for van der Waals contact between two interacting centers.

Marcus has shown [20] that the activation free energy ΔG^* can be related to the driving force ΔG° by

$$\Delta G^* = \frac{\lambda}{4}\left(1 + \frac{\Delta G^\circ}{\lambda}\right)^2 \tag{3}$$

Hence, substituting (2) and (3) into (1) gives

$$k_{ET} = \nu \exp\left[-\beta(d-3)\right]\exp\left[-\frac{\lambda}{4RT}\left(1 + \frac{\Delta G^\circ}{\lambda}\right)^2\right] \tag{4}$$

Equation (4) splits naturally into two parts, one describing the electronic contribution to the ET rate [the effects of distance (d) and medium (β)], and the other being the nuclear term [relating to changes in nuclear configuration around the

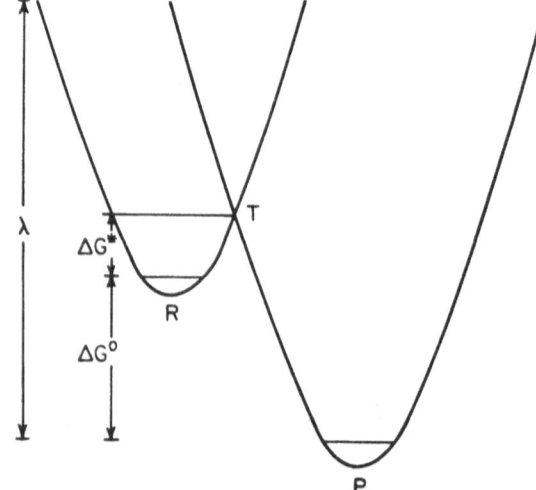

Fig. 2. Potential energy surfaces for reactants and products. The important energy parameters (λ, ΔG^*, and ΔG°), and the vibrational energy levels in each ground state and at the transition state T are shown.

redox sites (λ) and the relative changes in the ground-state energy levels of the reactant and product potential energy wells ($\Delta G°$)]. Each part, therefore, contains one of the two experimentally derivable parameters of interest namely, β and λ, describing the influence of medium and nuclear reorganization, respectively.

By writing the rate expression, $k_{ET} = \nu \exp[-\beta(d-3)] \exp(-\Delta G^*/RT)$, as a logarithmic function,

$$\ln k_{ET} = \ln \nu - \beta(d-3) - \frac{\Delta G^*}{RT} \qquad (5)$$

it is clear that there are two systematic experiments that should be particularly revealing. Plotting $\ln k_{ET}$ *vs. d* gives a line with a slope $-\beta$ and a 3 Å intercept $\ln \nu - \Delta G^*/RT$, whereas plotting $\ln k_{ET}$ *vs.* $1/T$ gives a line with slope $-\Delta H^*/R$. Knowing ΔG^* and ΔH^*, ΔS^* can be calculated. From equation (3) the reorganization energy λ can be determined by substituting appropriate values for ΔG^* and $\Delta G°$.

Results and discussion

Electron transfer

The rate for the intramolecular ET reaction, $a_5Ru^{3+}(His)Mb(^3MgP^*) \rightarrow a_5Ru^{2+}(His)Mb(^3MgP^+)$, was obtained for each of the four $a_5Ru^{3+}(His)Mb(MgP)$ derivatives from the quenching of the long-lived triplet state of the magnesium porphyrin, as depicted in Fig. 3:

$$k_{ET} = k_{obs} - k_d \qquad (6)$$

where k_{ET} is the intramolecular electron-transfer rate, k_{obs} is the observed decay rate, and k_d is the decay rate for native $Mb(^3MgP^*)$.

Although the decay profile for native $Mb(^3MgP^*)$ followed strict first-order behavior, biphasic plots were obtained for the modified $a_5Ru(His)Mb(^3MgP^*)$ derivatives [21]. The decay curve for the His-48 derivatives (Fig. 4) could, however, be fit to a first-order exponential with non-zero endpoint, since the main component was three orders of magnitude faster than the minor component. The longer-range derivatives were fitted to a biexponential function with zero endpoint (Fig. 5), and the faster components, which we ascribe to intramolecular electron transfer, are given in Table II. In the case of the long-range derivatives, the second component was generally within experimental error of the native decay rate and contributed less than 15% toward the observed kinetics. The origin of this minor component has

not yet been resolved. Under the conditions of the laser experiment, no significant degradation of the protein solution was observed; the absorption spectrum was the same before and after the experiment.

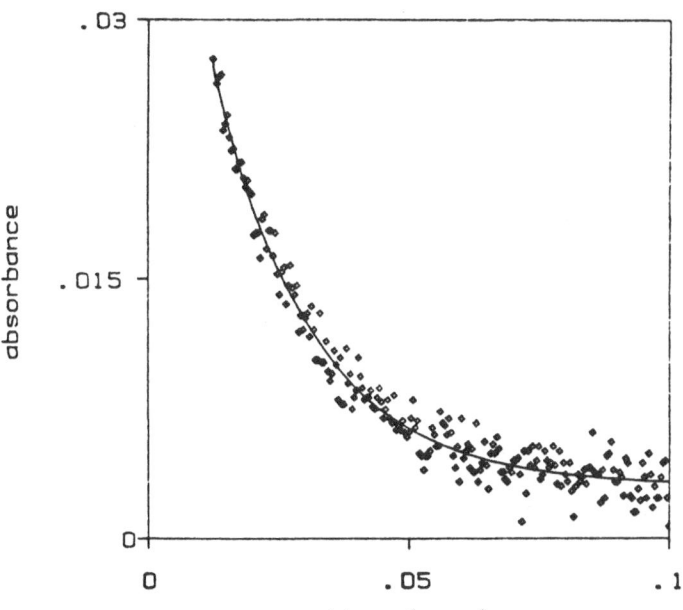

Fig. 4. Analysis of transient-absorption data obtained for $a_5Ru^{3+}(His-48)Mb(^3MgP^*)$ at 23 °C. The decay at 409 nm was fit by a monophasic curve with non-zero endpoint ($k = 57000$ s^{-1}). Procedures have been described previously [12].

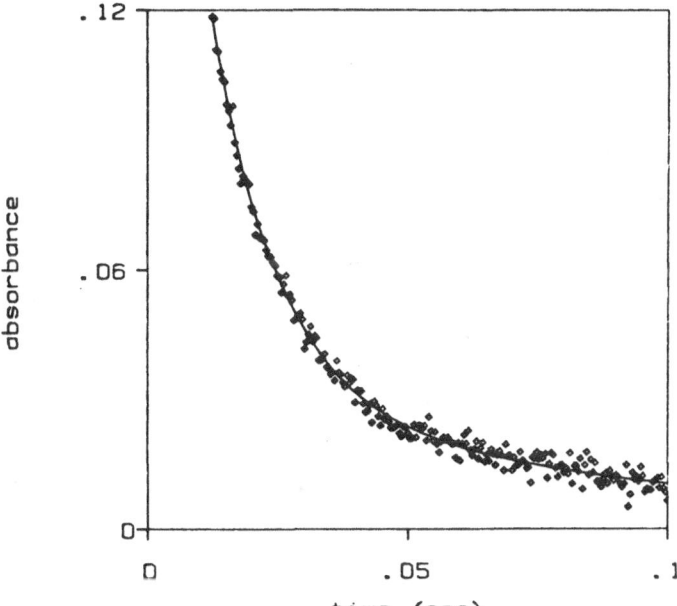

Fig. 5. Analysis of the transient-absorption data obtained for $a_5Ru^{3+}(His-116)Mb(^3MgP^*)$ at 23 °C. The decay at 409 nm was fit by a biphasic curve with zero endpoint ($k_1 = 69$ s^{-1}, $k_2 = 16$ s^{-1}). Procedures have been described previously [12].

Table II. *Intramolecular electron-transfer rates^a for $a_5Ru^{3+}(His)Mb(^3MgP^*)$*

Acceptor	d (Å)	k(s^{-1}) (P = diacid)	k(s^{-1}) (P = diester)
a_5Ru(His-48)	12.7	$57 \pm 4 \times 10^3$	$32 \pm 5 \times 10^3$
a_5Ru(His-81)	19.3	82 ± 6	48 ± 9
a_5Ru(His-116)	20.1	69 ± 10	49 ± 8
a_5Ru(His-12)	22.0	67 ± 4	39 ± 7

^{a} Measured at ambient temperature (23 °C).

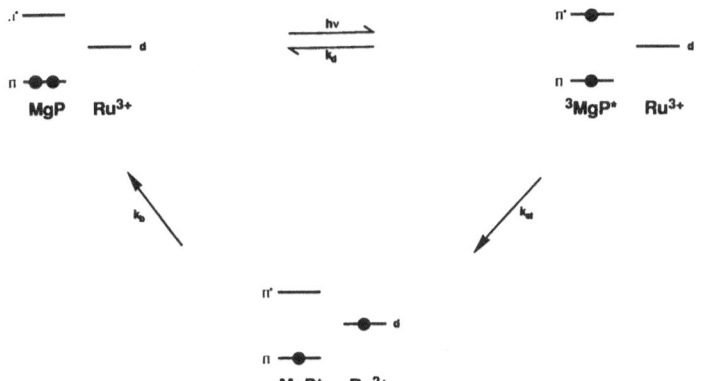

Fig. 3. Orbital diagrams illustrating the pathway for excited-state ET ($^3MgP^* \rightarrow Ru^{3+}$] in ruthenated myoglobin.

The natural decay rate for native Mb(^{3}MgP*)) was found to be 18 ± 3 s^{-1} for both the diacid and diester derivatives. This is in reasonable agreement with the value determined by Hoffman *et al.* [22] (24 s^{-1}) in their studies of energy transfer in magnesium and zinc modified hemoglobin and myoglobin. The long native lifetime allowed determination of the intramolecular electron-transfer rates for both the 13 Å and 19–22 Å derivatives (Table II). The decay rate of the a$_5$Ru(His-116)$^{2+}$Mb(^{3}MgP*) derivatives was also measured, the rate of 25 s^{-1} being in close agreement with k$_d$ for native Mb(^{3}MgP*). In addition, the measured rates in Table II were found to be essentially unchanged over a concentration range 15–100 μM, thereby confirming that the reaction in question is intramolecular.

A plot of ln k_{ET} *vs.* distance is shown in Fig. 6. A β value of *c.* 0.9 Å^{-1} was obtained and an activation free energy $\Delta G^* = 0.23 \pm 0.04$ eV (5.4 ± 0.9 kcal mol^{-1}) was calculated from the intercept at 3 Å. Variable temperature experiments with the diacid derivatives of a$_5$Ru(His)Mb(MgP) allowed the determination of ΔH^*, and so also, using the previously found value for ΔG^*, the activation entropy ΔS^*. The results are given in Table III and again strongly support the trends observed previously in the series of a$_5$Ru(His)Mb(ZnP) derivatives [12].

From the similarity in ground-state redox potentials and triplet energy levels for zinc and magnesium porphyrins [22, 23] the driving force for ET from ^{3}MgP* to a$_5$Ru^{3+}(His) is estimated to be the same as that for a$_5$Ru^{3+}(His)Mb(^{3}ZnP*) [*c.* 0.8 V] [12]. In going from the porphyrin diacid to diester, the driving potential will drop, due to the loss of electrostatic stabilization of the porphyrin cation radical by the carboxylate side-chains. This difference in driving force is clearly reflected in the intramolecular ET rates at all distances studied.

In the plot of ln k_{ET} *vs.* distance in Fig. 6, an average was taken for the three longer distance points (His-12, 81 and 116). In keeping with the previous discussion of ET in a$_5$Ru(His)Mb(ZnP) [12], the rate for the Ru(His-12) derivative may be enhanced by the intervening medium. A computer graphics analysis has indicated the presence of an aromatic residue (Trp-14) between the donor and acceptor sites [12]. Since there is an element of variability in the intersite distance due to the fluxional motion of the residues on the protein surface, we prefer to average over the long-range results at this stage, rather than treating His-12 as a special case, keeping in mind that our value for β may have to be altered in the future should we find additional evidence for specific rate enhancement by aromatic groups. The value determined for β was found to be the same for both the MgP diacid and diester derivatives. The rate-distance equation,

$$k_{ET} = 2.3 \times 10^8 \exp\left[-0.9\,(d-3)\right] s^{-1} \qquad (7)$$

was then derived and a ΔG^* of 0.20–0.27 eV (4.5–6.3 kcal mol^{-1}) was calculated from the intercept at 3 Å, where ν was allowed to vary between 10^{12} and 10^{13}.

By using the estimated $\Delta G°$ and the experimentally derived ΔG^* in equation (3), a reorganization energy $\lambda = 2.0$–2.5 eV was calculated for both the diacid and diester systems, where the range of possible values is a reflection of the uncertainty in ν and ΔG^*. This λ range accords closely with the reorganization energy determined by driving force variations in Ru(His-48)Mb derivatives [13].

The activation enthalpies in Table III fall into two distinct groups, with $\Delta H^* \sim 2.0$ kcal mol^{-1} for the His-48 derivative and ~ 6.0 kcal mol^{-1} for the more distant derivatives. Again, these results are in agreement with the trend found for a$_5$Ru(His)Mb(ZnP) [12]. Any implied distance dependence of ΔH^*, however, may be misleading. The His-48 side-chain is known to occur at the conformationally mobile CD helical corner of the protein backbone [24]. Recent theoretical studies have indicated that conformational fluctuations may play an important role in modulating the activation parameters. This has been most explicitly developed in the work by Beratan *et al.* [25]. These authors make a clear distinction between the approach to the transition state via a through-bond or through-space pathway. The former will depend on small changes in bond lengths and force constants, whereas the latter, which depends on the through-space overlap of orbitals, will show a variation dependent on the relative positioning of the protein backbone, i.e. it will depend on the dynamic motion of the protein at particular points in space.

Absorption spectroscopy

The absorption spectrum of Mb(MgP) is illustrated and compared with that of magnesium mesoporphyrin in Fig. 7. The porphyrin Q-bands are perturbed following insertion into the myoglobin pocket, with a splitting clearly observable in the band at *c.* 580 nm. This effect has also been noted in Mb(ZnP) [12]. Experiments with non-coordinating freebase,

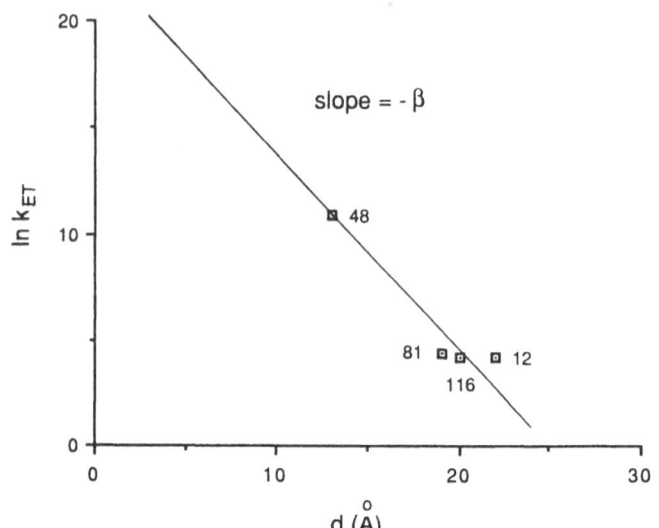

Fig. 6. Distance dependence of the ET rate (ln k_{ET}) in a$_5$Ru^{3+}(His)Mb (^{3}MgP*). P = mesoporphyrin IX diacid.

Table III. *Activation parameters for intramolecular electron transfer in* a$_5$Ru^{3+}(His)Mb(^{3}MgP*) *(P = mesoporphyrin IX diacid)*

Acceptor	ΔH^* (kcal mol^{-1})	ΔS^* (e.u.)
a$_5$Ru(His-48)	2.2 ± 0.3	-11 ± 7
a$_5$Ru(His-81)	5.8 ± 0.5	1 ± 6
a$_5$Ru(His-16)	5.2 ± 0.5	3 ± 7
a$_5$Ru(His-12)	5.2 ± 0.5	-1 ± 6

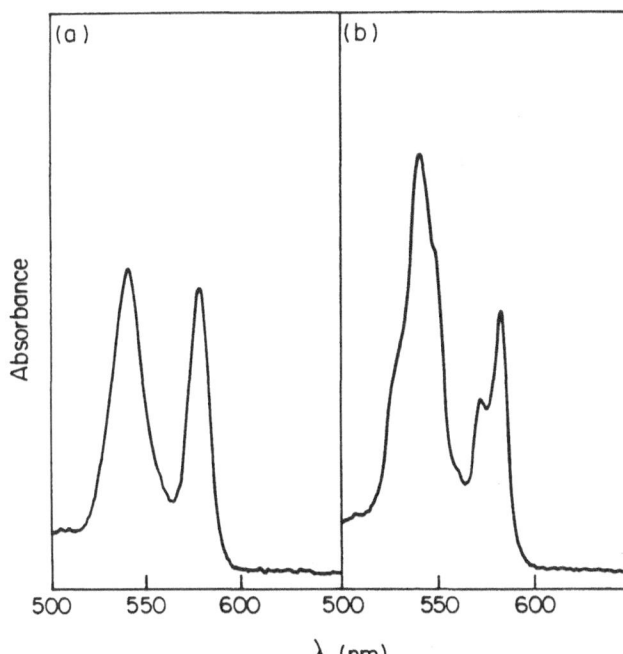

Fig. 7. Absorption spectra of (*a*) magnesium mesoporphyrin IX diester (in dichloromethane) and (*b*) Mb (magnesium mesoporphyrin IX) (in 0.1 M ionic strength phosphate buffer). A similar splitting has been observed in Mb (zinc mesoporphyrin IX) [12]. The absorbance scale for each spectrum is similar but not identical.

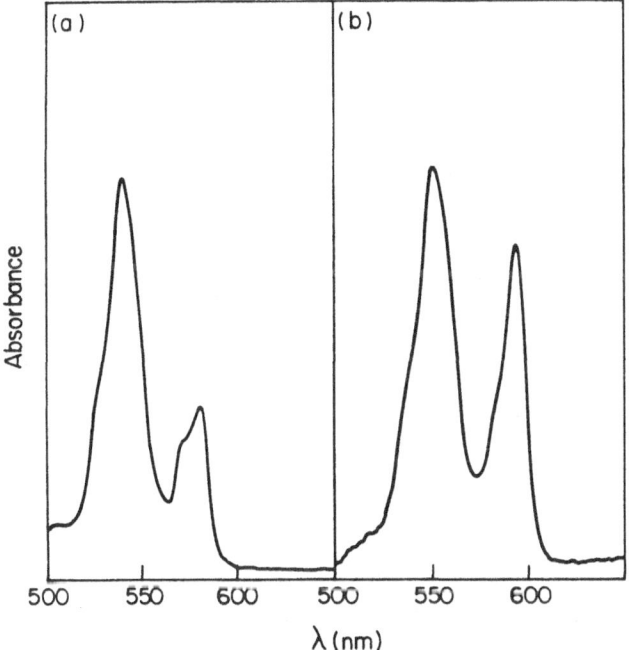

Fig. 8. Absorption spectra of (*a*) Mb (magnesium deuteroporphyrin IX) and (*b*) Mb (zinc protoporphyrin IX) (in 0.1 M ionic strength phosphate buffer). The absorbance scale for each spectrum is similar but not identical.

silver, copper, palladium, and platinum porphyrins did not show such effects, and therefore suggest coordination as the primary reason for the observed perturbation in the absorption spectrum. It is well known that the Q-bands are sensitive to the coordination state of the metal center [17], although it is important to restrict comparisons to cases in which the absorption spectra are dominated by porphyrin transitions.

Two reasonable explanations can be proposed to explain this behavior. First, the coordinating proximal histidine is known to adopt a fixed orientation relative to the porphyrin unit [26]. It is entirely possible that this might destroy the degeneracy of the Q_x and Q_y transitions in the porphyrin absorption spectrum (17), generating the split Q-band. Second, from NMR studies of metmyoglobin, the heme moiety is known to adopt two distinct diastereomeric orientations in the Mb pocket [27]; i.e. the proximal histidine can adopt two coordination geometries relative to the metal center, thereby providing a reasonable explanation for the perturbed Q-bands.

Detailed NMR studies by La Mar *et al.* have shown that with mesoporphyrin IX the diastereomeric forms exist in the ratio 5.5:1, whereas the ratio is *c.* 12:1 and 11:1 for deuteroporphyrin and protoporphyrin, respectively [27]. Reconstitution of myoglobin with either the magnesium or zinc derivatives of these porphyrins would then provide a means of distinguishing between these two models. From the latter model it would be predicted that one of the components of the split Q-band at *c.* 580 nm would become less intense, whereas no dramatic change would be expected were the former explanation to be correct. Following reconstitution of apomyoglobin with magnesium or zinc deuteroporphyrins or zinc protoporphyrin, the perturbations of the Q-bands were found to become much less prominent (Fig. 8), hence the latter explanation is preferred.

Related work

Cadmium porphyrins belong in the same general class of metalloporphyrins as the magnesium and zinc derivatives. The similarity in driving force for electron transfer from the triplet state is reflected in the rate ($k_{ET} = 63\,000$ s^{-1}) measured for a_5Ru^{3+}(His-48)Mb(^{3}CdP*). The faster native decay (*c.* 450 s^{-1}) prevented accurate measurement of the rates for the more distant residues. Variable temperature experiments gave $\Delta H^* = 2.8$ kcal mol^{-1}, which is in line with the corresponding results for a_5Ru(His-48)Mb(MgP) and a_5Ru(His-48)Mb(ZnP).

Zinc has assumed an important role in the study of modified cytochrome *c* [28]; the heme unit is covalently bound to the protein and zinc can be readily inserted into the freebase porphyrin while it resides in the protein pocket. We have found that cadmium is also very readily incorporated into freebase porphyrins in the pocket of myoglobin, and therefore represents an alternative to zinc for incorporation into the bound heme in cytochrome *c*, following removal of the iron center. As an NMR active nucleus [29], cadmium has the additional attraction of potentially allowing the active site to be probed directly. We are currently investigating this possibility.

Acknowledgement

We thank J. K. M. Sanders, A. W. Axup, M. Albin, and S. L. Mayo for helpful discussions. J. A. C. thanks SERC (United Kingdom) for a NATO postdoctoral fellowship. Our research on the electron-transfer reactions of myoglobin is supported by National Science Foundation Grants CHE85-18793 and CHE85-09637. This is contribution No. 7630 from the Arthur Amos Noyes Laboratory.

References

1. Hatefi, Y., *Annu. Rev. Biochem.* **54**, 1015 (1985); Dixit, B. P. S. N. and Vanderkooi, J. M., *Curr. Top. Bioenerg.* **13**, 159 (1984). *Antennas and Reaction Centers of Photosynthetic Bacteria* (ed. M. E. Michel-Beyerle), Springer-Verlag, Berlin (1985); Williams, G., Moore, G. R. and Williams, R. J. P., *Comments Inorg. Chem.* **4**, 55 (1985); Williams, R.-J. P. and Concar, D., *Nature* **322**, 213 (1986).

2. Pennington, D. E., in *Coordination Chemistry*, vol. 2 (ed. A. E. Martell), pp. 476–590, American Chemical Society (1978); Taube, H., *Angew. Chem. Int. Ed.* **23**, 329 (1984); Balzani, V., Bolletta, F., Gandolfi, M. T. and Maestri, M., *Topics in Current Chemistry* **75**, 1 (1968).

3. Marcus, R. A. and Sutin, N., *Biochim. Biophys. Acta* **811**, 265 (1985).

4. DeVault, D., *Quantum Mechanical Tunneling in Biological Systems*, 2nd ed. Cambridge University Press, Cambridge (1984).

5. Churg, A. K., Weiss, R. M., Warshel, A. and Takano, T., *J. Phys. Chem.* **87**, 1683 (1983); Onuchic, J. N., Beratan, D. N. and Hopfield, J. J., *J. Phys. Chem.* **90**, 3707 (1986); Hopfield, J. J., *Proc. Natl. Acad. Sci. USA* **71**, 3640 (1974); Bixon, M. and Jortner, J., *J. Phys. Chem.* **90**, 3795 (1986); Jortner, J., *Biochim. Biophys. Acta* **594**, 193 (1980); Buhks, E. and Jortner, J., *FEBS Lett.* **109**, 117 (1980); Jortner, J., *J. Chem. Phys.* **64**, 4860 (1976); Hush, N. S., *Coord. Chem. Rev.* **64**, 135 (1985); Larsson, S., *J. Chem. Soc., Faraday Trans. II* **79**, 1375 (1983); Kuki, A. and Wolynes, P. G., *Science* **236**, 1647 (1987).

6. Brunschwig, B. S., DeLaive, P. J., English, A. M., Goldberg, M., Gray, H. B., Mayo, S. L. and Sutin, N., *Inorg. Chem.* **24**, 3743 (1985); Gray, H. B., *Chem. Soc. Rev.* **15**, 17 (1986); Mayo, S. L., Ellis, W. R., Crutchley, R. J. and Gray, H. B., *Science* **233**, 948 (1986); Yocom, K. M., Shelton, J. B., Shelton, J. R., Shroeder, W. A., Worosila, G., Isied, S. S., Bordignon, E. and Gray, H. B., *Proc. Natl. Acad. Sci. USA* **79**, 7052 (1982); Winkler, J. R., Nocera, D. G., Yocom, K. M., Bordignon, E. and Gray, H. B., *J. Am. Chem. Soc.* **104**, 5798 (1982); Yocom, K. M., Winkler, J. R., Nocera, D. G., Bordignon, E. and Gray, H. B., *Chem. Scr.* **21**, 29 (1983); Kostic, N. M., Margalit, R., Che, C.-M and Gray, H. B., *J. Am. Chem. Soc.* **105**, 7765 (1983); Margalit, R., Kostic, N. M., Che, C.-M., Blair, D. F., Chiang, H.-J., Pecht, I., Shelton, J. B., Schroeder, W. A. and Gray, H. B., *Proc. Natl. Acad. Sci. USA* **81**, 6554 (1984); Nocera, D. G., Winkler, J. R., Yocom, K. M., Bordignon, E. and Gray, H. B., *J. Am. Chem. Soc.* **106**, 5145 (1984); Scott, R. A., Mauk, A. G. and Gray, H. B., *J. Chem. Educ.* **62**, 932 (1985); Crutchley, R. J., Ellis, W. R. and Gray, H. B., *J. Am. Chem. Soc.* **107**, 5002 (1985); Lieber, C. M., Karas, J. L. and Gray, H. B., *J. Am. Chem. Soc.* **109**, 3778 (1987); McLennon, G., Winkler, J. R., Nocera, D. G., Mauk, M. R., Mauk, A. G. and Gray, H. B., *J. Am. Chem. Soc.* **107**, 739 (1985).

7. Sykes, A. G., *Chem. Soc. Rev.* **14**, 283 (1985); Isied, S. S., Worosila, G. and Atherton, S. J., *J. Am. Chem. Soc.* **104**, 7659 (1982); Isied, S. S., Kuehn, C. and Worosila, G., *J. Am. Chem. Soc.* **106**, 1722 (1984); Bechtold, R., Gardineer, M. B., Kazmi, A., van Hemelryck, B. and Isied, S. S., *J. Phys. Chem.* **90**, 3800 (1986); Bechtold, R., Kuehn, C., Lepre, C., Isied, S. S., *Nature* **322**, 286 (1986); McGourty, J. L., Blough, N. V., Hoffman, B. M., *J. Am. Chem. Soc.* **105**, 4470 (1983); Ho, P. S., Sutoris, C., Liang, N., Margoliash, E. and Hoffman, B. M., *J. Am. Chem. Soc.* **107**, 1070 (1985); Peterson-Kennedy, S. E. McGourty, J. L., Kalweit, J. A. and Hoffman, B. M., *J. Am. Chem. Soc.* **108**, 1739 (1986).

8. Liang, N., Pielak, G. J., Mauk, A. G., Smith, M. and Hoffman, B. M., *Proc. Natl. Acad. Sci. USA* **83**, 1249 (1986); McLendon, G., Miller, J. R., *J. Am. Chem. Soc.* **107**, 7811 (1985).

9. Simolo, K. P., McLendon, G., Mauk, M. R. and Mauk, A. G., *J. Am. Chem. Soc.* **106**, 5012 (1984); McLendon, G., Guarr, T., McGuire, M., Simolo, K., Strauch, S. and Taylor, K., *Coord. Chem. Rev.* **64**, 113 (1985); Guarr, T., McLendon, G., *Coord. Chem. Rev.* **68**, 1 (1985); Cheung, E., Taylor, K., Kornblatt, J. A., English, A. M., McLendon, G. L. and Miller, J. R., *Proc. Natl. Acad. Sci. USA*: **83**, 1330 (1986); Conklin, K. T., McLendon, G., *Inorg. Chem.* **25**, 4804 (1986).

10. Crutchley, R. J., Ellis, W. R., Gray, H. B., in *Frontiers in Bioinorganic Chemistry* (ed. A. V. Xavier), VCH Verlagsgesellschaft, Weinheim (1986); Crutchley, R. J., Ellis, W. R., Shelton, J. B., Shelton, J. R., Schroeder, W. A. and Gray, H. B. (in preparation).

11. Oevering, H., Paddon-Row, M. N., Heppener, M., Oliver, A. M., Cotsaris, E., Verhoeven, J. W. and Hush, N. S., *J. Am. Chem. Soc.* **109**, 3258 (1987); Gust, D., Moore, T. A., Liddell, P. A., Nemeth, G. A., Makings, L, R., Moore, A. L., Barrett, D., Pessiki, P. J., Bensasson, R. V., Rougée, M., Chachaty, C., DeSchryver, F. C., Van der Auweraer, M., Holzwarth, A. R. and Connolly, J. S., *J. Am. Chem. Soc.* **109**, 846 (1987); Heiler, D., McLendon, G. and Rogalskyj, P., *J. Am. Chem. Soc.* **109**, 604 (1987); Leland, B. A., Joran, A. D., Felker, P. M., Hopfield, J. J., Zewail, A. H. and Dervan, P. B., *J. Phys. Chem.* **89**, 5571 (1985); Closs, G. L., Calcaterra, L. T., Green, N. J., Penfield, J. W. and Miller, J. R., *J. Phys. Chem.* **90**, 3673 (1986); Wasielewski, M. R., Niemczyk, M. P., *J. Am. Chem. Soc.* **106**, 5043 (1984); Fox, L. S., Marshall, J. L., Gray, H. B. and Winkler, J. R., *J. Am. Chem. Soc.* **109**, 6901 (1987).

12. Axup, A. W., Albin, M. A., Mayo, S. L., Crutchley, R. J. and Gray, H. B., *J. Am. Chem. Soc.* (submitted for publication); Axup, A. W., Ph.D. Thesis, California Institute of Technology (1987).

13. Karas, J. L., Lieber, C. M. and Gray, H. B., *J. Am. Chem. Soc.* (submitted for publication); Lieber, C. M., Karas, J. L., Mayo, S. L., Axup, A. W., Albin, M., Crutchley, R. J., Ellis, W. R. and Gray, H. B., in *Trace Elements in Man and Animals* (ed. B. Lonnerdal), Plenum, New York (1987).

14. Teale, F. W. J., *Biochim. Biophys. Acta* **35**, 543 (1959); Yonetani, T., *J. Biol. Chem.* **242**, 5008 (1967).

15. Magnesium mesoporphyrin IX diester was synthesized by standard procedures [16]. The diacid was obtained by base hydrolysis in THF/H_2O.

16. Buchler, J. W., in *The Porphyrins*, vol. 1 (ed. D. Dolphin), pp. 425–427. Academic Press, New York (1979).

17. Gouterman, M., in *The Porphyrins*, vol. 3 (ed. D. Dolphin), pp. 12–17. Academic Press, New York (1979).

18. These include zinc, platinum, and silver mesoporphyrin IX diacids and diesters.

19. Marcus, R. A., *J. Chem. Phys.* **24**, 966 (1956); Marcus, R. A., *J. Chem. Phys.* **26**, 867 and 872 (1957); Marcus, R. A., *Can. J. Chem.* **37**, 155 (1959); Marcus, R. A., *Discuss. Faraday Soc.* **29**, 21 (1960); Cave, R. J., Siders, P., Marcus, R. A., *J. Phys. Chem.* **90**, 1436 (1986).

20. Marcus, R. A., *Faraday Discuss. Chem. Soc.* **74**, 7 (1982).

21. The same observation has been made with a_5Ru^{3+}(His)Mb(^{3}ZnP*) derivatives [12].

22. Zemel, H. and Hoffman, B. M., *J. Am. Chem. Soc.* **103**, 1192 (1981).

23. Peterson-Kennedy, S. E., McGourty, J. L. and Hoffman, B. M., *J. Am. Chem. Soc.* **106**, 5010 (1984).

24. Ringe, D. and Petsko, G. A., *Prog. Biophys. Mol. Biol.* **45**, 197 (1985).

25. Beratan, D. N., Onuchic, J. N. and Hopfield, J. J., *J. Chem. Phys.* **83**, 5325 (1985); Beratan, D. N., Onuchic, J. N. and Hopfield, J. J., *J. Chem. Phys.* **86**, 4488 (1987).

26. Bertini, I., Luchinat, C., in *NMR of Paramagnetic Molecules in Biological Systems*, vol. 3 (ed. A. B. P. Lever and H. B. Gray), pp. 182. Benjamin/Cummings, Menlo Park (1986).

27. La Mar, G. N., Emerson, S. D., Lecomte, J. T. J., Pande, U., Smith, K. M., Craig, G. W. and Kehres, L. A., *J. Am. Chem. Soc.* **108**, 5568 (1986).

28. Elias, H., Chou, M. H. and Winkler, J. R., *J. Am. Chem. Soc.* (submitted for publication).

29. Armitage, I. M. and Boulanger, Y., in *NMR of Newly Accessible Nuclei*, vol. 2 (ed. P. Laszlo), p. 337. Academic Press, New York (1983).

Chemica Scripta 1988, **28A**, 27–31

Magnetic Interaction in Ribonucleotide Reductase

Anders Ehrenberg

Department of Biophysics, University of Stockholm, Arrhenius Laboratory, S-106 91 Stockholm, Sweden

Paper presented at the Nobel Conference 'Biophysical Chemistry of Dioxygen Reactions in Respiration and Photosynthesis', Fiskebäckskil, Sweden, 1–4 July, 1987

Abstract

Protein B2, one of the subunits of RNR (ribonucleotide reductase) from *E. coli*, has been shown to contain two centers with paramagnetic properties. One of the centers is an antiferromagnetically coupled pair of high-spin ferric ions ($J = -108$ cm^{-1}) connected by a μ-oxo-bridge and is similar to the iron center of hemerythrin. The other center is a remarkably stable tyrosyl radical. The radical and the iron center may be formed from radical-free apoprotein in a reaction with ferrous ions in the presence of dioxygen. Several RNRs, e.g. the mammalian enzyme, have been shown to contain similar centers. The features of the two centers are reviewed.

Magnetic interaction between the tyrosyl free radical and the iron center is indicated by the enhanced relaxation of the radical at temperatures above *c.* 30 K. Compared with the RNR from *E. coli*, the enhancement is more pronounced in the mammalian RNR, suggesting a stronger magnetic interaction in this case.

Apparently the two centers of each protein molecule are situated close enough so that magnetic interaction can take place between them. This interaction has both analogies and dissimilarities to magnetic interactions found in electron-transport systems in respiratory or photosynthetic systems.

A simulation of the EPR spectrum of the radical of the mammalian RNR is presented. A radical in the oxygen evolving complex of chloroplasts has been shown also to be a tryosyl radical (Barry, B. A. and Babcock, G. T., *Chemica Scripta* (in the press)). The spectrum of the latter may be obtained by simply reducing the major coupling of the RNR simulation. This change in coupling may be caused by a proton dissociation or rotation around the C_β-ring bond. The chloroplast radical EPR spectrum is not broadened by magnetic interaction even at room temperature.

Introduction

The biological systems for reduction of dioxygen to water or oxidation of water to dioxygen must be capable to store and transfer four electrons in close succession. For these purposes the active centers or aggregates contain more than one metal ion which in certain steps of the reaction cycle may become paramagnetic. Also semiquinones or other organic free radicals may be involved. Magnetic interactions and couplings have been demonstrated or hypothesized between those centers.

Several other redox-enzymes contain in each active center two or more units which in the resting enzyme are, or during the reaction may become, paramagnetic. In some of these enzymes, magnetic interactions have been demonstrated between the paramagnetic units. One such enzyme is the ribonucleotide reductase (RNR) found in *Escherichia coli*. This enzyme, in its active resting form, has been found to contain an unusually stable protein free radical in a tyrosine residue [1] and a binuclear iron center with two antiferromagnetically coupled ferric ions [2]. We have recently shown that there is a magnetic interaction between the free radical and the iron center of this enzyme [3]. Since there are analogies between this system and the redox centers brought

into focus at this conference it seems justified here to discuss also the properties of the paramagnetic centers of this class of RNRs and how the interaction between the centers is revealed. In addition, the radical and the iron center are formed in a reaction involving molecular oxygen.

General characteristics of the enzyme

RNR from *E. coli* is the most thoroughly studied representative of this class of enzymes. It catalyses the reduction of ribonucleoside diphosphates into deoxyribonucleoside diphosphates. The enzyme consists of two protein subunits, B1 (MW = 160000) and B2 (MW = 78000), both of which are peptide homodimers. The two proteins bind to each other, and enzyme activity is only obtained when both of them are present. Protein B1 carries the two binding sites for the substrates and two reducing centers, each with a dithiol, which in their oxidized disulphide form are substrates of thioredoxin or glutaredoxin [4]. In addition, protein B1 has effector sites for a very elaborate regulation of the activity level and specificity [4]. Protein B2 carries the iron center and the tyrosyl radical and is the subject of the present discussion.

Apo-protein B2, devoid of iron and free radical, may be prepared by treatment of the active protein with a strong iron-ion chelator [1]. From inactive apo-protein active holo-protein may be regained in a reaction with ferrous iron and dioxygen [2] (see Scheme 1).

$$\text{Apoprotein B2} + 2\text{Fe}^{2+} + \text{O}_2$$
$$\Downarrow$$
Active protein B2
(oxidized tyrosine + 2Fe^{3+})

Scheme 1

In this reaction both the iron center and the free radical are reformed. The details of this interesting reaction have still to be established.

The iron center

The iron center of active protein B2 does not give rise to an EPR signal [1]. Mössbauer spectra from protein B2 reconstituted with ^{57}Fe revealed two distinct and equally intense quadrupole doublets demonstrating that the two iron ions are non-equivalent [5]. The Mössbauer data [5], the paramagnetic susceptibility [2], and the light absorption [2] in combination showed that the iron center is very similar to those of oxyhemerythrin or methemerythrin and must be an antiferromagnetically coupled pair of ferric ions [2]. The necessary μ-oxo bridge has been nicely demonstrated by

Raman [6, 7] and EXAFS [8] results. While NMR observations on paramagnetically shifted proton resonances from protein B2 [9] suggest that histidines are coordinated to the iron center, the EXAFS results [8] indicate that the number of coordinated histidines is smaller in protein B2 than in the hemerythrin center.

As already mentioned, the Mössbauer results show that the two ions of the iron center are not identical [5]. An analysis of the binding requirements needed to create this situation leads to the conclusion that the two identical peptides of protein B2 must donate different sets of ligands to the two iron ions.

The tyrosyl radical

The free radical of protein B2 was discovered during a search for an eventual EPR response from the iron [1]. The radical shows a characteristic EPR signal with $g = 2.0047$ [1] (cf. Fig. 1), and a characteristic light absorption at 410 nm [1, 2]. By growing bacteria overproducing the enzyme in completely deuterated medium except for specific amino acids in their normal non-deuterated forms and measuring the EPR spectra of the bacteria themselves it was demonstrated that the radical site is a tyrosine residue [10]. By measurements on

bacteria grown on normal medium but with added tyrosine with specific deuterium substitutions the proton couplings involved could then be identified and measured and the spin-density distribution estimated [11]. This identified the radical as a tyrosyl radical, i.e. an electron has been removed from the tyrosine ring. Comparison with Hückel calculations of the spin density distribution [12] suggested that the phenolic group is still protonated in the radical of protein B2. However, this hypothesis needs confirmation since Hückel calculations are not all that accurate.

Hydroxylamine or hydroxyurea abolishes the radical by reduction [1]. At the same time the enzyme activity is lost. The radical and the activity may be regained after removal of the iron and reconstitution according to Scheme 1. In *E coli* there are also enzyme systems for removal and reintroduction of the radical [13, 14]. Apparently this enzymatic reintroduction of the radical includes reduction of the iron center and reoxidation in the presence of oxygen [15].

By taking difference spectra between radical-containing and radical-free protein B2 the light absorption of the radical has been determined [2]. This spectrum is practically identical to those determined for phenoxy radicals [2]. However, in protein B2 the long wavelength absorption at 600 nm is comparatively weak with $A_{600}/A_{410} \approx 0.05$ as compared with the model phenoxy radical with $A_{610}/A_{400} = 0.27$. Land *et al.* reported [16] that both the neutral, dissociated, and the cationic, protonated, forms of the phenoxy radical show a sharp absorption at about 400 nm but that the long wavelength band is much reduced in the protonated form. This supports the view that the tyrosyl radical of protein B2 is present in the protonated form or else is interacting with a positively charged group.

By comparison of the amino-acid sequences of several enzymes of this class the very tyrosine that is the radical site in protein B2 has been identified [17]. This position, Tyr 122 has also recently been nicely confirmed by site directed genetic change replacing this tyrosine by phenylalanine [18]. The amount of tyrosyl radical in protein B2 has been determined from the EPR signal [1, 2]. In some experiments slightly more than one radical per dimeric iron center has been observed [2]. However, consideration of the difficulties in achieving high accuracy in such measurements and the persistent clustering of data around the value of one radical per iron center, has convinced us that the best working hypothesis at present is that there is just one radical per dimeric iron center of fully active protein B2. This is in perfect agreement with recent results by Sjöberg *et al.* [19] using heterodimeric protein B2 containing one normal peptide and one defect peptide that still can form the radical. In the reconstituted heterodimer, the radical is formed with the same frequency in both peptides. Only the molecules with the radical in the normal peptide have enzymatic activity and only their radicals may be abolished by reaction with azido-CDP (see below). The hypothesis of one radical per fully active protein B2 furthermore motivates the stoichiometry of Scheme 1 which, however, does not account for the fourth oxidizing equivalent of dioxygen.

Several RNRs coded for in mammalian cells, by bacteriophages and by viruses are of the same class as the *E. coli* enzymes. They contain iron and carry a tyrosyl radical necessary for the activity. The EPR spectra are similar but not identical for these enzymes (see Fig. 1). The differences

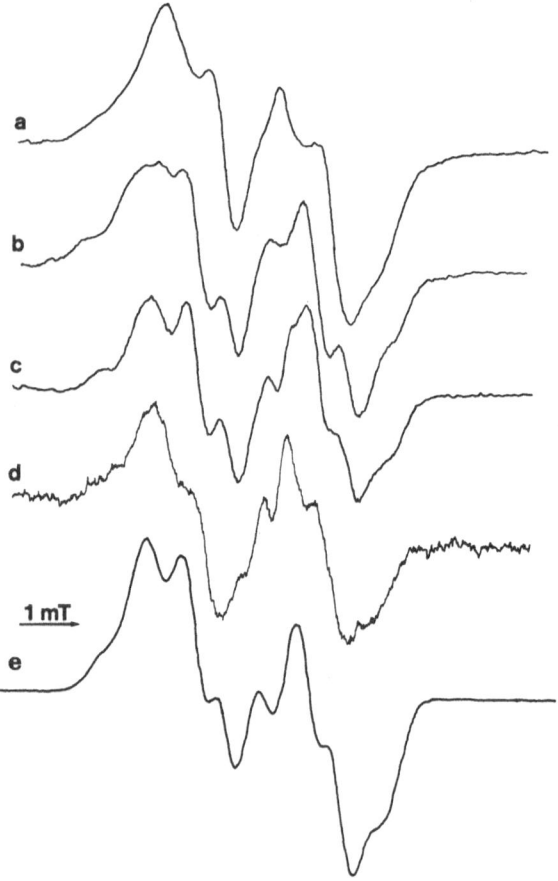

Fig. 1. EPR spectra of ribonucleotide reductases (RNR) of various origins recorded in frozen solutions at 9.5 GHz. (*a*) *E. coli* cells overproducing RNR, at 32 K. (*b*) Purified RNR from bacteriophage T4-infected *E. coli*, at 77 K [20]. (*c*) Hydroxyurea-resistant 3T6-HU11 cells overproducing active M2, the mammalian counterpart of *E. coli* B2, at 32 K [22,23]. (*d*) Pseudorabies virus-infected mouse L-cells, at 32 K [24]. (*e*) Computer simulation with parameters adjusted to fit spectrum c of mammalian M2. Parameters used: hyperfine tensors with units in mT: $a_{H\beta1} = (1.9, 2.1, 1.9)$ and $a_{H3} = a_{H5} = (0.35, 1.05, 0.70)$; g-tensor $g = (2.0080, 2.0030, 2.0030)$; and Gaussian linewidth 0.6 mT [21].

may be explained by small rotations around the bond from the tyrosine β-carbon to the ring [20].

In Fig. 1, a simulation of the EPR spectrum of the enzyme in hydroxyurea treated mouse cells is also included [21]. Such simulations will be helpful for confirmation and refinement of hyperfine coupling schemes for the radicals. In this connection the possibility to record EPR spectra at several microwave frequencies from 0.8 to 35 GHz (ongoing work in collaboration with J. S. Hyde, Milwaukee, USA) and to measure detailed ENDOR spectra (ongoing work in collaboration with G. Babcock, East Lansing, USA) will be of great importance since the number of parameters is large.

Enzymatic mechanism

In what way the iron center and/or the tyrosyl radical participate in or support the enzymatic reaction is a crucial question. A strong indication for a direct involvement of the radical comes from experiments by Sjöberg *et al.* using a suicidal substrate analog, where the 2'-hydroxyl group of the ribonucleotide diphosphate is replaced by an azido group [25]. When this inhibitor is added to the holoenzyme, the tyrosyl radical is eliminated and a new but unstable radical appears [25]. This transient radical was shown not to be located in the protein but must be in the inhibitor molecule, most likely in the sugar residue. This radical location was later confirmed by Ator *et al.* [26] using ^{15}N-substitution in the azido group of the inhibitor.

Based on these EPR observations and on results from experiments with specifically tritiated substrates [27, 28] a mechanism for the enzymatic reaction has been proposed [25, 27, 28]. The first steps involve oxidation of the substrate 3'-site by the tyrosyl radical and proton dissociation from the same site, followed by elimination of a hydroxyl ion from the substrate 2'-site. Thus a substrate cation radical extending over both sites is formed, which is finally reduced to the product by the B1 thiols and the tyrosine, so that the tyrosyl radical is reformed. Even if the proposed mechanism is supported by quite strong indirect evidence some more direct experimental verification would be desirable.

Several experimental attempts have been carried out in this laboratory to try to detect any change in the EPR signal during the enzymatic reaction. However, so far these efforts have been without success. The amount of change at any instance may be too small to be detected by the present techniques.

Spin-trapping experiments have been tried in attempts to reveal the presence and nature of the hypothetical substrate radical(s), but again without success [29]. The substrate radical(s) may be sterically protected from reaction with the spin-trapping agent.

The proposed radical transfer between tyrosine and substrate requires proximity between the two groups. Such a proximity between a paramagnetic group on a protein and a ligand group might be detectable by NMR, provided that the ligand has an exchange rate that is rapid enough. Some preliminary experiments with ^{1}H-NMR were carried out some years ago without any conclusive evidence. More decisive experiments are possible today and will be tried.

As long as the mode of involvement of the radical and the iron center in the enzymatic reaction is not definitely settled we must keep our minds open for other possibilities. The

transfer of the radical to the site of the 2'-azido group could be due to a capacity of the azido group itself to react with the tyrosyl radical and reduce it. However, the results of experiments with half active heterodimers [19] (see above) show that reaction with the 2'-azido group does not happen when the site is enzymatically incompetent. The importance of the iron center and the radical in the holoenzyme could be on the structural level, i.e. they could be required to produce the exact milieu at the active site, of which milieu they themselves could be a part, so that the B1 thiols and the substrate are activated.

The stoichiometry of one tyrosyl radical per one iron center and two substrate binding sites in the holoenzyme offers a particular problem. Proteins B1 and B2 could continuously dissociate and rebind so that both sites of the symmetric B1 molecule are activated.

Interaction between the two magnetic centers

Already the simultaneous formation of the iron center and the radical in the reaction according to Scheme 1 suggests that the two centers are situated in the neighborhood of each other in protein B2. Hence some interaction between the two magnetic centers could be expected. However, a careful analysis of the EPR lineshape of the tyrosyl radical in protein B2 reconstituted with [57]Fe did not reveal any hyperfine interaction between the iron nuclear spin and the electronic spin [1]. Conversely, no difference in the Mössbauer spectra was detected between active radical containing protein B2 and the same sample made radical free by treatment with hydroxyurea [5]. These two sets of results suggest that there is no direct orbital overlap between the tyrosyl radical and the iron center.

When recording EPR spectra of the tyrosyl radical of the *E. coli* enzyme at 77 K or below, it is necessary to use very low microwave power in order to avoid saturation of the signal [1]. When the corresponding radical in the mammalian enzyme was discovered it was noted that much higher microwave powers could be used [22, 23]. This suggested to us that there exists a magnetic interaction between the iron center and the radical which enhances the relaxation of the radical spin and that this enhancement is different in the two enzymes. In later experiments with progressive microwave saturation and linewidth analysis at several temperatures this assumption has been amply verified [3]. As shown in Fig. 2 the relaxation rate products $(T_1 T_2)^{-1}$ of the enzyme radicals increase much faster with temperature than for model radicals in frozen solution. If the iron centers are determining the relaxation rates of the enzyme radicals the observed behaviour is what would be expected considering the temperature dependence of the paramagnetic moment of the iron center [2, 3]. Obviously the interaction between iron center and radical is stronger in the mammalian than in the *E. coli* enzyme. This is also obvious from the differences in line broadening with increasing temperature observed for the two enzyme radicals (see Fig. 3) [3].

The stronger relaxation effects in the mammalian enzyme could be due to several causes [3]: A weaker antiferromagnetic coupling in the iron center leading to a stronger paramagnetism; a change in the relaxation of the iron center itself making the enhancement more effective; a smaller distance between iron center and radical; or a combination of these

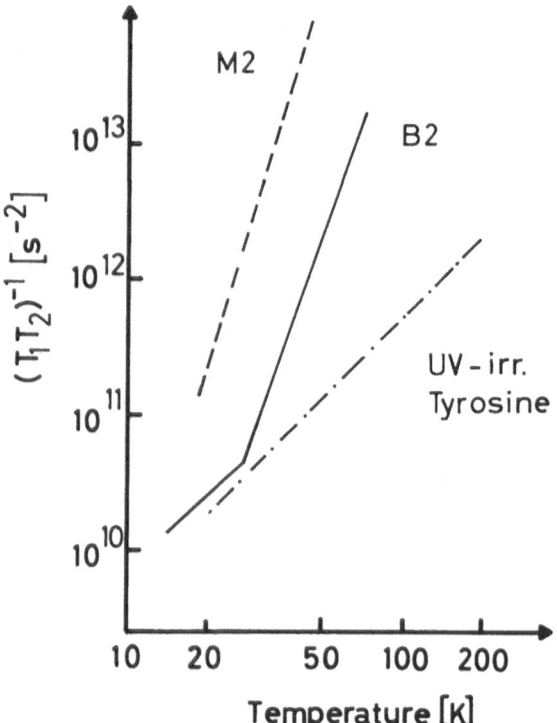

Fig. 2. Diagram showing the variation with temperature of the relaxation rate product $(T_1 T_2)^{-1}$ of the EPR at 9.5 GHz of the tyrosyl radical of purified mammalian protein M2 (–––), of purified protein B2 from *E. coli* (——), and in a model system obtained by UV-irradiation of tyrosine (–·–·–). The slopes of the lines in the log/log scale are 6.7, 5.7 and 2.0, respectively. Redrawn from [3].

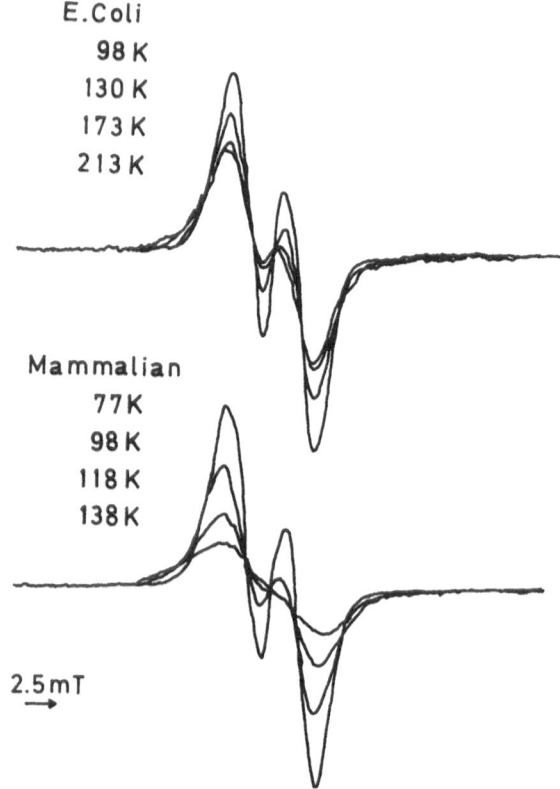

Fig. 3. EPR spectra of protein B2 from *E. coli* and protein M2 from mammalian cells at various temperatures. Spectra are taken at 9.5 GHz under non-saturating conditions and are normalized to the same double integral after correction for the Curie dependence [30].

effects. In the discussion so far it has been implicitly assumed that the interaction is dipolar in character. Possibly also a difference in type of interaction should be considered. The degree of exchange interaction could be negligible in the *E. coli* enzyme but more pronounced in the mammalian case [3].

It is hoped that further studies of the relaxation effects of the various enzymes will lead to a better understanding of the interaction between the centers and their functions. Obviously the radical is an EPR spectroscopic probe that is capable of giving information about the iron center which itself is EPR silent. Since no direct EPR information, such as relaxation properties, is available for the iron center the present data are not sufficient to permit calculation of a figure for the distance between the iron dimer and the radical.

Concluding comments

Both cytochrome oxidase and the oxygen-evolving center in photosynthesis very likely contain centers with antiferromagnetic couplings involved. They also contain other paramagnetic centers, and relaxation effects of the same type as in protein B2 are observed.

After reading the abstract to this conference by Bridgette and Babcock, who demonstrated that the component responsible for the slow radical signal appearing in the oxygen-evolving system of photosynthesis also is a tyrosyl radical, we have tried to simulate [21] the EPR spectrum of the slow radical spectrum [31, 32] (see Fig. 4). Simply by starting from the simulation of the EPR spectrum of the mammalian enzyme (shown in Fig. 1), and reducing the major doublet coupling, due to one of the β-hydrogens, by a factor of two, a spectrum very similar to that of the slow radical was obtained (Fig. 4). One possible explanation of this reduction of the hyperfine coupling could be dissociation of the phenolic group so that a neutral radical is formed, which according to Hückel calculations [12] should have a smaller spin density at the position causing this coupling. A second, and perhaps more likely mechanism, as also suggested by G. Babcock at this conference, is a mere rotation of *c.* 45°

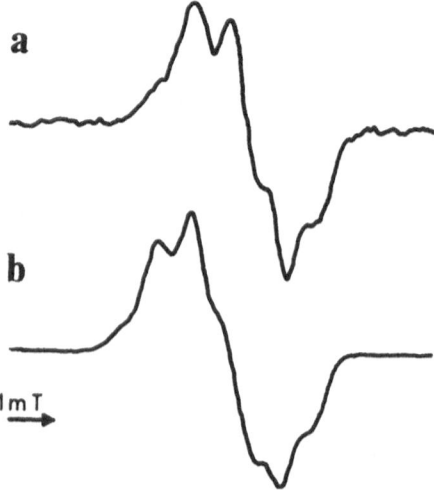

Fig. 4. (a) EPR spectrum at room temperature of the slow radical spectrum in the oxygen-evolving complex of chloroplasts, retraced from [31] (cf. also [32]); and (b) simulated spectrum, obtained by multiplying the hyperfine tensor $a_{H\beta 1}$ of spectrum (e) of Fig. 1 by 0.5 and keeping all other parameters unaltered.

around the bond axis from the β-carbon to the ring. It should be noted that the same linewidth is used in the two simulations, but while the RNR spectrum is taken at 32 K the chloroplast spectrum is recorded at room temperature. At room temperature all RNR spectra are broadened severely, because of the magnetic interaction with the iron center (cf. Fig. 3). It may be concluded that no similar line-broadening effect is operating in the case of the slow radical signal of the oxygen-evolving complex, as prepared by Hansson [31] or O'Malley *et al.* [32].

Acknowledgements

I wish to thank Astrid Gräslund for many important and helpful discussions. Thanks also to her and Torbjörn Astlind for collaboration in making the simulations. Haidi Astlind did the typing in a most professional way. This work was supported by grants from the Swedish Natural Science Foundation.

References

1. Ehrenberg, A. and Reichard, P., *J. Biol. Chem.* **247**, 3485 (1972).
2. Petersson, L., Gräslund, A., Ehrenberg, A., Sjöberg, B.-M. and Reichard, P., *J. Biol. Chem.* **255**, 6706 (1980).
3. Sahlin, M., Petersson, L., Gräslund, A., Ehrenberg, A., Sjöberg, B.-M. and Thelander, L., *Biochemistry* (in the press).
4. Thelander, L. and Reichard, P., *Annu. Rev. Biochem.* **48**, 133 (1979).
5. Atkin, C. L., Thelander, L., Reichard, P. and Lang, G., *J. Biol. Chem.* **248**, 7464 (1973).
6. Sjöberg, B.-M., Gräslund, A., Sanders Loehr, J. and Loehr, T. M., *Biochem. Biophys. Res. Commun.* **94**, 793 (1980).
7. Sjöberg, B.-M., Loehr, T. and Sanders Loehr, J., *Biochemistry* **21**, 96 (1982).
8. Bunker, G., Petersson, L., Sjöberg, B.-M., Sahlin, M., Chance, M., Chance, B. and Ehrenberg, A., *Biochemistry* **26**, 4708 (1987).
9. Sahlin, M., Ehrenberg, A., Gräslund, A. and Sjöberg, B.-M., *J. Biol. Chem.* **261**, 2778 (1986).
10. Sjöberg, B.-M., Reichard, P., Gräslund, A. and Ehrenberg, A., *J. Biol. Chem.* **252**, 536 (1977).
11. Sjöberg, B.-M., Reichard, P., Gräslund, A. and Ehrenberg, A., *J. Biol. Chem.* **253**, 6863 (1978).
12. Box, H. C., Budzinski, E. E. and Freund, H. G., *J. Chem Phys.* **61**, 2222 (1974).
13. Barlow, T., Eliasson, R., Platz, A., Reichard, P. and Sjöberg, B.-M., *Proc. Natl. Acad. Sci. USA* **80**, 1492 (1983).
14. Eliasson, R., Jörnvall, H. and Reichard, P., *Proc. Natl. Acad. Sci. USA* **83**, 2373 (1986).
15. Fontecave, M., Eliasson, R. and Reichard, P., *J. Biol. Chem.* (in the press).
16. Land, E. J., Porter, G. and Strachan, E. *Trans. Faraday Soc.* **57**, 1885 (1961).
17. Sjöberg, B.-M., Eklund, H., Fuchs, J. A., Carlson, J., Standart, N. M., Ruderman, J. V., Bray, S. J. and Hunt, T., *FEBS Lett.* **183**, 99 (1985).
18. Larson, Å and Sjöberg, B.-M., *EMBO J.* **5**, 2037 (1986).
19. Sjöberg, B.-M., Karlsson, M. and Jörnvall, H., *J. Biol. Chem.* **262**, 9736 (1987).
20. Sahlin, M., Gräslund, A., Ehrenberg, A. and Sjöberg, B.-M., *J. Biol. Chem.* **257**, 366 (1982).
21. Gräslund, A., Astlind, T. and Ehrenberg, A. (unpublished work).
22. Åkerblom, L., Ehrenberg, A., Gräslund, A., Lankinen, H., Reichard, P. and Thelander, L., *Proc. Natl. Acad. Sci. USA* **78**, 2159 (1981).
23. Gräslund, A. Ehrenberg, A. and Thelander, L., *J. Biol. Chem.* **257**, 5711 (1982).
24. Lankinen, H., Gräslund, A. and Thelander, L., *J. Virol.* **41**, 893 (1982).
25. Sjöberg, B.-M., Gräslund, A. and Eckstein, F., *J. Biol. Chem.* **258**, 8060 (1983).
26. Ator, M., Salowe, S. P., Stubbe, J., Emptage, M. H. and Robins, M. J., *J. Am. Chem. Soc.* **106**, 1886 (1984).
27. Stubbe, J. A. and Ackles, D., *J. Biol. Chem.* **255**, 8027 (1980).
28. Stubbe, J. A., Ator, M. and Krenitsky, T., *J. Biol. Chem.* **255**, 1625 (1983).
29. Gräslund, A., Sahlin, M. and Sjöberg, B.-M., *EHP, Environ, Health Perspect.* **64**, 139 (1985).
30. Sahlin, M., Petersson, L., Gräslund, A., Ehrenberg, A., Sjöberg, B.-M. and Thelander, L. (unpublished work).
31. Hansson, Ö., The role of manganese in photosynthetic production of oxygen from water, Ph.D. thesis, Chalmers Institute of Technology, Göteborg, Sweden, p. 12 (1986).
32. O'Malley, P. J., Babcock, G. T. and Prince, R. C., *Biochim. Biophys. Acta* **766**, 283 (1984).

Respiration

Chemica Scripta 1988, 28A, 35–40

From Indophenol Oxidase and Atmungsferment to Proton Pumping Cytochrome Oxidase aa_3 $Cu_A Cu_B (Cu_C?)ZnMg$

Helmut Beinert

Department of Biochemistry and National Biomedical ESR Center, Medical College of Wisconsin, Milwaukee, WI 53226, USA

Paper presented at the Nobel Conference 'Biophysical Chemistry of Dioxygen Reactions in Respiration and Photosynthesis', Fiskebäckskil, Sweden, 1–4 July, 1987

Abstract

Historical aspects of the development of research on and knowledge of cytochrome c oxidase are considered. Passages, relevant to the conference are quoted from pages written by Otto Warburg, Philip George, Juda H. Quastel and Max Rudolf Lemberg. Milestones in the history of research on cytochrome oxidase up until 1985, of experimental and conceptual nature, are then summarized in table form with references and are briefly discussed in the text.

Introduction

The roughly 60-year-old history of cytochrome oxidase provides a fairly faithful mirror of the respective state of development of techniques and concepts in biochemistry and adjacent fields of science, modulated by prevalent beliefs and attitudes of scientists as a group and, on occasion, as strong-minded individuals. Once hotly contested issues, after having consumed many man hours of debate and laboratory work, have quietly waned away, as they turned out to be non-issues with our advancing understanding of Nature's principles. An example of what I mean here, is the debate about the question as to whether a and a_3 are 'Siamese twins' or separate proteins, i.e. whether both hemes occur on the same protein or on separate proteins. With our understanding of the subunit structures of proteins and the observation that proteins with multiple prosthetic groups are quite often found in Nature, this issue could no longer excite anyone. In some cases, the stream of thought has swung around by 180 degrees. So I was almost stoned at one time when daring to suggest that CO might also bind to copper, not only to heme. Today this is accepted as interesting news. Or I might recall that I once had to wage a major battle for the two copper ions in cytochrome oxidase while I am now finding myself in the position wondering whether to accept the third copper ion which has been proposed to be part of the enzyme [1, 2].

Quotes from the historical record

At the occasion of preparing my mind for this lecture, I went back over the earlier and very early original literature and I would like to share with you a few remarks by some of the protagonists in the battles around cytochrome oxidase and of some critical observers of those days. It is probably fair to give Otto Warburg the first words. Early in his Nobel address

in 1931 [3] he said this (my translation from German): 'Of the two processes which are part of the utilization of oxygen, namely the oxidation of the iron in the enzyme and its reduction, the first one offers no problems.' Why did we then covene here today? It has been my own experience that 30–40 years ago almost nobody thought that there was such a problem: Iron goes from ferrous to ferric. The link of the reductive pathway to the oxidase received almost all of the attention. It was probably the merit of Philip George to have drawn attention to the non-triviality of the oxygen reaction in his incisive talk at the first meeting on oxidases and related redox systems in 1964 [4]. When considering the reaction of the enzyme with oxygen, he asked

What is the chemical nature of the oxidation states of the catalysts that participate at each stage of the reduction process if the four equivalents are not transferred to oxygen at once?...if one- or two-equivalent steps are involved..., then the thermodynamic data for the principal reduction pathways also have their place, and the question of oxidation states immediately becomes important.

Emphasizing the unique fitness of oxygen as principal biological oxidant he goes on to say:

Nevertheless, in looking over the electron transport chain, it is obvious that the trend toward a greater simplicity of reaction type is abruptly broken at the terminus. Two-equivalent organic substrates and coenzymes are followed by the one-equivalent cytochromes, but oxygen apparently is unable to respond equivalent by equivalent. It is almost as if the major part of the chain had evolved for a different terminal oxidant capable of reacting in this simple fashion, and when oxygen became available, the oxidase evolved as a special but very necessary appendage.

But let us return to Warburg; toward the end of his address [3], he alluded to his method of determining the absorption bands of his Atmungsferment: 'The procedure resembles... the spectral analysis of stars, and indeed the matter of which the enzyme is made up, even though close at hand, is out of reach for us as is that of the stars.' This statement by one of the greatest biochemists of his time, clearly puts before our mind the state of science at the time when this enzyme was discovered: the knowledge of protein structure and properties was essentially non-existent. And it is this lack of a general basis of biological chemistry that has made the definition of structure and function of cytochrome oxidase such a painfully slow and error-ridden process. Once more let me quote from

Warburg [5] from his book in the translation of A. Lawson. He says under the title: *The Oxygen Transporting Haem*: 'The discussions which took place with Willstätter, Wieland and Euler about oxygen transporting iron began in 1922 and ended in 1928.* They were brought to an end by a controversy with Keilin which started in 1929 and finished in 1939.' Concerning this controversy, J. H. Quastel, one of the contemporaries of Warburg and Keilin wrote this in his 1984 article on 'The development of biochemistry in the 20th century':

Two schools of thought were becoming current, one headed by Otto Warburg, that considered activation of oxygen by surface catalysts containing iron as the determining respiratory agent, and another headed by Wieland that considered activation of hydrogen, present in the substrates burned in the cells, as the more responsible factor....The two German schools were so bitter, the Cambridge school siding with Wieland..., that I personally felt that this matter could only be settled on the dueling ground, the protagonists armed with sharpened swords, unbuttoned, and ready to have at each other.

Into this tense arena strode David Keilin, all 5 foot 2 inches of him, bursting with intelligence and common sense. Keilin came into biochemistry, not from chemistry, not from medicine, but from biology. He came armed with cytochrome, whose absorption bands in the thoracic muscles of the blowfly and the honey bee he had noted when he was quite unaware of the earlier wrongly discredited work of Mac Munn on the histohematins. Our David, like another David who flourished 3000 years before, faced the modern Goliath, Warburg, with his cytochrome. Although Warburg was not demolished..., he retreated, still waving his banner 'Atmungsferment', and he began to turn his attention to other matters that became of the greatest value to biochemistry [6].

I have seen the commentary, somewhat similar to that voiced by Quastel, that Keilin, as a biologist, was more prone to taking the broader, flexible and more integrating view which was required to visualize the functioning of a 'respiratory chain', than was Warburg with the rigorous thinking of a physical biochemist and his demand for exact experimental proof. It was the concept of an enzyme as we have it today, that was entirely missing at that time, so that the ideas of a heme–protein catalyzing this reaction was not obvious before the CO-dissociation spectrum had been obtained. Quastel puts it this way:

Warburg himself missed major developments in intermediary metabolism because of his stubbornly held views on iron-catalysed respiration and on cancer metabolism. These views met with fierce opposition spurring on other workers to obtain results that led to modern ideas on cellular electron-transport systems. [6]

Let us leave the early days and hear from a colleague who stood in the middle between those days and us here, Max Rudolf Lemberg, who wrote one of the most comprehensive reviews [7] on cytochrome oxidase in 1969. I quote from his introductory remarks to a meeting on 'structure and function of cytochromes' at Osaka (1967):

The 1965 Symposium at Amherst ended on a rather dismal note – it appeared that everybody's oxidase preparation was not only different, but also better than everybody else's. While we have not yet overcome this state of affairs entirely...there is nevertheless, I feel, justification today for greater optimism. For one thing, the actual findings appear to converge more and more, even if their

*With the establishment of the spectrum of the CO compound of the Atmungsferment (present author's comment).

interpretations differ widely. Let us then remember, firstly that new tools will have to be applied to the field, in particular that we still lack any magnetochemical evidence, and secondly, that as much as our friends from protein chemistry and quantum physics can help us, they themselves do not yet have all the answers [8].

Leaving now pure history – or to some extent also psychology of research – we will move closer to the biochemical events and development of concepts that have remained as milestones from the path of history. I have assembled these into Tables I and II, somewhat similar to the table given by Wikström and his colleagues [9] but extending further into our days. Here the emphasis in choosing the time or time-span, as it may be, for certain events or developments is not so much on the mostly diffuse early inklings of a phenomenon rather than on the time of the experiment or exposition that made a certain point stick.

Milestones in the history of cytochrome oxidase
Early observations and preparative attempts.

Table I relates more to experimental contributions, while I have tried to point out important conceptual contributions in Table II. It is however, not always easy to clearly separate the two. I have also refrained from carrying the matter beyond 1981; this does not mean that I do not recognize important or exciting contributions that have emerged since then, but there is a certain distance that one has to allow for viewing the events and weighing their significance. Our discussions here today and tomorrow will contribute substantially in this process.

To build some background for those of you that aspire to making oxygen out of water rather than water out of oxygen we will go briefly through the decades and the events enumerated in the tables. The measurement of the CO photodissociation spectrum of the oxidase in crude, dilute suspensions of cells and the calculation from these measurements of the molecular weight of the protein was indeed a brilliant achievement if we consider the state of science and instrumentation of those days 60 years ago. The discovery of the missing link between substrate dehydrogenation and oxygen reduction, namely the cytochromes which led to the concept of a 'respiratory chain' was then the step which ushered in the period of 20th-century biochemistry through which we have lived during the past few decades. Viewed against this background all other 'milestones' may appear rather minor, although they did mean advances necessary to build our present picture. It does strike me by going over the old literature how far-reaching some of the ideas and discoveries of these pioneers were – in some admittedly rare instances – amidst almost total ignorance about most that we learn and know today about biochemistry and science altogether. Going back to the table: Cytochrome oxidase is a membrane protein and isolation procedures for such proteins were not available and had to be developed, initially mostly by trial and error, more recently by more rational approaches. Originally, bile salts were the most effective agents known, but their use led to heavy losses of activity. The availability of non-ionic detergents stimulated a second spurt of activity in this area in the late 1960s and we still are essentially in the wake of this period. These repeated preparative attempts deserve consideration, because it is only since decent preparations were available that our more

Table 1. *Significant events in the history of research on cytochrome* c *oxidase*

Year	Event	Ref.
1887, 1925	Discovery and rediscovery of cytochromes	[10–11]
1924	Warburg's experiments and ideas on the 'Atmungsferment'	[12]
1928	Photodissociation spectrum of the 'Atmungsferment'	[13]
1938	Requirement for cytochrome c to link substrate oxidation to indophenol oxidase	[14]
1939	Discovery of a_3 and its identity with the 'Atmungsferment'	[15]
1953	Direct measurement of difference and photodissociation spectra Final confirmation of the identity of a_3 with 'Atmungsferment'	[16]
1954	Oxidative phosphorylation accompanying e^- transfer through cytochrome oxidase	[17, 18]
1941–60	Early solubilization and purification procedures using bile salts	[19–22]
1959	Copper as constituent of cytochrome oxidase	[23]
1963	Flow–flash method for producing reduced oxidase for reoxidation experiments	[24]
1965	Two different intrinsic copper-sites in cytochrome oxidase	[25]
1964–65	$a:a_3$ ratio in cytochrome oxidase	[26–28]
1966	Individual electronic spectra of a and a_3	[29]
1966–74	Preparative advances by use of non-ionic detergents	[30, 31]
1963–75	Structure of Hemin a	[32, 33]
1967	Appearance of high-spin a_3^{3+} on reduction of oxidase	[34]
1967–74	Oxidoreductive titrations and measurement of oxidation-reduction potentials	[35–38]
1976–83	Subunit structure and organization of cytochrome oxidase	[44–47]
1970–72	Two-dimensional crystalline arrays of oxidase molecules	[53–55]
1979	Electron microscopy of these arrays	[56, 57]
1977	O_2-'pulsed' cytochrome oxidase	[64]
1979	Amino-acid sequence of subunits	[65]
1978	Triple trapping technique for studying O_2 reaction of reduced oxidase	[66]
1980	Comprehensive study of reoxidation by this technique	[67, 68]
1979–80	Bacterial aa_3-type oxidases with 2 or 3 subunits	[75–79]
1977–84	Cytochrome oxidase as proton pump	[80, 81]
1980–82	EPR signals for CU_B and early oxygen intermediate	[69, 70]
1981	Partial reversal of cytochrome oxidase reaction	[82]
1981–85	Histidine as ligand to a and a_3 and cysteine to Cu_A	[83–85]

Table II. *Significant conceptual contributions to research on cytochrome* c *oxidase*

Year	Contribution	Ref.
1928	Concept of measuring photodissociation spectrum and deriving various parameters from it	[13]
1930–40	Concept of a respiratory chain	[11, 14]
1965	Exposition of the fitness of oxygen as biological oxidant	[4]
1974–1979	Quantitative treatment of metal–metal interactions in protein	[86, 87]
1974–76	The 'neoclassical' model as explanation of the observed redox behavior of cytochrome oxidase	[88, 89]
1981	Synthesis of knowledge on cytochrome oxidase	[9]

incisive tools of spectroscopy and protein chemistry could come to fruition, and, I think, we have not reached the goal yet on this front, before we can produce homogeneous material, not only by criteria of protein chemistry but also of active site homogeneity.

The metal centers, redox-titrations, kinetics

Copper in cytochrome oxidase was not a new discovery. It was already known to Keilin, but it remained a controversial issue for many years until EPR came to the rescue. As usual, acceptance was slow. Despite the EPR experiments around 1960, there were still schemes for electron (e^-) transfer in the literature into the 1970s, where only hemes were involved and copper was totally ignored.

We hardly recall today that even the stoichiometry of the cytochromes a and a_3 was once a hot issue. Apparently a_3 was degraded during preparation and, in fact, some of the heterogeneities observed today are probably descendants of these earlier experiences.

The era in the 1960s and 1970s when redox-titrations of the enzyme became feasible, although not leading to a coherent picture of the events, nevertheless, laid the ground work for much of what has been done in the past 5 or 10 years. It was then when the number of electrons taken up by the enzyme was determined, when midpoint oxidation-reduction potentials were measured – often without knowledge to what component they were to be attributed – and when much of the complexity of the situation was discovered that we are still trying to sort out today. Simultaneous with these attempts were the rapid kinetic experiments of Gibson and Greenwood [39] and their associates by spectrophotometry and by us using EPR [40]. There are many more EPR signals than there are components in the enzyme, described in numerous publications through the years [40–42]. Gibson was also the first to make meaningful experiments on reoxidation of reduced oxidase by introducing the flow–flash technique which, in some form, has remained an important ingredient of many recent investigations. It was out of these kinetic studies that a emerged as the component reduced first by cytochrome *c*, although it is not considered entirely certain that the initial acceptor is not Cu_A (cf. [43]). It appeared also from the titration studies that *a* was the acceptor which retained the majority of reducing equivalents introduced initially (see however below under conceptual contributions, 'neoclassical' model), which seemed incompatible with the idea of *a* being the 'low potential' heme, as had been assumed by a number of investigators. It also became clear in the kinetic work that e⁻ transfer from *a* to Cu_A was fast, whereas to a_3 it was slow, much too slow to account for the turnover of the enzyme in multiple turnover experiments with the purified enzyme or in subcellular particles.

Protein structure, subunits, electron microscopy

It was in the 1970s when the tools and the know-how of protein chemistry made it feasible to attack the questions of primary and eventually quarternary structure, of subunits and their amino-acid sequences. While this has had no decisive influence yet on the studies of the reaction mechanism it, nevertheless, has provided us with an idea where the metals are most likely located and what distances separate them. Buse, to whom we owe most of the sequence data, has concisely summarized our knowledge of the composition of cytochrome oxidase as of 1984–5 in his contribution to the Caprarola meeting [48]. Calculated on the basis of amino-acid analyses, the monomer of beef heart cytochrome oxidase has 1793 amino acids and a protein weight of 202 787 Da. On the assumption of 2 hemes per monomer and of a 1:1:1 ratio of heme:Fe:Cu it contains 9.86 nmol heme *a*, 0.55 g Fe and 0.627 g Cu* per mg protein; it has 13 subunits of 12 different types,* 17 cysteine residues of which 6 are involved in disulfide formation. There is also a variable amount of phospholipid, 5–8% in Buse's preparation (but from 1 to 30% in others) and in addition detergent. The organization of the subunits in the molecule has also been explored and various authors have shown schemes which have evolved through the years to quite complicated edifices. I will refrain from showing any but will provide some references here [43, 46, 49–52]. These endeavors have been substantially aided by the fact that two-dimensional crystalline arrays of the enzyme could be produced under certain conditions, which led to a reasonable model of the quarternary structure, as shown in Fig. 1. All this information may be summarized in approximate terms in the following way: Depending on the method of preparation, the enzyme can be obtained as a monomer or a dimer (cf. Fig. 1). The molecule is inserted into the membrane measuring approximately 120 Å in the direction normal to the membrane plane and maximally 100 Å in the direction of this plane. It protrudes considerably into the cytosol space (55 Å) and much less into the matrix space (~20 Å) with two lobes. The three largest subunits (those coded in mitochondria [58–60]) seem to be those exclusively involved in the catalytic function and one or two of these contain all the metals, namely subunit I both hemes and Cu_B and subunit II possibly Cu_A. One has to consider the

---

*More recent work indicates that there are three copper atoms per molecule [2] and 13 different subunits [47*a*]. A yet different view on copper content, polypeptide composition and protein weight of the monomer is derived from an analysis of crystallized oxidase [47*b*].

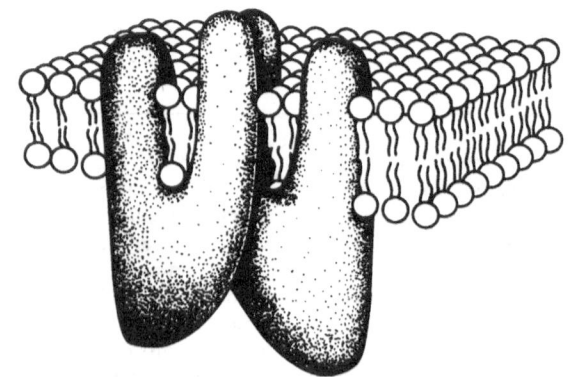

Fig. 1. A model for the structure of a cytochrome oxidase dimer showing 2-y-shaped monomers, interacting along their M_2 domain arms. (Reproduced with permission from ref. [57].)

possibility, however, that two subunits together may provide a single binding site. Cytochrome *c* approaches from the cytosolic side and is most likely bound to subunit II. It has always been some kind of surprise that the distances between the various active groups including cytochrome *c* when bound, appear to be considerable [49, 50, 61–63]. Thus, cytochrome *c* is ~25 Å from the nearest *a* heme, presumably cytochrome *a*; *a* and a_3 are separated by 12–16 Å, *a* and Cu_A by ~20 Å, *a* or Cu_A from Cu_B ~20 Å, only a_3 and Cu_B must be very close, namely 3.4 Å. The hemes of the cytochromes are oriented such that the heme plane is normal to the plane of the membrane [49].

The reoxidation reaction

As mentioned earlier, the emphasis for many years was on the reaction of the oxidase with cytochrome *c* or other reductants, not the reoxidation of reduced oxidase. The main reason for this was the lack of tools for the study of this very rapid reaction. The flow–flash technique [24] provided a means to rapidly produce the bare reduced enzyme; it was now necessary to slow the reaction with oxygen down so that observation of intermediates would become possible. This was accomplished by the 'triple-trapping' technique at cryogenic temperatures of Chance *et al.* [66] which has led to elaborate and comprehensive studies of the reoxidation reaction by optical and EPR spectroscopies ([67, 68], cf. [43]). Details of these will certainly be the subject of discussion at this meeting and I prefer not to anticipate any of this.

In the course of this work, finally, EPR signals for 'undetectable' copper, Cu_B were discovered in a few specific but rarely reached states of the enzyme [69, 70] which supported previous conclusions that the two copper components are very different and that Cu_B interacts magnetically with another component in the enzyme, which was presumed to be a_3. This notion has been rather generally accepted.

Active site heterogeneity

A difficulty in the interpretation of many of the studies on cytochrome oxidase was that the metal centers of the enzyme showed spectral and catalytic properties that varied depending on such conditions as presence of oxygen and/or cytochrome *c* or previous redox cycling, etc. It has recently become clear that there are two principal causes for this.

First, the enzyme can, without losing its integrity or changing its oxidation state, exist in a number of states, depending on the presence of substrates or other effectors or following a redox cycle. To this category belongs the 'pulsed' enzyme, originally described by Antonini *et al.* [64]. However, superimposed on this existence of different but, apparently naturally occurring states, there often is an active site heterogeneity that can be traced to a modification, usually irreversible, occurring in the course of isolation. While the first type of heterogeneity can often be eliminated by reduction and reoxidation or addition of ligands such as, e.g. cyanide, the second type of preparation-derived heterogeneity is more difficult to combat. It seems from more recent work that it is easier to avoid conditions that lead to this heterogeneity than to eliminate the unwanted species once they have been formed [71].

The introduction of various ligands or mixtures of ligands such as NO, azide, cyanide, or NO plus fluoride helped to sort out the various species present in enzyme as isolated [72]. It seems that all preparations extensively used in recent years contained several species, often dependent on the particular batch made [71, 73, 74]. Any claims to superiority of one over another preparation, in this respect, seems therefore meaningless at this point.

Bacterial oxidases; proton pumping

An important development coming to the aid of structural studies and attempts to locate the metals in the enzyme was the isolation of cytochrome oxidases from lower organisms which have only two or three subunits (cf. [79]). Interestingly, they show essentially the same properties as the 13 subunit enzyme, and are even able to perform proton pumping. This brings us to the area attracting most recent interest, namely the link of e^- transfer to energy conservation via vectorial proton transfer.

Conceptual contributions

This is as far as I want to carry the discussion of experimental contributions and we may now take a look at Table II, where I have assembled some of the more decisive conceptual contributions to the field. It is much shorter than Table I and one reason probably is that many contributions of this kind are in some way already built into the events shown in Table I. It is obvious that the idea and design of the procedure for measuring a CO photodissociation spectrum is as basic to the whole field as the concept of a respiratory chain. The elaboration on the fitness of oxygen I referred to above. It was in contributions from Göteborg in the 1970s that attention was drawn to and the basis was laid for quantitative treatment of site–site interactions influencing the oxidation-reduction potentials [86, 87]. These considerations fostered the acceptance of the 'neoclassical model' in which these interactions play a paramount role. This model provides a consistent explanation for the spectral (optical and EPR) observations and measurements of midpoint oxidation-reduction potentials in oxidoreductive titrations (Fig. 2). In essence the model implies that both hemes are of equal potential at the outset, but as one becomes reduced the potential of the other heme drops substantially so that at the halfway point there is a mixture of $a^{2+}\ a_3^{3+}$ and $a^{3+}\ a_3^{2+}$. The

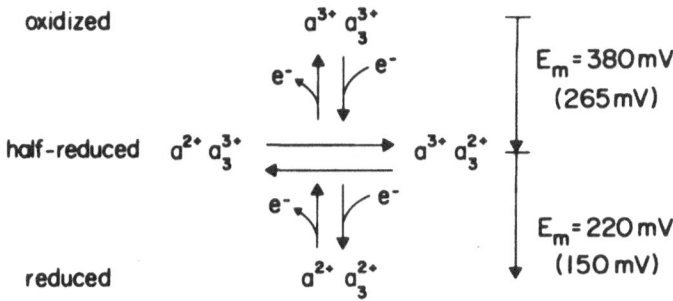

Fig. 2. The neo-classical model of cytochrome c oxidase. The scheme only refers to the behavior of the two hemes, not the copper atoms. Midpoint potentials in parentheses refer to the values found at a high phosphate potential. (Reproduced with permission from ref. [89].)

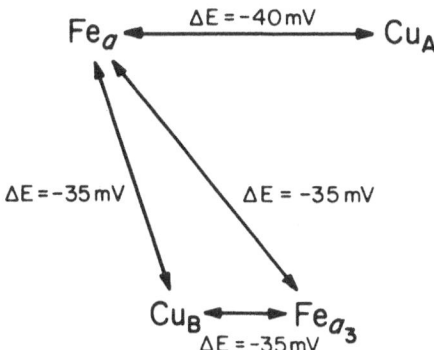

Fig. 3. Proposed interaction scheme to account for available spectro-electrochemical data on native, CO-inhibited, and cyanide-inhibited cytochrome c oxidase. The magnitudes of the various interactions involving the cytochrome a_3/Cu_B site are difficult to assess precisely, but the interaction magnitudes indicated are not expected to be very much (more than about 15 mV) in error. All of the interactions suggested are anti-cooperative and thus exert a buffering effect on the redox state of the protein. (Reproduced with permission from ref. [74].)

theme of site–site interactions have recently been elaborated on by Chan and his colleagues [74]. They find interactions as shown in Fig. 3. Finally, I think, the book by Wikström, Krab and Saraste [9] should also be mentioned here because, contrary to most other reviews on the subject, it was the first comprehensive attempt to bring all the then available knowledge together, weigh the evidence critically and interpret the large body of experimental results in a unifying fashion.

I am certain that many of the ideas and concepts laid down in this book will be the subject of discussion here at the present meeting.

Acknowledgements

I am indebted to Dr J. H. Quastel and Dr P. George for the permission to quote from their writings, to Dr T. G. Frey for kindly furnishing Fig. 1, and to Drs M. Wikström and S. I. Chan for permission to use Figs 2 and 3, respectively, from their publications.

References

1. Bombelka, E., Richter, F.-W., Stroh, A. and Kadenbach, B., *Biochem. Biophys. Res. Commun.* **140**, 1007 (1986).
2. Steffens, G. C. M., Biewald, K. and Buse, G., *Eur. J. Biochem* **164**, 295 (1987).
3. Warburg, O., *Zeitschr. Ang. Chem.* **45**, 1 (1932).
4. George, P. The fitness of oxygen, in *Oxidases and Related Redox Systems*, p. 3 (ed. T. E. King, H. S. Mason and M. Morrison), Wiley & Sons, New York (1965).

5. Warburg, O., in *Heavy Metal Prosthetic Groups and Enzyme Action*, p. 166. Oxford, Clarendon Press (1949).
6. Quastel, J. H., *J. Biochem. and Cell Biol*, **62**, 1103 (1984).
7. Lemberg, M. R., *Physiol. Revs.* **49**, 48 (1969).
8. Lemberg, M. R., in *Structure and Function of Cytochromes* (ed. K. Okunuki, M. D. Kamen, and I. Sekuzu, p. 3. University of Tokyo Press (1968).
9. Witkström, M., Krab, K. and Saraste, M., *Cytochrome Oxidase. A Synthesis*, Academic Press, London (1981).
10. Mac Munn, C. A. *J. Physiol.* **8**, 51 (1887).
11. Keilin, D., *Proc. Roy. Soc. Lond. B***98**, 312 (1925).
12. Warburg, O., *Bioch. Zeitschr.* **152**, 479 (1924).
13. Warburg, O. and Negelein, E., *Bioch. Zeitschr.* **193**, 339 (1928).
14. Keilin, D. and Hartree, E. F., *Proc. Roy. Soc. Lond.*, B**125**, 171 (1938).
15. Keilin, D. and Hartree, E. F., *Proc. Roy. Soc., Lond.*, B**127**, 167 (1939).
16. Chance, B., *J. Biol. Chem.* **202**, 383, 397, 407 (1953).
17. Maley, G. F. and Lardy, H. A., *J. Biol. Chem.* **210**, 903 (1954).
18. Lehninger, A. L., Ul Hassan, M. and Sudduth, H. C., *J. Biol. Chem.* **210**, 911 (1954).
19. Yakushiji, E. and Okunuki, K., *Proc. Imp. Acad.* **17**, 38 (1941).
20. Straub, F. B., *Z. Physiol. Chem.* **268**, 227 (1941).
21. Wainio, W. W., Cooperstein, S. J., Kollen, S. and Eichel, B., *J. Biol. Chem.* **173**, 145 (1948).
22. Yonetani, T., *J. Biol. Chem.* **235**, 845 (1960).
23. Sands, R. H. and Beinert, H., *Biochem. Biophys. Res. Commun.* **1**, 175 (1959).
24. Gibson, Q. H. and Greenwood, C., *Biochem. J.* **86**, 541 (1963).
25. Beinert, H., in *The Biochemistry of Copper* (ed. J. Peisach, P. Aisen, and W. E. Blumberg), p. 213. Academic Press, New York (1966).
26. van Gelder, B. F. and Muijsers, A. O., *Biochem. Biophys. Acta* **81**, 405 (1964).
27. Vanneste, W. H., *Biochem. Biophys. Res. Commun.* **18**, 563 (1965).
28. Gibson, Q. H., Palmer, G. and Wharton, D. C., *J. Biol. Chem.* **240**, 915 (1965).
29. Vanneste, W. H., *Biochemistry* **5**, 838 (1966).
30. Jacobs, E. E., Andrews, E. C., Cunningham, W. and Crane, F. L., *Biochem. Biophys. Res. Commun.* **25**, 87 (1966).
31. Jacobs, E. E., Kirkpatrik, F. H. Jr., Andrews, E. C., Cunningham, W. and Crane, F. L., *Biochem. Biophys. Res. Commun.* **25**, 96 (1966).
32. Grassl, M., Coy, U., Seyffert, R. and Lynen, F., *Biochem. Z.* **338**, 771 (1963).
33. Caughey, W. S., Smythe, G. A, O'Keefe, D. H., Maskasky, J. and Smith, M. L., *J. Biol. Chem.* **250**, 250, 7602 (1975).
34. van Gelder, B. F., Orme-Johnson, W. H., Hansen, R. E., and Beinert, H., *Proc. Natl. Acad. Sci. USA* **58**, 1073 (1967).
35. Wilson, D. F., Lindsay, J. G. and Brocklehurst, E. S., *Biochem, Biophys. Acta* **256**, 277 (1972).
36. Heineman, W. R., Kuwana, T. and Hartzell, C. R., *Biochem. Biophys. Res. Commun.* **49**, 1 (1972).
37. Tiesjema, R. H., Muijsers, A. O. and van Gelder, B. F., *Biochim. Biophys. Acta* **305**, 19 (1973).
38. Leigh, J. S. Jr., Wilson, D. F. and Owen, C., *Arch. Biochem. Biophys.* **160**, 476 (1974).
39. Greenwood, C. and Gibson, Q. H., *J. Biol. Chem.* **242**, 1782 (1967).
40. van Gelder, B. F. and Beinert, H., *Biochim. Biophys. Acta* **189**, 1 (1969).
41. Hartzell, C. R. and Beinert, H., *Biochim Biophys. Acta* **368**, 318 (1974).
42. Aasa, R., Albracht, S. P. J., Falk, K.-E., Lanne, B. and Vänngård, T., *Biochim. Biophys. Acta* **422**, 260 (1976).
43. Hill, B. C., Greenwood, C. and Nicholls, P., *Biochim. Biophys. Acta* **853**, 91 (1986).
44. Downer, N. W., Robinson, N. C. and Capaldi, R. A., *Biochemistry* **15**, 2930 (1976).
45. Steffens, G. J. and Buse, G., *Z. Physiol. Chem.* **357**, 1125 (1976).
46. Capaldi, R. A., Malatesta, F. and Darley-Usmar, V. M., *Biochim. Biophys. Acta* **726**, 135 (1983).
47. (a) Kadenbach, B., Jarausch, J., Hartmann, R., and Merle, P., *Anal. Biochem.* **129**, 517 (1983). (b) Yoshikawa, S., Tera, T., Takahashi, Y.,

Tsukihara, T. and Caughey, W. S., *Proc. Natl. Acad. Sci. USA* (in the press).
48. Buse, G. Meinecke, L. and Bruch, B., *J. Inorg. Biochem.* **23**, 149 (1985).
49. Ohnishi, T., LoBrutto, R., Salerno, J. C., Bruckner, R. C. and Frey, T. G., *J. Biol. Chem.* **257**, 14821 (1982).
50. Blair, D. F., Martin, C. T., Gelles, J., Wang, H., Brudvig, G. W., Stevens, T. H. and Chan, S. I., *Chem. Scr.* **21**, 43 (1983).
51. Jarausch, J. and Kadenbach, B., Eur. J. Biochem., **146**, 219 (1985).
52. Malmström, B. G., *Biochim. Biophys. Acta* **549**, 281 (1979).
53. Seki, S., Hayashi, H. and Oda, T., *Arch. Biochem Biophys.* **138**, 110 (1970).
54. Vanderkooi, G., Senior, A. E., Capaldi, R. A. and Hayashi, H., *Biochim. Biophys. Acta* **274**, 38 (1972).
55. Wakabayashi, T., Senior, A. E., Hatase, O., Hayashi, H. and Green, D. E., *J. Bioenergetics* **3**, 339 (1972).
56. Fuller, S., Capaldi, R. A. and Henderson, R., *J. Mol. Biol.* **134**, 305 (1979).
57. Frey, T. G. and Chang, T., *Ann. New York Acad. Sci.* **483**, 120 (1986).
58. Mason, T. L. and Schatz, G., *J. Biol. Chem.* **248**, 1355 (1973).
59. Rubin, M. S. and Tzagoloff, A., *J. Biol. Chem.* **248**, 4275 (1973).
60. Hare, J. F., Ching, E., and Attardi, G., *Biochemistry* **19**, 2023 (1980).
61. Vanderkooi, J. M., Landesberg, R., Hayden, G. W. and Owen, C. S., *Eur. J. Biochem.* **81**, 329 (1977).
62. Brudvig, G. W., Blair, D. F. and Chan, S. I., *J. Biol. Chem.* **259**, 11001 (1984).
63. Kornblatt, J. A. and Luu, H. A., *Eur. J. Biochem.* **159**, 407 (1986).
64. Antonini, E., Brunori, M., Colosimo, A., Greenwood, C. and Wilson, M. T., *Proc. Natl. Acad. Sci. USA* **74**, 3128 (1977).
65. Steffens, G. and Buse, G., *Z. Physiol. Chem.* **360**, 613 (1979).
66. Chance, B., Saronio, C. and Leigh, J. S., *J. Biol. Chem.* **250**, 9226 (1975).
67. Clore, G. M., Andréasson, L.-E., Karlsson, B., Aasa, R. and Malmström, B. G., *Biochem. J.* **185**, 139 (1980).
68. Clore, G. M., Andréasson, L.-E., Karlsson, B., Aasa, R. and Malmström, B. G., *Biochem. J.* **185**, 155 (1980).
69. Karlsson, B., Aasa, R., Vänngård, T. and Malmström, B. G., *FEBS Lett.* **131**, 186 (1981).
70. Hansson, O., Karlsson, B., Aasa R., Vänngård, T. and Malmström, B. G., *EMBO J.* **1**, 1295 (1982).
71. Baker, G. M. Noguchi, M., and Palmer, G., *J. Biol. Chem.* **262**, 595 (1987).
72. Brudvig, G. W., Stevens, T. H. Morse, R. H. and Chan, S. I., *Biochem.* **20**, 3912 (1981).
73. Scott, R. A., Schwartz, J. R. and Cramer, S. P., *Biochemistry* **25**, 5546 (1986).
74. Blair, D. F., Ellis, W. R., Jr., Wang, H., Gray, H. B. and Chan, S. I., *J. Biol. Chem.* **261**, 11524 (1986).
75. Sone, N., Ohyama, T., and Kagawa, Y., *FEBS Lett.* **106**, 39 (1979).
76. Ludwig, B. and Schatz, G., *Proc. Natl. Acad. Sci. USA* **77**, 196 (1980).
77. Yamanaka, T. and Fujii, K., *Biochim Biophys. Acta* **591**, 53 (1980).
78. Fee, J. A., Choc, M. G., Findling, K. L. Lorence, R. and Yoshida, T., *Proc. Natl. Acad. Sci. USA* **77**, 147 (1980).
79. Fee, J. A., Kuila, D., Mather, M. W. and Yoshida, T., *Biochim. Biophys. Acta* **853**, 153 (1986).
80. Wikström, M. K. F. and Saari, H. T., *Biochim Biophys. Acta* **462**, 347 (1977).
81. Wikström, M. and Krab, K., *Biochim Biophys. Acta* **549**, 177 (1979).
82. Wikström, M., *Proc. Natl. Acad. Sci. USA* **78**, 4051 (1981).
83. Stevens, T. H. and Chan, S. I., *J. Biol. Chem.* **256**, 1069 (1981).
84. Stevens, T. H., Martin, C. T., Wang, H., Brudvig, G. W., Scholes, C. P. and Chan, S. I., *J. Biol. Chem.* **257**, 12106 (1982).
85. Martin, C. T., Scholes, C. P. and Chan, S. I., *J. Biol. Chem.* **260**, 2857 (1985).
86. Malmström, B. G., *Quart. Rev. Biophys.* **6**, 389 (1973).
87. Lanne, B. and Vänngård, T., *Biochim. Biophys. Acta* **501**, 449 (1978).
88. Nicholls, P. and Petersen, L. C., *Biochim. Biophys. Acta* **357**, 462 (1974).
89. Wikström, M. K. F., Harmon, H. J., Ingledew, W. J. and Chance, B., *FEBS Lett.* **65**, 259 (1976).

Chemica Scripta 1988, **28A**, 41–46

The Rapid and Slow Forms of Cytochrome Oxidase

Graham Palmer, Gary M. Baker and Masato Noguchi

Department of Biochemistry, Rice University, P.O. Box 1892, Houston, TX 77251, USA

Paper presented by Graham Palmer at the Nobel Conference 'Biophysical Chemistry of Dioxygen Reactions in Respiration and Photosynthesis', Fiskebäckskil, Sweden, 1–4 July, 1987

Abstract

Small modifications to the Hartzell–Beinert procedure for the isolation of solubilized cytochrome oxidase leads to an enzyme preparation which reacts rapidly and homogeneously with cyanide. This 'rapidly-reacting' form of the enzyme can be converted quantitatively to a 'slowly-reacting' form by at least two, simple manipulations. The reaction of these two forms of the enzyme with fluoride, azide, hydrogen peroxide, nitric ooxide, carbon monoxide, and formate has been compared. It appears that the rapidly reacting form of the enzyme reacts readily with all of these reagents while the slowly reacting form does not. The magnetic susceptibility (20–200 K) together with the EPR, optical, MCD, and resonance Raman spectra of the two forms have also been compared. The data show that, following the conversion to the slow form, both hemes retain their basic electronicc properties (oxidation state, spin state and coordination number), that there appears to be a lowering in symmetry at cytochrome a_3 probably due to subtle changes in axial ligation, additional structural changes occur at the periphery of cytochrome a_3' and that solvent accessibility to cytochrome a is markedly reduced.

It is now generally accepted that purified preparations of cytochrome oxidase are inhomogeneous. With most proteins it is the case that variations in the polypeptide are the explicit source of the problem; however, with cytochrome oxidase it is inhomogeneity in the structure of one or more of the redox-active centers that is of special concern.

The first clear evidence for this active-center inhomogeneity is to be found in the work of van Buren [1] who observed that the rate of reaction of the enzyme with the heme ligand, cyanide, yielded complicated kinetics which could be resolved into several exponential processes. Subsequently, Brudwig *et al.* [2] discovered that the isolated enzyme reacted spontaneously with nitric oxide leading to the formation of variable amounts of a high spin $g = 6$ heme electron paramagnetic resonance (EPR) signal. The production of this signal was explained as due to the reaction of NO with Cu_B which was then rendered diamagnetic.* As a result the antiferomagnetic coupling between Cu_B and cytochrome a_3 was eliminated and the intrinsic EPR of this heme produced. Brudwig *et al.* [2] noted that the size of this EPR signal varied with preparation and proposed that this was due to some form of heterogeneity of the binuclear center which led to a variable extent of reaction with NO.

More recently, Kumar *et al.* [3] examined the cyanide kinetics of a number of preparations of cytochrome oxidase and found that most enzymes exhibited several kinetic phases of reaction and that the several phases of reaction varied in extent from preparation to preparation. While some samples were dominated by rapidly reacting material, in others the rapid phase of reaction was small and the kinetics were dominated by a much more slowly reacting component. Kumar *et al.* [3] attempted to correlate this behaviour with differences that they oberved in the copper EXAFS of these preparations but they could not draw any convincing correlations.

Because much of our understanding of the properties of the redox centers of this enyme stems from the interpretation of experiments, notably physical measurements, which assumes that the redox centers have homogeneous structural and functional properties the discovery of this inhomogeneity was both a blessing and a curse. On the positive side one now had an objective and plausible explanation why (nominally) identical experiments performed in different laboratories often led to apparently different results. At the same time the recognition of this inhomogeneity raised the fundamental and crucial question as to which of these experimental results reflected the property of the enzyme in its physiological state.

Recently my colleagues Gary Baker and Masato Noguchi studied the cyanide-binding kinetics and spectroscopic properties of a number of popular preparations of cytochrome oxidase and made the observation that there appeared to be a correlation between the amplitude of the slow phase of reaction with cyanide and the intensity of an EPR signal present at low magnetic fields, the so-called $g' = 12$ signal [4] (compare Fig. 1, top left and bottom). They observed that those enzyme preparations that were dominated by the fast phase of reaction gave only a small $g' = 12$ signal while the intensity of this signal was large in those preparations in which the reaction with cyanide was primarily slow. Baker and Noguchi subsequently showed [4] that this correlation was maintained in a quantitative way in a variety of experiments in which the proportion of enzyme reacting slowly with cyanide had been changed systematically, and we thus became convinced that these two parameters (i.e. the extent of slow reaction with cyanide and the size of the $g' = 12$ signal) were manifestations of a common property of the enzyme.

We attempted to observe the behaviour of the $g' =$ signal during the isolation of the enzyme by the modified Volpe–Caughey procedure [5] and found that it could not be detected in submitochondrial fragments or in samples taken during the early stages of the preparation (Fig. 1, top right, traces A–D); this signal could only be detected during the final steps of the purification (traces E–H) when its intensity

---

* This could be the consequence of either of the two procedures: (i) a redox reaction between the NO and the Cu leading to the formation of Cu(I); or (ii) an antiferromagnetic interaction between the NO and the Cu(II) which, as both are $S = \frac{1}{2}$ species, leads to a $S = 0$ magnetic ground state for the copper–nitrosyl adduct.

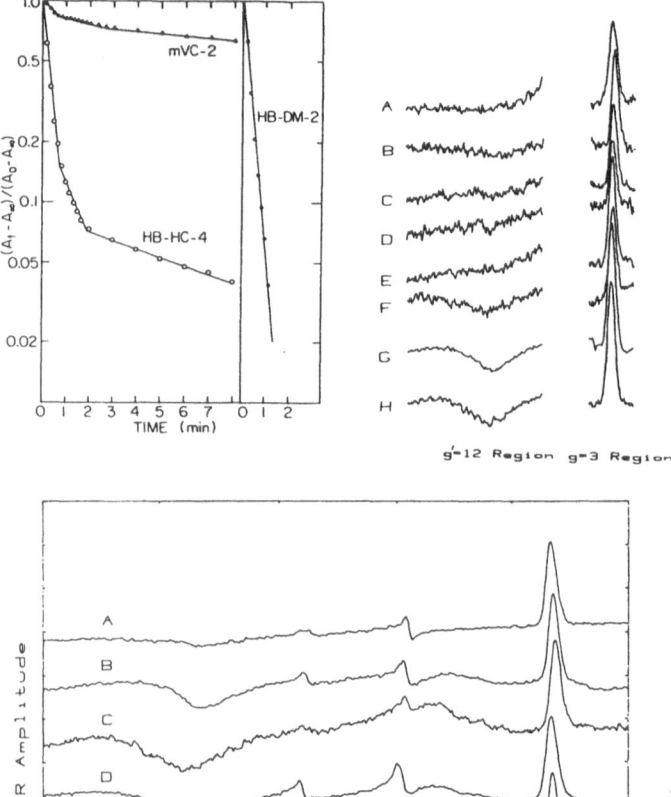

Fig. 1. Top left: the kinetics of cyanide binding to several preparations of cytochrome oxidase. (A) The reaction of the preparations prepared according to Volpe and Caughey (mVC-2, Δ) and Hartzell and Beinert (HB-HC-4, ○). (B) The reaction of enzyme (HB-DM-2) prepared using our modified procedure [4]. *Top right*: the development of the $g' = 12$ EPR signal during the isolation of mVC-3. (A) Keilin–Hartree particles. (B) After cholate extraction. (C) After the third ammonium sulfate step. (D) After the fourth ammonium sulfate step; the enzyme is introduced into Tween 20 at this point. (E) After fifth ammonium sulfate step. (F) After storage at 4 °C overnight and after the sixth ammonium sulfate step (25–35%). (G) After the seventh ammonium sulfate step (25–33%). (H) After dialysis and concentration. *Bottom*: low-field EPR spectra of various cytochrome c oxidase preparations. (A) HB-HC-4; (B) mVC-2; (C) Yonetani; (D) Yu and King; (E) HB-DM-2.

increased rapidly. As expected, the final product exhibited an EPR intensity consistent with the cyanide reactivity of the enzyme preparation.

We thus became persuaded that the presence of the $g' = 12$ EPR signal is a consequence of some manipulation that occurs during the enzyme purification, and proceeded to modify the Hartzell–Beinert procedure [6] so that the final product exhibited exclusively rapid reaction with cyanide (Fig. 1, top left, HB-DM-2) and had no detectable $g' = 12$ EPR (Fig. 1, bottom, trace E). These modifications required only small changes to the original protocol.

We later proceeded to examine what factors might be responsible for the production of the $g' = 12$ form of the enzyme and discovered that exposure of g12-less enzyme to low pH and also incubation in a relatively dilute form both led to the formation of the $g' = 12$ containing species. No differences in sensitivity to detergents could be identified providing pH and enzyme concentration were controlled.

With this information we were able to define a protocol for the quantitative conversion of rapidly reacting g12-less enzyme into the slowly reacting $g' = 12$-containing form using incubation overnight at pH 6.8 at 100 μM (or less) heme a. The conversion proceeded with a blue-shift of the Soret maximum, from 424 to 417 nm. However, this optical change appears not to be directly linked to the rapid-to-slow conversion, for it can be reversed by subsequently raising the pH, even though the cyanide reactivity and epr characteristics of the original enzyme are not restored.

Comparison of the reactivity of the two forms with common reagents

Addition of 0.1 M fluoride to rapid enzyme results in only small changes in its optical properties, the most conspicuous being a slight blue-shift in the 655 nm band most readily seen in the magnetic circular dichroism (MCD) spectrum (Fig. 2, right). There is also a small blue-shift in the Soret, from 424 to 421 nm which is accompanied by small changes in the MCD (Fig. 2, left) though the strong feature due to cytochrome a^{3+} is unchanged. These changes are present within 10 min and are stable overnight. With slow enzyme no changes are apparent after 10 min and only small changes occur with further incubation, the 655 nm band shifting slightly to the blue and the Soret shifting 11 nm to the red. The changes in the EPR spectrum parallel the optical changes. With rapid enzyme the formation of the fluoride signal at $g = 5$ and 3.2 [2] is maximal in 10 min; the EPR spectrum of slow enzyme is unchanged after 10 min reaction but after incubation overnight the EPR resembles that of the 10 min rapid sample.

0.1 M azide immediately shifts the Soret of fast enzyme from 424 to 426 nm and produces a small red-shift in the α-band which might be the origin of an apparent red-shift also observed in the 655 nm band. At the same time the rate of reaction with cyanide is modified. For example with 2 mM azide about one-third of the absorbance change occurs some three times more rapidly than is normally found with rapid enzyme. With slow enzyme the rapid red shift in the Soret is also observed but no other changes are observed in the absorbance spectrum.

The reaction with nitric oxide is complicated. Rapid enzyme yielded about 0.2 equiv. of high-spin heme (per aa_3) within 10 min. Slow enzyme gave very little high-spin signal at this time though about 0.1 equiv was found after 90 min. Interestingly, the optical spectrum of the 10 min rapid form resembles that obtained after complete reaction with cyanide with a 4 nm red-shift and intensification of the Soret, and a large decrease in the 655 nm band. The extent of these spectrophotometric changes was consistent with complete reaction of cytochrome a_3 even though only a small fraction of the enzyme exhibited a high-spin EPR spectrum. Prolonged incubation with NO led to additional high-spin material only in the case of the slow form. However, rapid enzyme exhibited further changes in EPR with the appearance of the signal characteristic of the ferrous $a_3 \cdot$NO compound superimposed upon that of Cu_A; the amount of nitrosyl formed accounted for 0.1 equiv. of cytochrome a_3.

A partial explanation for these observations can be made by extending the hypothesis of Brudwig *et al.* [2] for the formation of the $g = 6$ EPR signal and by drawing on our

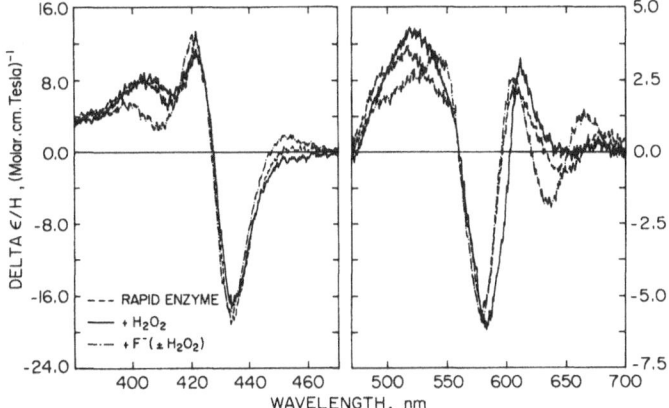

Fig. 2. MCD spectra of the rapid form of cytochrome oxidase together with its fluoride and hydrogen peroxide derivatives. The enzyme concentration was 18 μM, fluoride was at 0.1 M and hydrogen peroxide at 100 μM. Data was obtained immediately after mixing. The spectra obtained when fluoride was added after hydrogen peroxide are not significantly different from those shown with fluoride alone (M. Noguchi, G. M. Baker and G. Palmer, unpublished data).

Table I. *Comparison of reaction with several ligands*

Ligand	Fast enzyme	Slow enzyme
Potassium cyanide	100% rapid	0% rapid, 80–100% slow
Potassium fluoride	EPR and optical changes are rapid	Requires overnight for complete reaction
Sodium azide	Small optical changes, reaction with cyanide enhanced	No obvious effect
Nitric oxide	0.2 h.s. heme immediately	Requires overnight for complete reaction
Hydrogen Peroxide	Rapid optical changes to give compound C	No effect[a]
Carbon monoxide	Mixed-valence compound formed within 10 min	Requires overnight for complete reaction
Sodium formate	Immediate blue-shift of Soret	No reaction

[a] A small excess of H_2O_2 gives no reaction while a large excess destroys the heme.

understanding of the reaction of the enzyme with the cyanide anion and with dioxygen, both of which are believed to form a bridge between the heme iron of a_3 and Cu_B.* Thus it seems that about 0.2 equiv. of the enzyme has a defect at the copper site which allows direct reaction with NO. In the remainder of the enzyme we propose that NO functions as a bridge between the two metal ions leading to a structure of the form [Fe–NO–Cu]. As a consequence the heme is converted to the low-spin state. The only complicating feature is that NO contains an unpaired electron and thus the formation of a new EPR-active species might be expected; this has not been observed. Three alternatives suggest themselves: (i) this resonance is located at $g = 2$ and has a short relaxation time so that its EPR is obscured by the NO matrix signal which is very intense at liquid helium temperatures; (ii) Cu_B is reduced in this adduct so that the interaction between the NO and the heme again leads to an even-spin system; (iii) the three electrons couple in a complicated way with both ferromagnetic and antiferromagnetic interactions which could result in a very anisotropic, and hence hard to detect, EPR signal. In a small fraction of the enzyme the bridged center undergoes reduction resulting in the formation of the classical heme-nitrosyl derivative which is observed.

Addition of a small excess of hydrogen peroxide leads immediately to a red-shift of the Soret band, a small intensification of the α-band and no obvious decrease in intensity of the 655 nm band; the maximum changes are produced by about 5 equiv. of ligand. Slow enzyme does not exhibit any optical changes under these conditions. The MCD spectra of the product are almost unchanged from that of the native enzyme the most significant difference being an increase in intensity at 600 nm and a change in the appearance of the 655 band (Fig. 2, right). Subsequent addition of fluoride reverses these optical changes and results in spectra almost identical to those obtained upon direct addition of fluoride to rapid enzyme (Fig. 2). It would thus appear that hydrogen peroxide does not convert cytochrome a_3 to the low-spin state and is readily displaced by fluoride. It is worth recalling that Carter *et al.* [7] found the Raman core-size

marker of this derivative to be present at 1591 cm^{-1}, a slightly higher frequency than that of the low-spin cytochrome a, and raised the possibility that cytochrome a_3 was $S = \frac{3}{2}$ in the hydrogen peroxide treated enzyme. Recent magnetic susceptibility studies on this derivative† are consistent with this idea provided that the heme is still antiferromagnetically coupled to the copper.

Incubation of rapid enzyme with carbon monoxide resulted in the immediate (10 min) formation of the mixed-valence carbon monoxide adduct; with slow enzyme complete reaction required overnight incubation.

0.1 M formate immediately shifts the Soret of rapid enzyme to 417 nm with a small decrease in intensity of the α-band and accentuation of the 655 nm band. Formate has no clear effect on the optical properties of slow enzyme.

These reactions with ligands are summarized in Table I.

Comparison of some physical properties of the two forms.

We have already noted only physical property that is clearly different in the two forms of the enzyme, namely the presence of the $g' = 12$ EPR signal in the slow form. Although it is generally agreed that this EPR signal arises from the $S = 2$ binuclear center the specific explanation for the resonance is currently debated with Brudwig *et al.* [8] claiming that it is due to transitions between the ± 1 levels of the $S = 2$ configuration while Hagen [9] believes that it is due to transitions between the ± 2 levels of the same electronic configuration. (It should be noted that, with appropriate values for certain ligand-field and exchange parameters, both of these transitions can occur [8]; thus distinguishing between the two alternatives has been quite difficult.)

A second apparent difference between the two forms is the wavelength of the Soret maximum. However, while it is true that this maximum shifts to the blue on formation of the slow form, raising the pH causes the maximum to shift back to the red *without any loss in $g' = 12$ signal or recovery of the rapidly reacting form*. It thus seems that this blue-shift in the Soret is not an obligatory characteristic of the slow form.

* In the case of oxygen, formation of this bridge is accompanied by a redox reaction so that the actual bridging species appears to be peroxide.

† D. Lee, V. Chunplan, L. Wilson, M. Noguchi, G. M. Baker and G. Palmer and Z. K. Barnes, J. Dye and G. T. Babcock, unpublished data.

Chemica Scripta 28A

The Soret mcd spectra of the two forms are indistinguishable. As the MCD spectrum is dominated by cytochrome *a* this establishes that the absorbance changes just described are due to cytochrome a_3. The 655 nm band undergoes a red-shift on conversion to the slow form; this is best seen in the MCD spectrum but can be visualized in the absorbance spectrum as improved resolution of the 655 nm feature. The continued presence of this band is evidence that cytochrome a_3 is still high spin and this conclusion is supported by magnetic susceptibility measurements.

The magnetic moments of the two forms of the enzyme only differ by 3%, which is within the uncertainty of the method, and reinforces the conclusion that cytochrome a_3 is still high spin and the binuclear center magnetically unchanged. Further evidence for this conclusion is to be found in the Raman spectra (see below).

Robert Scott and his colleagues at the University of Illinois have examined the copper edge and copper EXAFS of the two forms of the enzyme (Fig. 3). The edge spectra (Fig. 3, inset) are comparable to those described as 'pulsed' and 'resting' by Powers *et al.* [10] except that the assignments appear to be inverted, for our *rapid* form exhibits an edge spectrum similar to the *resting* form of Powers *et al.* and vice versa. However, the actual differences are small and may not be significant.

The Fourier transform of the copper EXAFS also show differences, the most conspicuous being the increase in resolution of the two main peaks found with the slow form of the enzyme (Fig. 3). However, curve-fitting to the data lead to two shells, one for Cu–(N,O), the other for Cu–S, which are not significantly different for the two forms of the enzyme and it is probable that, once more, these differences are within experimental error. Scott [11] had previously noted the preparation dependent variability in the FT resolution in this region and concluded that this variability was not necessarily significant.

Additional differences between the fast and slow forms are found in their resonance Raman (RR) spectra [112]. With both forms of the enzyme the oxidation state marker is found at 1371 cm^{-1}, while the characteristic spin-state markers of *a* and a_3 are unchanged in the two forms (Fig. 4). Thus cytochrome a_3 is ferric, high-spin‡ and six-coordinate in both forms of the enzyme. One clear difference in this high-frequency region is that there is significantly more intensity at *c.* 1620 cm^{-1} in the slow form of the enzyme than there is in the fast form. The vibrations in this region have been claimed to reflect partial reduction of the heme (caused by laser photons) [13]. However, this mode develops in the slow form without any corresponding increase in intensity at 1359 cm^{-1}, the vibration associated with *ferrous* heme, and it thus appears that the presence of Raman intensity in the 1620 cm^{-1} region is not unequivocal evidence for photoreduction.

The most significant difference in the low-frequency region (metal–ligand modes) is the shift of a line present at 223 cm^{-1} in fast enzyme to 220 cm^{-1} in slow enzyme. This mode is reasonably interpreted as being the cytochrome a_3-N(His) stretch, though we should note that previous attempts [14] to observe such modes in *ferric* heme–imidazole model compounds have been unsuccessful.

‡ Because the behaviour off the spin-state (or core-size) marker is determined principally by the occupation of d$_{x^2-y^2}$, the Raman spectrum demonstrates that neither form of cytochrome a_3 has the $S = \frac{3}{2}$ state.

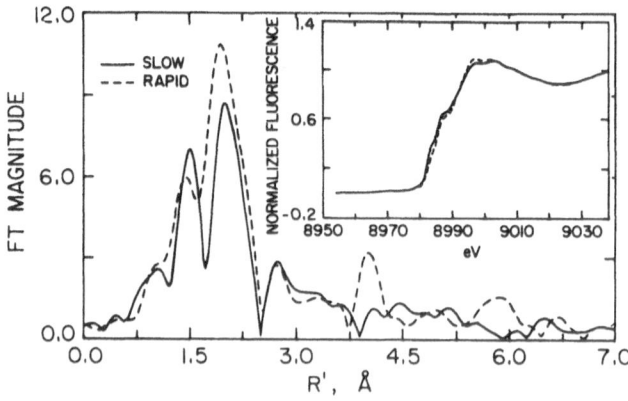

Fig. 3. Comparison of the copper X-ray edge (inset) and EXAFS of the fast and slow forms of cytochrome oxidase. This experiment was performed by Robert Scott, Richard J. Sullivan and Marly Eidsness of the University of Illinois.

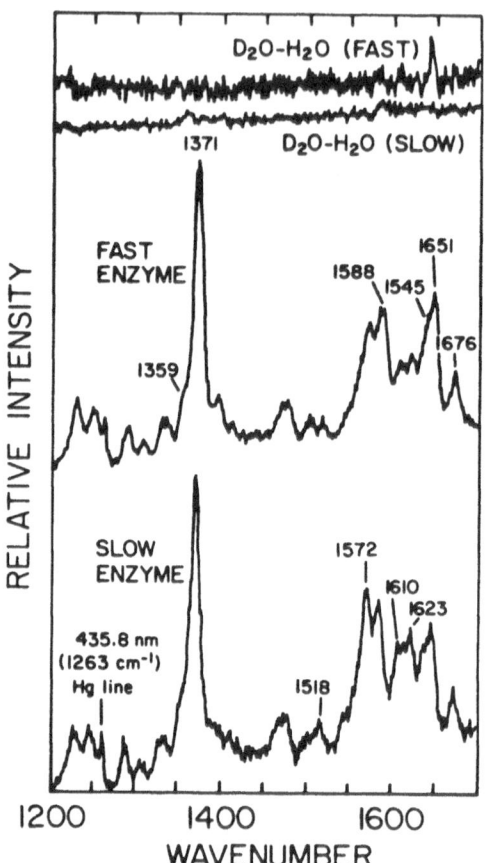

Fig. 4. Comparison of the resonance Raman spectra of the fast (upper) and slow (lower) forms of cytochrome oxidase. The noisy lines at the top of the panel are difference spectra (D$_2$O–H$_2$O) of the two forms of the enzyme. (From Schoonover *et al.* [12].)

Upon transferring enzyme from H$_2$O to D$_2$O small changes are noted. First the iron-histidine stretch just described shifts by 3 cm^{-1} to lower energy in rapid enzyme but not in slow enzyme. This sensitivity to deuteration suggests that N(1) of the histidine is hydrogen bonded, and that the contributory hydrogen atom is exchangeable in fast enzyme but not in slow enzyme. In the high frequency region exchange into D$_2$O leads to shift of the mode believed to be the –C=O stretch of the formyl group of cytochrome *a*R (Fig. 4, upper traces). This effect was first observed by Copeland and Spiro [15] and interpreted by them as resulting from a strengthening of a hydrogen bond to the formyl group when a proton is replaced by a deuteron. We were surprised to find, however, that slow enzyme did not undergo this shift even if the exposure to D_2O was increased from 6 to 24 h. Thus the accessibility of the formyl of cytochrome a, and possibly the imidazole of cytochrome a_3, is markedly reduced when the enzyme is transformed into the slow form.

Discussion

It now appears that the conformational change responsible for the fast-to-slow transition is sufficiently extensive that both cytochrome a and cytochrome a_3 are affected. The involvement of cytochrome a is deduced from the absence of any effect of D_2O on the carbonyl stretch of its formyl substituent ($1651 cm^{-1}$) in the slow form even though this formyl group is easily accessible in the fast species. It should be apparent that this does not necessarily mean that there are structural changes immediately at cytochrome a; either the accessibility to D_2O has been eliminated, or the residue which is hydrogen bonded to the carbonyl group can no longer exchange its proton.

The changes to cytochrome a_3 are more substantial: a general loss in ligand reactivity, a change in magnetic properties (i.e. the appearance of the $g' = 12$ resonance) and an apparent weakening of the bond of the proximal histidine to the heme. At the same time the spectroscopic data make it clear that a_3 maintains its high-spin ferric electronic configuration and is still 6-coordinate.

Although the loss in ligand reactivity might be explained by a severe decrease in accessibilty to cytochrome a_3 due to a closing of a channel from the exterior to the binuclear center, we prefer to interpret the result in terms of a change in properties of the ligand which bridges a_3 and Cu_B. Reaction of a reagent with the heme iron requires initial dissociation of this intrinsic ligand, a process which is controlled by the strength of the iron–ligand bond. A change in bridging ligand leading to an increase in the strength of this bond would undoubtedly slow down the reaction with added reagents. However, this change need not require replacement of the ligand for a modification in iron–ligand geometry might well accomplish the same result. Increasing the iron–ligand bond-strength would weaken the iron––histidine bond and lead to a lowering of the Raman frequency. The impact of these changes on the magnetic properties depends upon the origin of the $g' = 12$ resonance. If, as has been proposed by Hagen [9], the $g' = 12$ signal is a $\Delta m = 4$ transition then a suitable value of the cubic component of the ligand field could enable this EPR signal. On the other hand, if it is a $\Delta m = 2$ signal, as Brudwig believes [8], then a low symmetry component of the ligand field (or the exchange interaction) would have to change. We do not have enough information to evaluate these alternatives.

Even though conversion to the slow form is permanent it is possible to convert the slow form back to the rapid form for a short time. This is achieved via the pulsing protocol [16] whereby the enzyme is subjected to a cycle of reduction and reoxidation. The product of this reaction reacts rapidly with cyanide and lacks a $g' = 12$ EPR signal [4]. However, over a period of hours it spontaneously reverts to the slow form. This memory of the enzyme for its origin suggests that the fundamental process underlying the fast to slow transition is the modification of an amino acid. One possibility for this modification is the hydrolysis of an asparagine residue to aspartate, a reaction which would lead to the appearance of a negative charge and the loss of a group capable of hydrogen bonding; such changes may well lead to alterations in conformation.

There is circumstantial evidence for the appearance of a negative charge close to cytochrome a_3 for the conversion to slow enzyme is accompanied by a blue-shift of the Soret absorbance. This shift cannot be ascribed to changes in oxidation state, spin-state or coordination number nor can it be attributed to changes in the orientation of the formyl substituent relative to the porphyrin.* Thus it appears to us that the blue-shift of the Soret band can have only one of two origins: (i) a massive change in dielectric about the heme – this we feel is quite unlikely; (ii) the appearance of a negative charge at the perimeter of the porphyrin. As the electronic transitions of the porphyrin are associated with the displacement of electron density from the center to the periphery placing a negative charge at the periphery will oppose such transitions and lead to a blue-shift of the spectrum.†

However, to integrate this idea with the reversal of the optical changes produced by raising the pH and with the transient formation of pulsed enzyme which occurs on re-oxidation requires invoking *ad hoc* chemical and conformational changes for which there is no specific evidence. We are thus left with only a hint as to the origin of the fast-to-slow transition but the basic idea is sufficiently reasonable as to merit further experimentation. Certainly a number of obvious possiblities have been eliminated and further spectroscopic data should define the process more clearly. RR spectroscopy will be particularly useful in this effort.

Acknowledgements

The work described in this paper was undertaken with support provided by the National Institutes of Health (GM-21337) and the Robert A. Welch Foundation (C-636). We express our appreciation to Dr Robert Scott for permission to publish Fig. 3.

References

1. Van Buuren, K. J. H., Nicholls, P. and Van Gelder, B. F., *Biochim. Biophys. Acta* **256**, 258–276 (1972).
2. Brudwig, G. W., Stevens, T. H., Morse, R. and Chan, S. I., *Biochemistry* **20**, 3912–3921 (1981).
3. Naqui, A., Kumar, C., Ching, Y.-C., Powers, L. and Chance, B., *Biochemistry* **23**, 6222–6227 (1984).
4. Baker, G. T., Noguchi, M. and Palmer, G., *J. Biol. Chem.* **262**, 596–604 (1987).
5. Yoshikawa, S., Choc, M. G., O'Toole, M. C. and Caughey, W., *J. Biol. Chem.* **252**, 5498–5509 (1977)
6. Hartzell, C. R. and Beinert, H., *Biochim. Biophys. Acta* **368**, 318–338 (1974).
7. Carter, K. R., Antalis, T. A., Palmer, G., Ferris, N. S. and Woodruff, W. H., *Proc. Natl. Acad. Sci. USA* **78**, 1652–1655 (1981).

* Because of the constancy in frequency ($1676 cm^{-1}$) of the Raman mode associated with the formyl group of the heme of a_3 it follows that this formyl group remains in conjugation with the porphyrin. It is the interaction between the π^*-orbitals of the carbonyl and the porphyrin that is mainly responsible for raising the degeneracy of the porphyrin LUMO's. This change in energy leads to a shift in wavelength and so movement of the formyl in and out of the prophyrin plane would modulate the Soret wavelength.

† A loss of a positive charge would have a similar effect; the positive charge would assist the electronic transition in the rapid species.

8. Brudwig, G. W., Morse, R. H., and Chan, S. I., *J. Mag. Res.* **67**, 189–201 (1986).
9. Hagen, W. R., *Biochim Biophys. Acta* **708**, 82–98 (1982).
10. Powers, L. and Chance, B. J. *Inorg. Chem.* **23**, 207–217 (1985).
11. Scott, R. A., Schwartz, J. R. and Cramer, S. P. *Biochemistry* **25**, 5546–5555 (1986).
12. Schoonover, J.R., Dyer, R. B., Woodruff, W. H., Baker, G. M., Noguchi, M. and Palmer, G. (in review).
13. For a review see Babcock, G. T. in *Biological Applications of Raman Spectroscopy* (ed. T. G. Spiro), vol. III (in the press).
14. Mitchell, M. L., Canmbell, D. H., Traylor, T. G. and Spiro, T. G., *Inorg. Chem.* **24**, 967–971 (1985).
15. Copeland, R. A. and Spiro, T. G. *FEBS Lett* **97**, 239–243 (1986).
16. Antonini, E., Brunori, M., Greenwood, C. and Wilson, M. T. *Proc. Natl. Acad. Sci. USA* **74**, 3128–3132 (1975).

Chemica Scripta 1988, **28A**, 47–50

Electron-Transfer Reactions in Cytochrome *c* Oxidase

Ron Wever, L. B. Vlegels, H. Dekker and A. C. F. Gorren

Laboratory of Biochemistry, University of Amsterdam, P.O. Box 20151, 1000 HD Amsterdam, The Netherlands

Paper presented by Ron Wever at the Nobel Conference 'Biophysical Chemistry of Dioxygen Reactions in Respiration and Photosynthesis', Fiskebäckskil, Sweden, 1–4 July, 1987

Abstract

The effect of steady-state illumination on the redox state of cytochrome a_3 in mixed-valence carboxy-cytochrome *c* oxidase ($a^{3+}a_3^{2+} \cdot CO$) has been studied. It is shown that the pH markedly affects the resulting difference spectra (light–dark) and that more cytochrome a_3 gets oxidized at higher pH values. A model is proposed in which the apparent midpoint potential of cytochrome $a_3^{2+} \cdot CO$ decreases upon illumination leading to electron-transfer (ET) reactions from cytochrome $a_3^{2+} \cdot CO$ to Cu_A and cytochrome a_3 in the partially reduced enzyme. Biphasic recombination kinetics of cytochrome a_3 with CO are observed in the mixed-valence enzyme above pH 8.0. The first phase corresponds to normal recombination of CO with cytochrome a_3. The amplitude of the second phase was a function of the period of illumination and it was prominent at higher pH values. It is suggested that during illumination under those conditions a hydroxyl ion is bound to cytochrome a_3 which stabilizes the oxidized state of this redox site. In the dark, this ligand is slowly expelled by CO from its binding site.

Introduction

Cytochrome *c* oxidase (ferrocytochrome *c*:oxygen oxido-reductase EC 1.9.3.1) is the terminal enzyme of the mitochondrial respiratory chain. The enzyme contains four metal centres as prosthetic group: two haem *a* groups (of cytochrome *a* and of cytochrome a_3) and two copper atoms (Cu_A and Cu_B) [1]. Two of these centres, the haem *a* group of cytochrome *a* and Cu_A, are involved in accepting electrons from cytochrome *c*, whereas the other two, the haem *a* group of cytochrome *a* and Cu_B form a binuclear reaction site, where oxygen reduction takes place [2, 3]. When electrons are donated by cytochrome *c* to the oxidase in the absence of oxygen the rate of internal electron transfer (ET) from cytochrome *a* to the a_3–Cu_B is slow (0.6–5 s^{-1}). However, when the reduced enzyme is reoxidized by oxygen, cytochrome *a* and Cu_A are oxidized at rates of 700–6000 s^{-1} [4, 7], respectively. These values are in line with the turnover number which at low pH (4.6–5.2) reaches a value of 630–1000 s^{-1} [8, 9]. Similarly, as demonstrated by Boelens *et al.* [10], when CO is ligated to the a_3^{2+}–Cu_B site the rate of internal ET between cytochrome a_3 and Cu_A is fast (7000 s^{-1}). The acceleration in rate of internal ET is probably caused by an increase of redox potential difference between cytochrome *a* and the a_3–Cu_B pair, when CO or O_2 are ligated to this redox site [11, 12].

As shown in [10, 13, 14] it is possible to modulate the redox potential of the cytochrome $a_3^{2+} \cdot CO$ compound by light and as a consequence it is possible to induce ET reactions in mixed-valence carboxy-cytochrome *c* oxidase. There is some agreement [15, 16] that there is a coupling between ET reactions in the enzyme and the proton-pumping properties of the membraneous enzyme. Therefore, we have studied the effect of pH on the extent of the ET reactions induced by

light. The experiments show that at pH values above neutral, more cytochrome a_3 gets oxidized. Part of this work appeared in a preliminary form [17].

Experimental

Bovine cytochrome *c* oxidase was isolated according to refs. [18] and [19]. The oxidase concentration was determined spectrophotometrically using an absorbance coefficient (redox) of 24.0 mM^{-1} cm^{-1} at 605 nm [20]. Chemicals were mainly from British Drug Houses (Analar grade). CO was from Matheson Gas Products containing less than 10 ppm oxygen, and helium was from Hoek Loos. Phosphate buffers and carbonate buffers in 0.1 M concentrations were used in the pH range 6.5–8.8 and 8.7–10.8, respectively. As a detergent 1.0 % Tween 80 was used. The experiments were carried out in the absence of oxygen in anaerobic Thunberg cells which could be filled with CO of variable pressures. The samples were made anaerobic as described before [14]. The fully reduced cytochrome *c* oxidase CO compound was prepared by adding CO to samples which were reduced by dithionite. The mixed-valence CO compound was prepared by incubation of oxidized cytochrome *c* oxidase under CO (100 kPa) at 20 °C for 3 h [14]. The absorbance changes were studied with a Hewlett-Packard 8451A diode array spectrophotometer. Photodissociation of cytochrome $a_3^{2+} \cdot CO$ was achieved by steady-state illumination with a 150 W xenon lamp (Oriel) filtered through an orange filter (OG 570) as described [13]. A blue filter (BG 12) was placed in front of the diode array.

Results

When fully reduced carboxy-cytochrome *c* oxidase is photo-dissociated a peak is formed at 445 nm in the difference spectrum (light minus dark) corresponding to formation of unliganded cytochrome a_3^{2+} and a trough at 428 nm that can be assigned to photodissociation of the cytochrome $a_3^{2+} \cdot CO$ complex (results not shown) in line with refs. [13] and [17]. Mixed-valence carboxy-cytochrome *c* oxidase ($a^{3+}a_3^{2+} \cdot CO$) yielded a light-induced difference spectrum that was similar to that seen in the fully reduced carboxy-cytochrome *c* oxidase ($a^{2+}a_3^{2+} \cdot CO$). Notable differences were however present between the light-induced spectra of both enzyme species (cf. ref. [13], Fig. 3). For example the ratio $\Delta A_{445}/\Delta A_{428 \text{ nm}}$ changed from 1.4 in the fully reduced enzyme to 0.8 in mixed-valence carboxy-cytochrome *c* oxidase (Table I). There was also a shift in the isosbestic point from 435.9 nm in fully reduced enzyme to 437.0 nm in the mixed-valence enzyme. At pH values higher than 8 the ratio $\Delta A_{445}/\Delta A_{428 \text{ nm}}$

Table I. *Effect of pH on the light-induced difference spectra of carboxy-cytochrome c oxidase*

Oxidation state	pH	$\Delta A_{445}/\Delta A_{428\,nm}$	Isosbestic[a] point (nm)
Fully reduced	7.4	1.41	435.9
	10.2	1.40	436.0
Mixed valence	6.4	0.84	—
	7.4	0.85	437.0
	7.8	0.76	—
	8.3	0.78	437.2
	8.5	0.59	—
	8.9	0.50	437.5
	9.4	0.57	437.7
	10.2	0.43	438.1

[a] The ratios and isosbestic points were calculated from the spectra obtained after 4 s of illumination.

Experimental conditions 5–8 μM cytochrome c oxidase; p_{CO}, 2.5–30 kPa.

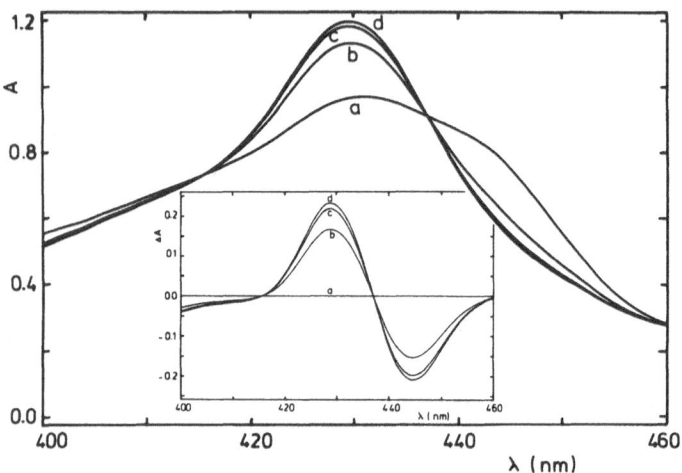

Fig. 1. Recombination of mixed-valence carboxy-cytochrome c oxidase with CO after illumination (10 s). Experimental conditions: 4.7 μM cytochrome c oxidase; 100 mM phosphate buffer (pH 7.4) in 1.0% Tween 80, p_{CO} = 2.5 kPa. (*a*) During illumination (10 s); (*b*) 0.4 s; (*c*) 0.8 s; and (*d*) 3 s after illumination, respectively. The inset shows the difference spectra of the recombination reaction. The spectrum obtained during illumination (*a*) was taken as a reference.

decreased further, whereas the ratio was not affected by pH in the fully reduced enzyme. Similarly, there was a further shift in the isosbestic point to 438.1 nm; this shift was not observed in fully reduced enzyme.

In addition, in the mixed-valence carboxy-cytochrome c oxidase a small absorbance increase was present around 410 nm, which was not seen in the fully reduced enzyme. Previous studies [21, 22] using EPR have already shown that upon illumination of mixed-valence cytochrome c oxidase, oxidation of cytochrome a_3 will occur. The oxidation of cytochrome a_3 will result in a less intense light-induced band at 445 nm. Thus, as judged from the ratio $\Delta A_{445}/\Delta A_{428\,nm}$ (Table I) cytochrome a_3 becomes progressively more oxidized at pH values higher than 8.3.

Recombination experiments

The recombination of CO with mixed-valence carboxy-cytochrome c oxidase was followed by measuring optical absorption spectra during continuous illumination and immediately after illumination. As Fig. 1 shows the recombination of CO with the mixed-valence enzyme at pH 7.4

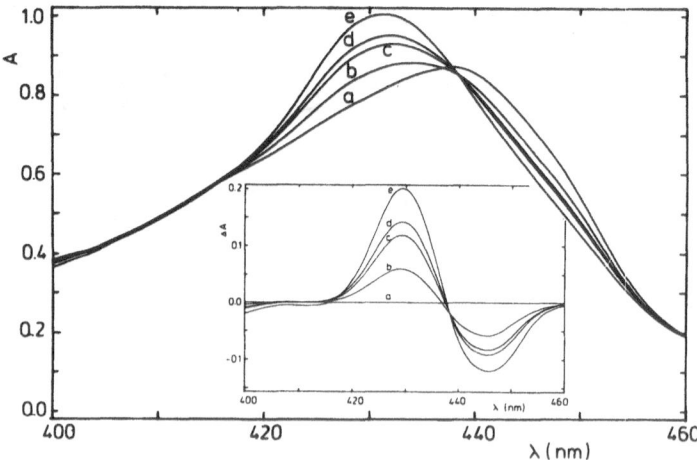

Fig. 2. Recombination of mixed-valence carboxy-cytochrome c oxidase with CO after illumination. Experimental conditions as in Fig. 1, except that 100 mM sodium carbonate (pH 10.1) was used. (*a*) During illumination; (*b*) 0.4 s; (*c*) 2.0 s; (*d*) 4.8 s; and (*e*) 12 s after illumination, respectively. The inset shows the difference spectra of the recombination reaction. The spectrum obtained during illumination (*a*) was taken as a reference.

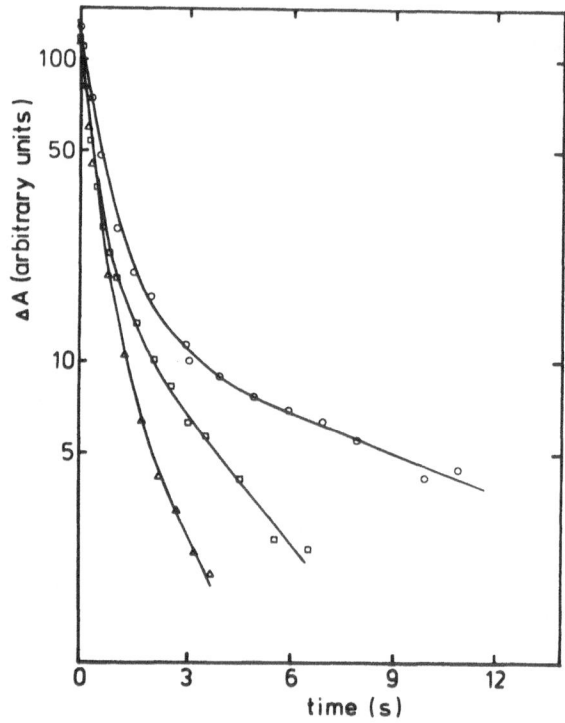

Fig. 3. Time course of the recombination of CO with mixed-valence cytochrome c oxidase. Experimental conditions 4.7 μM cytochrome c oxidase in 100 mM sodium carbonate (pH 9.4), 1% Tween 80, p_{CO} = 2.5 kPa. △—△, Recombination after 1 s illumination; □—□, after 4 s illumination; ○—○, after 10 s illumination.

is fast and complete within 3 s. The inset shows the difference spectra of the recombination reaction with reference to the spectrum obtained during continuous illumination. From the time course of this reaction it was possible to calculate a rate constant for the recombination reaction of 6×10^4 M^{-1} s^{-1} in good agreement with [10].

When the recombination reaction was followed at pH values higher than 8.0 the rate of this reaction slowed down considerably (Fig. 2) and a shift was observed in the isosbestic point. The reaction is clearly biphasic as is illustrated in Fig. 3. This figure shows the time course of the reaction at 428 nm. Two phases were observed, a fast phase which corresponded to the normal recombination of CO with cytochrome c oxidase and a much slower phase. This phase was a non-exponential, the rate of which decreased from 0.4^{-1} to $0.2\ \text{s}^{-1}$ when the illumination time was increased from 1 to 10 s. Clearly, also the amplitude of the slow phase increased when the illumination time was increased. The slower phase was nearly absent when the sample was illuminated for only 1 s. Illumination of the sample for a longer period than 10 s had no further effect on the amplitude and the rate of the second phase.

Discussion

In mixed-valence carboxy-cytochrome c oxidase it is possible to modulate the redox state of the various redox centres by light. These changes are related to the decrease in the apparent midpoint potential of the $a_3^{2+}\cdot\text{CO}–\text{Cu}_\text{B}$ site induced by illumination. Important information regarding the rate of internal electron transfer has been obtained [10] using a dye laser to dissociate the cytochrome $a_3^{3+}\cdot\text{CO}$ compound in the mixed-valence enzyme.

Scheme 1 gives a general mechanism for the light-driven redox changes in cytochrome c oxidase.

$$a^{3+}a_3^{2+}\cdot\text{CO}\underset{\text{CO}}{\overset{h\nu}{\rightleftharpoons}} a^{3+}a_3^{2+}\rightleftharpoons a^{2+}a_3^{3+}\overset{\text{L}}{\rightleftharpoons} a^{2+}a_3^{3+}\cdot\text{L}$$

Scheme 1

In the dark the equilibrium will shift to the left, whereas upon illumination the equilibria are shifted to the right. In the presence of a ligand L which has a high affinity for oxidized cytochrome a_3 the equilibria will shift completely to a state in which cytochrome a is completely reduced [13, 14]. The resulting light-induced effects on the optical absorption spectra are not easy to interpret. Upon dissociation of the cytochrome $a_3^{3+}\cdot\text{CO}$ compound (optical absorption band at 428 nm) by light, unligated reduced cytochrome a_3 will be formed (optical absorption band at 445 nm). As a result a peak at 445 nm and a trough at 428 nm will be found in the light-induced difference spectrum.

The ensuing oxidation of cytochrome a_3 will result in a loss of intensity of the band at 445 nm, which is, however, partly compensated by the synchronous reduction of cytochrome a which also absorbs at this wavelength. The trough at 428 nm will be caused not only by dissociation of cytochrome $a_3^{2+}\cdot\text{CO}$, but also by disappearance of the band of oxidized cytochrome a. According to Vanneste [23], the Soret peak of oxidized cytochrome a_3 is at 412 nm and this band is probably responsible for the small absorbance increase which is observed at this wavelength region [13]. However, when cytochrome a_3 is converted into a low-spin state its Soret peak will move towards higher wavelengths and may coincide entirely with the trough at 428 nm. Therefore, the resulting changes should be interpreted cautiously. However, it is obvious that as judged from the ratio $\Delta A_{445}/\Delta A_{428\ \text{nm}}$ cytochrome a_3 becomes partly oxidized upon illumination of the mixed-valence enzyme. Further, the increase of the ratio at higher pH values shows that more cytochrome a_3 gets oxidized at higher pH values and thus the equilibria in Scheme 1 will shift to the right.

The recombination experiments of CO with mixed-valence cytochrome c oxidase suggest that during illumination a ligand is bound to the partially reduced enzyme which upon recombination is slowly expelled from its binding site by CO. It is tempting to speculate that OH^- is involved in this process, since there is evidence from spectroscopic data that at pH 8.4 a hydroxyl ion can bind to oxidized cytochrome a_3 as the sixth ligand to the haem iron [24, 25] to yield a low-spin state of cytochrome a_3.

Steady-state experiments [8, 9] have indicated that a site is present in cytochrome c oxidase with a pK_a around 8 which when deprotonated inhibits cytochrome c oxidase. It is conceivable in line with [8] that binding of OH^- to cytochrome a_3 is responsible for the observed inhibition of the enzyme.

However, it is clear that more steps and intermediates are involved than binding of OH^- alone. For example, the recombination of CO is a non-exponential curve and the pH dependence of the ratio $\Delta A_{445}/\Delta A_{428\ \text{nm}}$ does not follow a single acid-base curve. Alternatively it is possible that a slow rate of internal electron transfer from cytochrome a and Cu_A to cytochrome a_3 governs the recombination kinetics of CO with cytochrome a_3. However, the values reported [4–6] for ET in such a system are faster (0.6–$5\ \text{s}^{-1}$) than the value of $0.2\ \text{s}^{-1}$ found by us in this study.

Acknowledgements

The authors wish to thank Dr R. Boelens for his help in the early stage of this investigation. This study was in part supported by grants from the Netherlands Organisation for the Advancement of Pure Research (ZWO) under the auspices of the Netherlands Foundation for Chemical Research (SON).

References

1. Van Gelder, B. F. and Muijsers, A. O., *Biochim. Biophys. Acta* **118**, 47 (1966).
2. Greenwood, C. and Gibson, Q. H., *J. Biol. Chem.* **242**, 1782 (1967).
3. Chance, B., Leigh, J. S., Jr. and Waring, A., in *Structure and Function of Energy-Transducing Membranes* (eds. K. van Dam and B. F. van Gelder), p. 1, Elsevier, Amsterdam (1977).
4. Gibson, Q. H. and Greenwood, C., *J. Biol. Chem.* **240**, 888 (1965).
5. Brunori, M., Colosimo, A., Rainoni, G., Wilson, M. T. and Antonini, E., *J. Biol. Chem.* **254**, 10769 (1979).
6. Antalis, T. M. and Palmer, G., *J. Biol. Chem.* **257**, 6194 (1982).
7. Greenwood, C. and Gibson, Q. H., *J. Biol. Chem.* **240**, 2694 (1965).
8. Wilms, J., van Rijn, J. L. M. L. and van Gelder, B. F., *Biochim. Biophys. Acta* **593**, 17 (1980).
9. Thörnström, P.-E., Soussi, B., Arvidsson, L. and Malmström, B. G., *Chem. Scr.* **24**, 230 (1984).
10. Boelens, R., Wever, R. and van Gelder, B. F., *Biochim. Biophys. Acta* **682**, 264 (1982).
11. Marcus, R. A. and Sutin, N., *Biochim. Biophys. Acta* **811**, 265 (1985).
12. Tollin, G., Meyer, T. E. and Cusanovich, M. A., *Biochim. Biophys. Acta* **853**, 29 (1986).
13. Boelens, R. and Wever, R., *Biochim. Biophys. Acta* **547**, 296 (1979).
14. Boelens, R. and Wever, R., *FEBS Lett.* **116**, 223 (1980).
15. Wikström, M., Krab, K. and Saraste, M., *Cytochrome Oxidase*. Academic Press, London (1981).
16. Brzezinsky, P. and Malmström, B. G., *Proc. Natl. Acad. Sci. USA* **83**, 4282 (1986).
17. Wever, R., Boelens, R. and Gorren, A. C. F., in *Frontiers in Bioinorganic Chemistry* (ed. A. Xavier), p. 714. VCH Verlagsgesellschaf, Weinheim (1985).
18. van Buuren, K. J. H., Ph.D. Thesis, University of Amsterdam, Gerja, Waarland (1972).
19. Hartzell, C. R., Beinert, H., van Gelder, B. F. and King, T. E., *Methods Enzymol.* **53**, 54 (1978).

20. Van Gelder, B. F., *Biochim. Biophys. Acta* **118**, 36 (1966).
21. Wever, R., van Drooge, J. H., van Ark, G. and van Gelder, B. F., *Biochim. Biophys. Acta* **374**, 215 (1974).
22. Wever, R. and van Gelder, B. F., *Biochim. Biophys. Acta* **368**, 311 (1974).
23. Vanneste, W. H., *Biochemistry* **5**, 838 (1966).
24. Shaw, R. W., Hansen, R. E. and Beinert, H., *Biochim. Biophys. Acta* **504**, 187 (1978).
25. Lanne, B., Malmström, B. G. and Vänngård, T., *Biochim. Biophys. Acta* **545**, 205 (1979).

Chemica Scripta 1988, 28A, 51–56

The Dioxygen Chemistry of Cytochrome *c* Oxidase

Sunney I. Chan*, Stephan N. Witt and David F. Blair

Noyes Laboratory of Chemical Physics 127-72, California Institute of Technology. Pasadena, CA 91125, USA

Paper presented by Sunney I. Chan at the Nobel Conference 'Biophysical Chemistry of Dioxygen Reactions in Respiration and Photosynthesis', Fiskebäckskil, Sweden, 1–4 July, 1987

Abstract

Dioxygen is reduced to water at the binuclear site of cytochrome *c* oxidase during cellular respiration. Low temperature kinetic experiments indicate that the O–O bond is cleaved at the three-electron level of reduction. The observations are consistent with spontaneous cleavage of a cupric Cu_B/ferrous heme a_3-hydroperoxide to give an oxy-ferryl heme a_3/cupric Cu_B intermediate. This species, which is EPR-silent above 1.8 K, reacts rapidly with CO even at low temperatures to form a product with a magnetically isolated rhombic Cu_B EPR signal. The chemistry is consistent with oxidation of CO to CO_2 by the oxy-ferryl heme a_3 followed by formation of a ferrous heme a_3–O_2 (or CO) adduct adjacent to the EPR-detectable cupric Cu_B. In corroboration of these findings, when partially reduced cytochrome *c* oxidase samples are re-oxidized with dioxygen, an intermediate, in which dioxygen is at the three-electron level of reduction, is trapped at the dioxygen reduction site in a subpopulation of the enzyme molecules. This intermediate exhibits novel spectral features at 580 and 537 nm in the re-oxidized-minus-resting difference spectrum. It also reacts rapidly with CO at 277–298 K to give the rhombic Cu_B EPR signal with concomitant abolition of the 580/537 nm features and intensity increase at 600–604 nm. A homogeneous population of the same oxy-ferryl heme a_3/cupric Cu_B species is produced, as judged by the intensity of the absorption bands at 580 and 537 nm in the difference spectrum, when *excess* H_2O_2 is added to either 'pulsed' or reduced cytochrome *c* oxidase. It is proposed that after re-oxidation of the fully reduced enzyme with H_2O_2, an equivalent of H_2O_2 binds to the pulsed enzyme to produce a peroxidic adduct, namely compound C, and the excess H_2O_2 can then act as a one-electron reductant to reduce the peroxy adduct to yield the trapped oxy-ferryl heme a_3/cupric Cu_B couple and superoxide. A species with the identical spectroscopic signatures as the oxy-ferryl heme a_3/cupric Cu_B intermediate is also formed in the initial step of the partial reversal of the dioxygen reduction reaction [Wikström, M., *Proc. Natl. Acad. Sci. USA* **78**, 4051–4054 (1981)].

Introduction

Cytochrome *c* oxidase mediates the transfer of electrons from ferrocytochrome *c* to molecular oxygen in the mitochondrion, reducing dioxygen to water:

$$4 \text{ Ferrocytochrome } c + O_2 + 4H^+ \xrightarrow{\text{Enzyme}}$$

$$4 \text{ Ferricytochrome } c + 2H_2O$$

Because the enzyme carries out this task rapidly and efficiently, essentially without side reactions, the chemistry that Nature has devised to accomplish the dioxygen cleavage in this reaction is of considerable interest. In the mitochondrion, cytochrome *c* oxidase assumes even greater significance because the enzyme also serves as a free energy transducer and converts part of the redox free energy between ferrocytochrome *c* and molecular oxygen into a trans-membrane protonomotive force ($\Delta\mu_{H^+}$) across the inner membrane where the protein resides. The protein ac-

---

* To whom correspondence should be addressed.

complishes this magnificent feat in two ways. First, the electrons from ferrocytochrome *c* enter the protein on the cytosol side of the inner mitochondrial membrane and the protons that are consumed in the reaction (scalar protons) are taken up from the mitochondrial matrix. In this manner, membrane sidedness is exploited to achieve charge separation across the inner membrane of the mitochondrion. Secondly, additional protons, called vectorial protons, are actively pumped from the matrix to the cytosol as electrons are transferred from ferrocytochrome *c* to the dioxygen anchored at the dioxygen reduction site [1]. Up to four protons are ejected from the matrix per dioxygen molecule reduced, i.e. the H^+/e^- stoichiometry approaches unity in the redox-linked proton translocation [2]. Since four scalar protons are also consumed per dioxygen molecule reduced, the overall reaction catalyzed by cytochrome *c* oxidase in the mitochondrion may be written as follows:

$$4 \text{ Ferrocytochrome } c + O_2 + (n+4) H^+ \text{ (matrix)}$$

$$\xrightarrow{\text{Enzyme}} 4 \text{ Ferricytochrome } c + 2H_2O + nH^+ \text{ (cytosol)}$$

where $0 \leqslant n \leqslant 4$.

The molecular machine that Nature has created to accomplish the above reaction is elegant indeed. Ferrocytochrome *c* does not reduce dioxygen directly. In the enzyme-mediated reaction, the electrons are first intercepted by the two 'low-potential' metal centers of the protein, namely heme *a* and Cu_A, and these two redox centers regulate the flow of electrons to the dioxygen molecule. Heme *a* is a bis-imidazole heme A [3–5]. The ligand structure of Cu_A is unusual, with two cysteines and one (possibly two) histidines [6–8]. The dioxygen that is reduced is in fact bound elsewhere, to a second pair of metal ions located some 15–20 Å away [9], a binuclear site consisting of a myoglobin-like heme ion (heme a_3) [10, 11] and a copper ion (Cu_B), whose ligand structure is still unsettled. Possibly three histidine nitrogens [12] and one methionine sulfur are coordinated to Cu_B [8]. The Cu_B site seems to be conformationally flexible, and it is possible that there is an additional coordination site available for binding of exogenous ligands, including H_2O and OH^-.

Dioxygen does not bind to oxidized cytochrome *c* oxidase. But the reaction of dioxygen with the reduced binuclear site is rapid (second-order rate constant of $10^8 \text{ M}^{-1} \text{ s}^{-1}$) [13]. Oxygen binding to the binuclear center raises the reduction potential of the site by some 500 mV at neutral pH, and this increase in the reduction potential is reflected in the electron transfer (ET) rates between the low potential centers and the dioxygen binding site. This ET is sluggish in the resting oxidase (0.5 s^{-1}) [14] and even in the pulsed state of the oxidized enzyme (~ 2.5 s^{-1}) [15], when compared with the rate of electron entry into the protein or the rate of electron equilibration between heme a and Cu$_A$ (20 s^{-1}) [16]. However, when dioxygen is bound to the binuclear site, the ET from the low potential sites becomes significantly more facile, with rate enhancements approaching 10^3-fold [13]. These observations are consistent with the known dependence of ET rates on the exergonicity ($-\Delta G°$) of the ET reaction, other factors being equal [17], and underscore the role of dioxygen in the terminal step of cellular respiration and in biological energy transduction. In the light of these considerations, we would expect Nature to devise a mechanism of dioxygen reduction that activates and cleaves the dioxygen molecule in steps that apportion the total free energy for the overall reaction in fairly even increments. A high ET exergonicity is particularly essential in the transfer of the last two electrons to the dioxygen molecule anchored at the binuclear site. It seems necessary to compensate for the impeding effects of the large structural reorganizations expected at the binuclear site during the bond cleavage steps on the ET rates.

The O–O bond is cleaved at the three-electron level of reduction

The reaction of reduced cytochrome c oxidase with dioxygen is extremely rapid. Accordingly, it is necessary to monitor the reaction employing rapid kinetic techniques or follow the formation of the intermediates at low temperatures.

In rapid kinetic experiments, the flow–flash method is typically used to photolyze CO from the fully reduced CO-inhibited enzyme in the presence of dioxygen. The rapid ensuing reaction with dioxygen is then monitored spectro-photometrically or by resonance Raman (RR) spectro-scopy. Babcock *et al.* [18, 19] recently combined the rapid-mixing flash photolysis technique with time-resolved RR spectroscopy to study the early phases of the dioxygen reaction at room temperature. Evidence was obtained for the formation of an initial dioxygen adduct (compound A) at the heme a_3 site within the first 50 μs. This intermediate is characterized by oxidation-state and spin-state vibrational marker bands at frequencies similar to those of the dioxygen adducts of myoglobin and hemoglobin, and is photolabile as expected. Non-photolabile intermediates were apparently not formed until well after 100 μs of the initiation of the reaction. Ogura *et al.* [20], in a similar rapid-mixing experiment using a low power continuous wave laser, instead of the high-power pump pulse and the delayed lower intensity probe pulse from a YAG laser employed by Babcock *et al.*, reported formation of an intermediate with a RR spectrum distinct from that of the dioxygen adduct within 450 μs of the initiation of the reaction at 277 K. It is likely that in this intermediate, electron redistribution has taken place between the dioxygen and the metal centers of the binuclear site to give the peroxy compound, i.e. compound C. Another possibility is that some ET has occurred between heme a/Cu$_A$ and the binuclear site so that the spectrum arises from an admixture of compound C and a three-electron-reduced dioxygen intermediate in this study.

In triple-trapping experiments [21], the fully reduced CO-inhibited enzyme is photolyzed in the presence of dioxygen at low temperatures and the reaction with dioxygen is followed over the slower timescales of seconds and minutes. Typically,

the formation of the various intermediates of dioxygen reduction is monitored by optical spectroscopy and electron paramagnetic resonance (EPR) spectroscopy. When the enzyme is incubated at 173 K following the photolysis, compound A, the dioxygen adduct of ferrous heme a_3, is formed within 20 s; the binding of dioxygen to heme a_3 takes place with a bimolecular rate constant of 81 M^{-1} s^{-1} at this temperature [22]. There is no reoxidation of either heme a or Cu$_A$ during the formation of this intermediate. Neither is compound C, the peroxy intermediate, observed under these conditions. In contrast, compound C is produced in high yield, even at 173 K, upon prolonged incubation of aerated mixed-valence CO-inhibited enzyme [23]. It is not clear whether the failure to detect compound C in the reaction with the fully reduced enzyme, is an artifact of the low temperature triple-trapping method of initiating the reaction. Fiamingo *et al.* [24] have shown by FTIR spectroscopy that the CO migrates to the cuprous Cu$_B$ following photolysis, thus it is conceivable that the involvement of Cu$_B$ in the reduction may be inhibited by the bound CO under these low temperature conditions. Alternatively, it could be that electron transfer from the low potential metal centers is greatly accelerated once compound A has been converted to compound C, and it is difficult to trap a sufficient concentration of the peroxy species for spectroscopic detection despite its characteristic intense α-band at 607 nm.

Clore *et al.* [22] and Blair *et al.* [25] have followed by combined optical and EPR spectroscopy the oxidation of heme a and Cu$_A$ as a result of ET from these sites to the partially reduced dioxygen intermediate(s) formed at the binuclear site over the temperature range 166–186 K. The rates are approximately 2×10^{-3} s^{-1} at 173 K. The Arrhenius activation parameters for Cu$_A$ oxidation are $\Delta H^\ddagger \sim 13$ kcal mol^{-1}, $\Delta S^\ddagger \sim 3$ cal mol^{-1} K^{-1} [25]. The kinetic branching ratio, as reflected by the level of heme a oxidation relative to Cu$_A$ oxidation during this early phase of the reaction, do not vary significantly over the temperature range examined, indicating that the activation enthalpies for the heme a and Cu$_A$ oxidations are similar.

The initial transfer of an electron from either Cu$_A$ or heme a to the dioxygen reduction site produces a reduced dioxygen intermediate at the three-electron level of reduction. Since this intermediate contains an odd number of electrons at the dioxygen reduction site, the site should yield an EPR signal. Such an EPR signal has indeed been observed [25, 26], with the rate of its appearance paralleling the oxidations of heme a/Cu$_A$ as expected. This EPR signal is unusual, however. It is very difficult to saturate, even at 9 K and only one of its g-components ($g \sim 2.26$) is evident in the spectrum. This EPR signal exhibits features characteristic of Cu hyperfine interaction. Accordingly it has been assigned to a hydroperoxide-bridged cupric Cu$_B$/ferrous heme a_3 intermediate. It has been proposed [25] that in this intermediate, the hydroperoxide coordination to copper may be stronger than its coordination to iron, so that the electronic structure of the intermediate may be viewed as a cupric hydroperoxide super-exchange or magnetic-dipolar coupled to a high spin ($S = 2$) or intermediate spin ($S = 1$) ferrous heme a_3. A number of cupric hydroperoxides have been synthesized and characterized in recent years [27], with similar EPR parameters as those obtained for the cytochrome oxidase intermediate. The super-exchange interaction with the high spin or intermediate spin ferrous heme a_3 is expected to shift the remaining components of the EPR spectrum to lower g-values and these components may be heterogeneously broadened by 'J strain' to be beyond detection.

Blair *et al.* [25] subsequently discovered that two dioxygen intermediates are in fact formed at the binuclear site at the three-electron level of reduction. These workers attributed the second intermediate to an O–O bond cleavage product of the first intermediate. When the hydroperoxide-bridged cupric Cu_B/ferrous heme a_3 intermediate is incubated above 183 K, the unusual Cu_B EPR signal due to this intermediate decays. This decay occurs without significant further oxidation of heme a or Cu_A (Fig. 1). From a temperature study of this decay process over the temperature range 183–207 K, Blair *et al.* deduced that the process was thermally activated ($\Delta H^{\ddagger} \sim 18$ kcal mol^{-1}) but entropically driven ($\Delta S^{\ddagger} \sim +23$ cal mol^{-1} K^{-1}). These results are consistent with a bond cleavage process with a highly ordered reactant state and a dissociative transition state. On this basis, Blair *et al.* proposed that the O–O bond in dioxygen is cleaved by cytochrome c oxidase at the three-electron level of reduction.

Nature of the second intermediate

If the Blair *et al.* [25] proposal proves correct, known chemistry suggests formation of an oxy-ferryl species at the heme a_3 site in the second intermediate of dioxygen reduction at the three-electron level. It could be either an oxy-ferryl heme a_3/cupric Cu_B species *or* an oxy-ferryl heme a_3 porphyrin π-radical cation/cuprous Cu_B species. The oxy-ferryl heme a_3 in the first instance would be similar to the porphyrin heme in the compound II species of horseradish peroxidase, and the latter would correspond to compound I. Although, given the known reduction potential of Cu_B (~ 400 mV), the electronic distribution of the site described by the second alternative would be unexpected, the two possibilities should be distinguishable by EPR spectroscopy since the site contains an odd number of electrons. Unfortunately, this intermediate appears to be EPR-silent and our efforts to observe its EPR spectrum at temperatures

as low as 1.8 K have proved unsuccessful. Our failure to record an EPR spectrum of this intermediate remains puzzling.

On the other hand, we have been successful in obtaining other spectroscopic signatures of the EPR-silent intermediate. It is possible to trap this intermediate in a subpopulation of the enzyme molecules by re-oxidation in the presence of dioxygen of a sample of the protein containing approximately only three reducing equivalents at the beginning of the reaction [28]. The intermediate so trapped is also EPR-silent, but it is sufficiently stable even at ambient temperatures to allow an optical difference spectrum to be recorded. This difference spectrum exhibits novel spectral features at 580 and 537 nm, that are not assignable to the peroxy adduct or the pulsed enzyme (Fig. 2). On the other hand, these bands are at the expected spectral positions for an oxy-ferryl heme a_3, i.e. a neutral Fe(IV)=O intermediate of heme A, based on comparison with the optical spectra of horseradish peroxidase

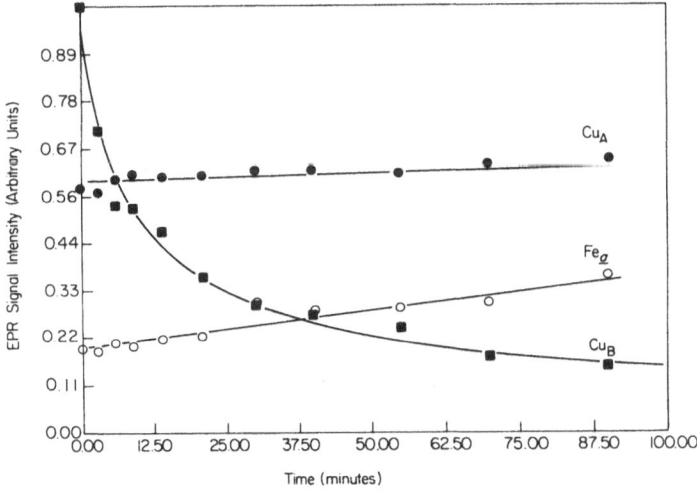

Fig. 1. Intensities of the EPR signals due to Cu_A, Fe_a (heme a), and Cu_B during incubation of a cytochrome c oxidase sample at 191 K following a 2 h preincubation of the fully reduced enzyme in the presence of dioxygen at a lower temperature (177 K) to produce the intermediate (proposed cupric hydroperoxide/ferrous heme a_3) which exhibits the Cu_B EPR signal. (Taken from ref. [25].)

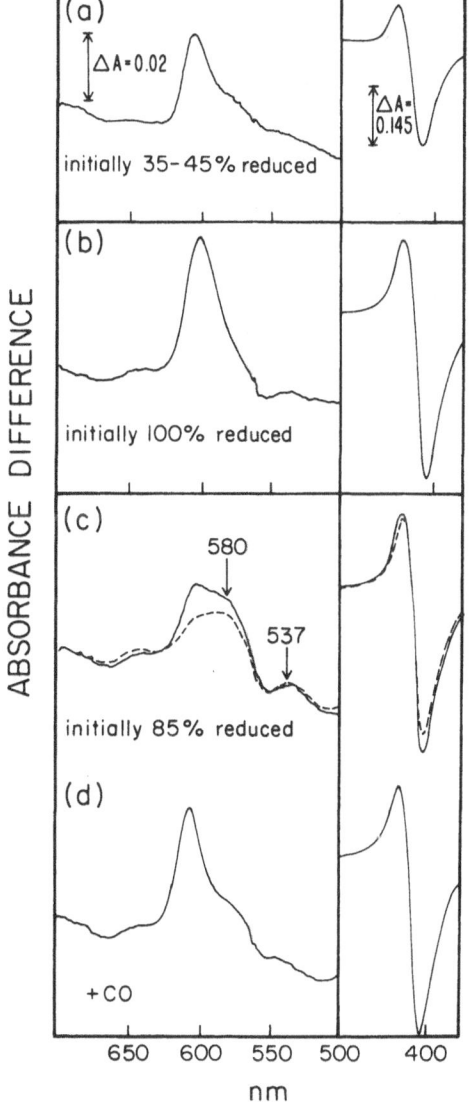

Fig. 2. Optical difference spectra (re-oxidized – resting) for re-oxidized (by dioxygen) cytochrome c oxidase samples with varying initial levels of reduction: (a) initially 35–45% reduced; (b) initially 100% reduced; (c) solid difference spectrum was obtained immediately after re-oxidation of a sample that was initially 85% reduced; dashed difference spectrum obtained 23 min of continued incubation at 277 K; (d) difference spectrum obtained 2–3 min after 1 atm of CO was admitted to sample (c) 28 min following re-oxidation. (Taken from ref. [28].)

compound II and of the ES compound (or the Fe(IV) peroxidase) in cytochrome c peroxidase [29].

An intermediate with identical optical signatures as well as other properties can be prepared by activation of pulsed cytochrome c oxidase or the reduced enzyme by excess H_2O_2 at ambient temperatures (see below) [30]. This method of preparation allows the intermediate to be produced in significantly greater yields and homogeneity so that its RR spectrum can be recorded. The RR spectrum in the high-frequency region exhibits an oxidation state marker ($\nu_4 = 1377$ cm^{-1}) characteristic of an oxy-ferryl species, i.e. an Fe(IV) = O porphyrin. A RR band is also identified at 811 cm^{-1}, that has tentatively been assigned to the Fe(IV) = NO stretch in the oxy-ferryl heme a_3 [31].

Perhaps, the best indication that the EPR-silent intermediate might be an oxy-ferryl heme a_3 is provided by its unusual reactivity toward CO. Blair *et al.* found that when the intermediate is incubated at 211 K or higher, a new EPR signal with g-values and hyperfine splittings characteristic of a magnetically isolated rhombic cupric ion appears. They attributed these new observations to the onset of an oxo-transfer reaction, wherein the ferryl oxo has been transferred from the oxy-ferryl heme a_3 to CO to give CO_2. Since CO is a two-electron reductant, the oxo-transfer reaction should leave an odd number of electrons at the binuclear site. Blair *et al.* proposed that a low-spin ferrous heme a_3 is stabilized by CO or O_2 binding when these ligands are present in excess. Thus

$$[Fe_a^{III}Cu_A^{II}Fe_{a_3}^{IV} = O\,Cu_B^{II}] + CO_2$$
$$\longrightarrow [Fe_a^{III}Cu_A^{II}Fe_{a_3}^{II}Cu_B^{II}] + CO_2$$
$$\xrightarrow{L} [Fe_a^{III}Cu_A^{II}Fe_{a_3}^{II}-L\,Cu_B^{II}] + CO_2 \quad (L = O_2, CO).$$

species with rhombic Cu_B^{II} EPR signal

The formation of a CO– or O_2– adduct at heme a_3 would lead to a magnetically isolated rhombic cupric EPR signal for Cu_B, as observed. In accordance with this explanation, the rhombic Cu_B EPR signal should be replaced by a low-spin hydroxy-ferric heme a_3 EPR signal at pH $\sim$ 9. Since the redox potential of Cu_B is higher than heme a_3 under these conditions, redistribution of the electrons should occur to give the hydroxy-ferric heme a_3/cuprous Cu_B species.

$$[Fe_a^{III}Cu_A^{II}Fe_{a_3}^{II}-L\,Cu_B^{II}] + OH^-$$
$$\longrightarrow [Fe_a^{III}Cu_A^{II}Fe_{a_3}^{III}-OH\,Cu_B^{I}] + L.$$
species with low-spin hydroxy-ferric heme a_3
EPR signal

The reactivity of the second three-electron-reduced dioxygen intermediate toward CO is tantalizing. Although a similar reaction between CO and the resting and pulsed forms of the oxidized enzyme has been recognized for some time [32–35], the products are different. With pulsed or resting oxidase, the product is the CO-mixed valence compound $[Fe_a^{III}Cu_A^{II}Fe_{a_3}^{II}-CO\,Cu_B^{I}]$ rather than $[Fe_a^{III}Cu_A^{II}Fe_{a_3}^{II}-L\,Cu_B^{II}]$, $L = O_2$, CO; i.e. the product formed with the three-electron-reduced dioxygen intermediate contains one additional oxidation equivalent. The binuclear site in the latter exhibits the magnetically isolated rhombic Cu_B EPR signal mentioned earlier, whereas that of the CO-mixed-valence species is EPR-silent. The reactivities of the various species

toward CO are also different: the rate of reaction is relatively slow in the case of the resting enzyme ($t_\frac{1}{2} \sim 400$ min at 277 K); the reaction proceeds much more rapidly with the pulsed enzyme ($t_\frac{1}{2} \sim 4$ min at 277 K); on the other hand, the reaction with the three-electron reduced dioxygen intermediate is extremely facile under comparable conditions. Moreover, the CO oxidation reaction is catalytic in the presence of either O_2 or H_2O_2 with the resting or pulsed enzyme, whereas the oxo-transfer reaction with the oxy-ferryl heme a_3/cupric Cu_B intermediate is catalytic only in the presence of H_2O_2. Despite these notable differences, there must exist an important common feature in the reaction mechanisms for the two processes with respect to activation of the CO prior to the oxo-transfer. We suspect that Cu_B plays a role in this CO activation. At this time, it is unclear whether cupric Cu_B alone acts in this activation, or whether some electron redistribution from heme a_3 to Cu_B must occur to form the transition state.

Activation by hydrogen peroxide

It is well known that hydrogen peroxide can replace dioxygen as the electron acceptor in the cytochrome c oxidase reaction. However, it is less appreciated that the *oxidized* enzyme can also be activated by H_2O_2. A homogeneous preparation of the peroxidic adduct compound C is formed at the binuclear site when one equivalent of H_2O_2 is added to the pulsed enzyme [36, 37]. In the presence of excess H_2O_2, however, the excess H_2O_2 can act as an one-electron reductant to produce the oxy-ferryl heme a_3/cupric Cu_B derivative and $HO_2^\bullet$ under ambient temperatures [30].

$$[Fe_a^{III}Cu_A^{II}Fe_{a_3}^{III}-O-O-Cu_B^{II}] + H_2O_2$$
$$\longrightarrow (Fe_a^{III}Cu_A^{II}Fe_{a_3}^{IV} = O\,Cu_B^{II}-OH] + HO_2^\bullet$$

Photoreduction of compound C also leads to the same identical species with heme a_3 one oxidation equivalent above the ferric state.

When fully reduced (or pulsed) cytochrome c oxidase is treated with excess hydrogen peroxide at room temperature and trace amounts of catalase are quickly added to remove the excess hydrogen peroxide, the resultant enzyme species exhibits an α-band at 596 nm, significantly blue-shifted relative to that in the resting enzyme, and a Soret maximum at 428 nm (Fig. 3 a, b). From the re-oxidized-minus-resting difference spectrum (Fig. 3 c), the α- and β-bands of the activated heme a_3 have blue-shifted to 582 and 533 nm, respectively. No EPR resonances associated with either Cu_B or heme a_3 are observed for this derivative, while Cu_A and heme a show their typical EPR signals with their expected intensities (Fig. 3 f). There is striking similarity between the absorption difference spectrum of the H_2O_2-treated enzyme, i.e. the 428/580 nm species (Fig. 3 c), and that arising from the subpopulation of the three-electron-reduced dioxygen species trapped at the binuclear site following reoxidation of partially reduced cytochrome c oxidase (Fig. 2). The derivative produced by the H_2O_2-treatment also catalyzes the rapid oxidation of CO to CO_2 (Fig. 3 d, e), with the concomitant formation of the magnetically isolated rhombic Cu_B EPR signal seen in the low temperature experiments. In fact, the rhombic Cu_B EPR signal is produced in substantial greater yields under these conditions (Fig. 3 g, h).

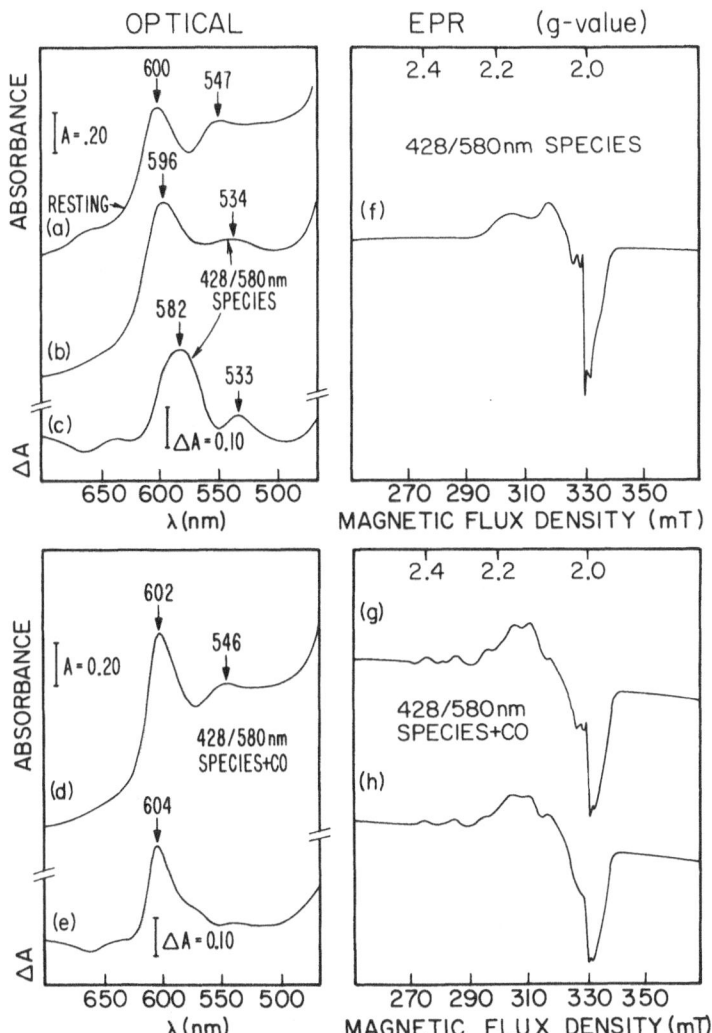

OPTICAL   EPR   (g-value)

Fig. 3. Optical absorption and EPR spectra of the 428/580 nm component of H_2O_2-treated cytochrome c oxidase: (*a*) optical absorption spectrum of resting enzyme; (*b*) optical absorption of the H_2O_2-treated enzyme (reduced enzyme was reoxidized with a 24-fold molar excess of H_2O_2, followed in 105 s by the addition of 0.1% catalase); (*c*) re-oxidized-minus-resting difference spectrum assigned to the 428/580 nm component of H_2O_2-treated enzyme; (*d*) optical absorption spectrum of the enzyme after incubating re-oxidized sample under 1 atm of CO for 20 s at approximately 283 K; (*e*) optical difference spectrum of the H_2O_2-treated enzyme after CO-incubation *vs.* resting enzyme; (*f*) X-band EPR spectrum of the H_2O_2-treated enzyme at 77 K; (*g*) 77 K X-band EPR spectrum of the H_2O_2-treated enzyme following CO-incubation; (*h*) re-recording of the X-band EPR spectrum at 77 K following thawing of sample in (*g*) to obtain the optical spectra in (*d*) and (*e*). All optical spectra were recorded at ice temperature. (Taken from ref. [30].)

These observations demonstrate that hydrogen peroxide can activate the fully reduced (or pulsed) cytochrome c oxidase to form the same EPR-silent oxy-ferryl heme a_3/ cupric Cu_B species produced with dioxygen in the low-temperature single-turnover experiments discussed earlier. Based on EXAFS measurements, which show a similarity in the 'iron–oxygen' bond length between the H_2O_2-treated enzyme and horseradish peroxidase compound I, Chance and co-workers [38] proposed that the species activated by excess hydrogen peroxide contains an oxy-ferryl heme a_3 porphyrin π-radical cation adjacent to a cupric Cu_B at the binuclear site and referred to the intermediate as 'pulsed peroxide I'. This conclusion is unlikely to be correct, since the activated heme a_3 is only *one*, not two, oxidation equivalent(s) above the ferric heme a_3. The oxy-ferryl heme a_3 porphyrin π-radical cation/cupric Cu_B structure proposed

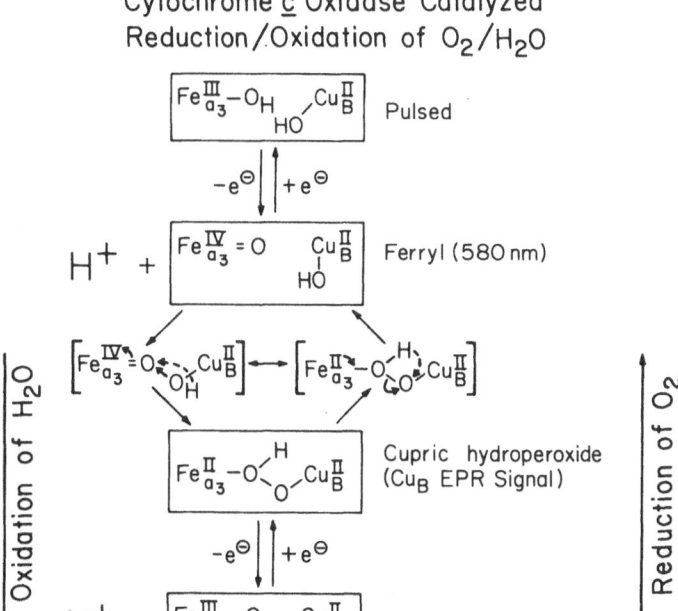

Cytochrome c Oxidase Catalyzed
Reduction/Oxidation of O_2/H_2O

Fig. 4. A possible scheme depicting the reaction mechanism for the forward and the reverse reactions in the final two electron-transfer steps of dioxygen reduction in cytochrome c oxidase.

corresponds to a peroxide-heterolytic-bond-cleavage product of compound C. In our view, it is unlikely that a porphyrin π-radical cation would be produced in any activated forms of cytochrome c oxidase, as the powerful electron withdrawing effect of the formyl substituent would render the heme A difficult to be oxidized. Instead, all the available evidence points to a heme a_3 in the H_2O_2-activated enzyme that is more akin to the oxy-ferryl protoporphyrin heme in horseradish peroxidase compound II.

Partial reversal of the dioxygen reduction reaction

In a classical set of experiments, Wikström [39] demonstrated that partial reversal of the dioxygen reaction is possible in mitochondria. When mitochondria are poised in a highly oxidized state with ferricyanide, the addition of high concentrations of ATP induces the sequential formation of two intermediates, which are similar, if not identical, to two of the intermediates that are formed when dioxygen reacts with fully reduced cytochrome c oxidase. The first intermediate Wikström observed upon stepwise reversal of the dioxygen reduction reaction was a 580 nm species, which Wikström suggested was a one-electron oxidation product of the ferric heme a_3/cupric Cu_B couple. It exhibits the same spectroscopic signatures as the oxy-ferryl heme a_3/cupric Cu_B species described earlier. The second intermediate displayed intense absorption at 607 nm and is spectroscopically indistinguishable from the peroxidic adduct compound C. Thus, a partial reversal of the dioxygen reduction reaction had occurred. It is evident that the 580/537 nm intermediate, which we have designated as the oxy-ferryl heme a_3/cupric Cu_B species, may be produced either by reducing dioxygen by three electrons *or* by reversing the dioxygen reaction by a single ET. A possible scheme depicting the reaction mechanism for the forward and the reverse reactions is suggested in Fig. 4. Subsequent potential measurements over a range of pH's indicate that the reduction potentials of the 580/530 nm species as well as compound C are *c.* 800 mV *vs.* NHE and that $2H^+$ are released into the matrix during each of the one-electron oxidation steps (M. Wikström, private communication). Only one H^+ is shown for each of these steps in Fig. 4. It is suggested that the remaining two protons originate from deprotonation of nearby amino acid residues of the catalytic site.

Concluding remarks

In this review we have attempted to summarize our current understanding of the dioxygen chemistry of cytochrome *c* oxidase. Although good progress has been made in recent years, our knowledge remains fairly rudimentary. Several lines of experimental evidence are presented here to indicate that the O–O bond is cleaved at the three-electron level of dioxygen reduction in this enzyme; the final oxidizing equivalent is stored in the form of oxy-ferryl species at the heme a_3 site. Thus, Nature exploits variants of the same basic chemistry in the peroxidases, cytochrome P-450 as well as in cytochrome *c* oxidase. Future work will concentrate on further structural characterization of the various intermediates formed during turnover, determination of the rate constants for the various electron transfer and protonation steps under coupled as well as uncoupled conditions, and identification of those steps that are linked to redox-linked proton translocation.

Acknowledgments

This work was supported by grant GM 22432 from the National Institute of General Medical Sciences, United States Public Health Service, and by BRSG grant RR 07003 awarded by the Biomedical Research Support Grant Program, Division of Research Resources, National Institutes of Health. D. F. B. was a recipient of a National Research Service Award (5T32GM-07616) from the National Institute of General Medical Sciences. S. I. C. is a Fogarty Scholar-in-residence at the National Institutes of Health, Bethesda, MD 20892. This article is contribution 7648 from the Division of Chemistry and Chemical Engineering, California Institute of Technology.

References

1. Wikström, M., *Nature* **266**, 271–273 (1977).
2. Wikström, M., Krab, K. and Saraste, M., *Cytochrome c Oxidase: A Synthesis.* Academic Press, New York (1981).
3. Babcock, G. T., Callahan, P. M., Ondrias, M. R. and Salmeen, I., *Biochemistry* **20**, 959–966 (1981).
4. Peisach, J. and Mims, W. B., *Isr. J. Chem.* **21**, 59–60 (1981).
5. Martin, C. T., Scholes, C. P. and Chan, S. I., *J. Biol. Chem.* **260**, 2857–2861 (1985).
6. Stevens, T. H., Martin, C. T., Wang, H., Brudvig, G. W., Scholes, C. P. and Chan, S. I., *J. Biol. Chem.* **257**, 12106–12113 (1982).
7. Martin, C. T., Scholes, C. P. and Chan, S. I. (manuscript submitted to *J. Biol. Chem.*).
8. Li, P. M., Gelles, J., Chan, S. I., Sullivan, R. J. and Scott, R. A., *Biochemistry* **26**, 2091–2095 (1987).
9. Brudvig, G. W., Blair, D. F. and Chan, S. I., *J. Biol. Chem.* **259**, 11001–11009 (1984).
10. Stevens, T. H. and Chan, S. I., *J. Biol. Chem.* **256**, 1069–1071 (1981).
11. Blair, D. F., Martin, C. T., Gelles, J., Wang, H., Brudvig, G. W., Stevens, T. H. and Chan, S. I., *Chem. Scr.* **21**, 43–53 (1983).
12. Cline, J., Reinhammer, B., Jensen, P., Venters, R. and Hoffman, B. M., *J. Biol. Chem.* **258**, 5124–5128 (1983).
13. Greenwood, C. and Gibson, Q. H., *J. Biol. Chem.* **242**, 1782–1787 (1967).
14. Gibson, Q. H., Greenwood, C., Wharton, D. C. and Palmer, G., *J. Biol. Chem.* **240**, 888–894 (1965).
15. Brunori, M., Colosimo, A., Rainori, G., Wilson, M. T. and Antonini, E., *J. Biol. Chem.* **254**, 10769–10775 (1979).
16. Halaka, F. G., Barnes, Z. K., Babcock, G. T. and Dye, J. L., *Biochemistry* **23**, 2005–2011 (1984).
17. Marcus, R. A. and Sutin, N., *Biochim. Biophys. Acta* **811**, 265–322 (1985).
18. Babcock, G. T., Jean, J. M., Johnson, L. N., Palmer, G. and Woodruff, W. H., *J. Am. Chem. Soc.* **106**, 8305–8311 (1984).
19. Babcock, G. T., Jean, J. M., Johnston, L. N., Woodruff, W. H. and Palmer, G., *J. Inorg. Biochem.* **23**, 243–251 (1985).
20. Ogura, T., Yoshikawa, S. and Kitagawa, T., *Biochim. Biophys. Acta* **832**, 220–223 (1985).
21. Chance, B., Saronio, C. and Leigh, J. S., Jr., *J. Biol. Chem.* **250**, 9226–9237 (1975).
22. Clore, G. M., Andréasson, L. E., Karlsson, B., Aasa, R. and Malmström, B. G., *Biochem. J.* **185**, 139–154 (1980).
23. Clore, G. M., Andréasson, L. E., Karlsson, B., Aasa, R. and Malmström, B. G., *Biochem. J.* **185**, 155–167 (1980).
24. Fiamingo, F. G., Altschuld, R. A., Moh, P. P., and Alben, J. O., *J. Biol. Chem.* **257**, 1639–1650 (1982).
25. Blair, D. F., Witt, S. N. and Chan, S. I., *J. Am. Chem. Soc.* **107**, 7389–7399 (1985).
26. Karlsson, B., Aasa, R., Vänngård, T. and Malmström, B. G., *FEBS Lett.* **131**, 186–188 (1981).
27. Thompson, J. S., *J. Am. Chem. Soc.* **106**, 4057–4059 (1984).
28. Witt, S. N., Blair, D. F. and Chan, S. I., *J. Biol. Chem.* **261**, 8104–8107 (1986).
29. Ho, P. S., Hoffman, B. M., Kang, C. H. and Margoliash, E., *J. Biol. Chem.* **258**, 4356–4363 (1983).
30. Witt, S. N. and Chan, S. I., *J. Biol. Chem.* **262**, 1446–1448 (1987).
31. Witt, S. N., Manthey, J., Kean, R., Centano, J., Babcock, G. T. and Chan, S. I. (unpublished results).
32. Young, L. and Caughey, W., *Fed. Proc.* **39**, 2090 (Abstr. 2562) (1980).
33. Bickar, D., Bonaventura, C. and Bonaventura, J., *J. Biol. Chem.* **259**, 10777–10783 (1984).
34. Brzezinski, P. and Malmström, B. G., *FEBS Lett* **187**, 111–114 (1985).
35. Morgan, J. E., Blair, D. F. and Chan, S. I., *J. Inorg. Biochem.* **23**, 295–302 (1985).
36. Wrigglesworth, J. M., *Biochem. J.* **217**, 715–719 (1984).
37. Bickar, D., Bonaventura, J. and Bonaventura, C., *Biochemistry* **21**, 2661–2666 (1982).
38. Chance, B. and Powers, L., *Curr. Topics in Bioenergetics* **14**, 1–19 (1985).
39. Wikström, M., *Proc. Natl. Acad. Sci. USA* **78**, 4051–4054 (1981).

Chemica Scripta 1988, **28A**, 57–61

The Reactions between the Carbon Monoxide Complex of Cytochrome *c* Oxidase and Oxygen: An Investigation of Electron Transfer between the Functional Units of Cytochrome *c* Oxidase

M. T. Wilson

Department of Chemistry, University of Essex, Wivenhoe Park, Colchester, Essex, United Kingdom

M. Brunori and P. Sarti

Department of Biochemical Sciences and Consiglio Nazionale delle Ricerche, Centre of Molecular Biology, University of Rome, 'La Sapienza', Rome, Italy

and G. Antonini and F. Malatesta

Department of Experimental Medicine and Biochemical Sciences, University of Rome 'Tor Vergata', Rome, Italy

Paper presented by Michael T. Wilson at the Nobel Conference 'Biophysical Chemistry of Dioxygen Reactions in Respiration and Photosynthesis', Fiskebäckskil, Sweden, 1–4 July, 1987

Abstract

The oxidation of the carbon monoxide adduct of cytochrome *c* oxidase by oxygen has been studied by stopped-flow spectrophotometry, The complex time-courses for the oxidation of cytochrome *a* and Cu_A have been used to monitor electron transfer (ET) between functional units. The rate of oxidation of cytochrome *a* in monomeric enzyme (subunit III-less or shark) exhibited a strong dependence on the enzyme concentration, the dimeric form of the enzyme very much less so, if at all. This result together with experiments on partially CO-saturated enzyme and on the effect of cytochrome *c* lead us to conclude that ET between monomers within a dimer can occur and that it is this reaction which dictates the shape of the time course for the dimer at low enzyme concentration. The inter-monomer ET rate is salt concentration-dependent with phosphate being particularly effective at stimulating this rate.

The results are discussed in terms of models describing the system and attention is drawn to the fact that experiments of this type provide a relatively simple method of distinguishing the monomeric and dimeric forms of the enzyme.

Introduction

Cytochrome *c* oxidase (EC $1 \cdot 9 \cdot 3 \cdot 1$), the terminal electron acceptor of the mitochondrion, is a transmembrane protein complex. It has been shown to exist, both in the inner mitochondrial membrane [1] and when reconstituted at high protein-to-lipid ratio into the phospholipid membrane of large vesicles [2], in a dynamic equilibrium comprising a relatively mobile fraction and an aggregated fraction. In such systems it may be possible that during the catalytic cycle of the enzyme and the accompanying electron flux through the enzyme electron transfer (ET) occurs between functional units of the oxidase (i.e. units containing one set of the redox active metals Cyt *a*, Cu_A, Cu_B and Cyt a_3). Any such ET would presumably be strongly catalyzed by cytochrome *c*.

ET between oxidase functional units has, to date however, received relatively little attention although it has been invoked to explain the slow re-oxidation of cytochrome *a* following the enzyme's reaction with ferrocytochrome *c* in the presence of excess oxygen [3] and the differences in the spectrum of cytochrome oxidase following oxygen displacement of CO either in the dark or by photolysis [4]. In order to investigate inter-oxidase ET reactions more fully we have followed the earlier work of Gibson and Greenwood [5] who studied the reactions of the carbon monoxide complex of fully reduced cytochrome *c* oxidase with molecular oxygen. The surprising result these workers obtained was that the oxidation of cytochrome *a* was autocatalytic and complete before the full oxidation of cytochrome a_3, which itself was oxidized, as expected, at a rate limited by the dissociation rate (~ 0.02 s^{-1}) of CO from the complex. As oxidation of cytochrome *a* proceeds via cytochrome a_3 these results were interpreted in terms of ET between cytochrome a moieties in different functional units leading to a single oxidized cytochrome a_3 eventually accepting electrons from a number of cytochromes *a*. Provided this ET is faster than CO dissociation this mechanism is capable of accommodating the observed kinetic results.

In this study we have used the maximum value of the observed rate constant for the oxidation of cytochrome *a* (and Cu_A) as a measure of the inter-oxidase ET rate and have examined the dependence of this rate constant on the oxidase concentration both for dimeric and monomeric forms of the enzyme. From these studies it appears that, at low enzyme concentration, ET within the dimer plays an important role in determining the oxidation profile for cytochrome *a*. The exact shape and rate of this oxidation is however rather variable from preparation to preparation and we have noted that it has a strong dependence on the ionic composition of the medium, with phosphate having the most marked effect.

At present we do not have available a full explanation in quantitative terms for the shape of the kinetic progress curves but some possibilities are discussed below. The fact, however, that the dimer plays an apparently important role has allowed us to devise a simple kinetic method for distinguishing dimers from monomers. This method has been applied not only to the enzyme in solution but to the enzyme reconstituted into vesicles [6].

Experimental

Enzyme and chemicals

Cytochrome c oxidase was prepared from beef heart by the method of Yonetani [7]. Enzyme concentrations (functional units) were determined spectrophotometrically using ϵ_{605} (redox) $= 22$ mM^{-1} cm^{-1}. Monomeric beef enzyme was prepared by removal of subunit III using the method of Puettner *et al.* [8]. Monomeric shark (*Sphyrna lewini*) enzyme was the kind gift of Dr D. Bickar. Reconstitution of the enzyme into vesicles was performed using the method of Hinkle *et al* [9] as modified by Sarti *et al.* [10].

The carbon monoxide derivative of the fully reduced enzyme was prepared by addition of a solution of dithionite (prepared in previously de-oxygenated buffer) to a final concentration of 500 μM and CO-equilibrated buffer to give 100 μM CO to degassed nitrogen-equilibrated enzyme preparations. Dilution of this complex was carried out as required using a degassed buffer containing 100 μM-CO and 500 μM dithionite.

The aggregation state of the enzyme and its derivatives was determined by measurement of the sedimentation coefficient using a Beckman Model E Analytical Ultracentrifuge.

Hepes (4-(2-hydroxyethyl)-1-piperazine-ethane sulphonic acid], cytochrome c (Type VI) and Soy bean phospholipids (L-α-phosphatidylcholine Type II-S), were purchased from Sigma Chemical Company.

Results

Aggregation state of the enzyme

The sedimentation coefficients of the native and subunit III-depleted enzyme were determined between pH 7.0 and pH 7.4, over the concentration range 3.5–40 μM enzyme and either in low ionic strength buffer (5 μM Hepes) or at higher salt (> 25 mM phosphate). In addition to the oxidized 'resting' form of the enzyme, both the reduced form and the reduced carbonmonoxy complex were examined. The enzyme was found in all cases to consist of a single peak comprising approximately 90 % of the total protein, > 95 % for the subunit III-less enzyme. The native enzyme, irrespective of protein concentration and redox state, gave S_{20} values (uncorrected for bound lipid) ranging between 11.3 and 12.9 S with a mean value of 12.1 ± 0.6 S. This value is characteristic of the dimer [11–13]. The subunit III-less enzyme, again irrespective of the conditions yielded S_{20} values in the range 5.7–6.9 S with a mean value of 6.1 ± 0.4 S, value characteristic of the monomer [11–13].

Stopped-flow experiments

On mixing the carbonmonoxy adduct of fully reduced cytochrome c oxidase with molecular oxygen, complex wavelength-dependent progress curves resulted. These have been analysed by Gibson and Greenwood [5] in terms of the sum of two component reactions, namely (i) the exponential time-course for the oxidation of cytochrome $a_3^{2+} \cdot$ CO by molecular oxygen, rate limited at the CO dissociation rate constant (~ 0.02 s^{-1}) and (ii) the autocatalytic oxidation of cytochrome a^{2+}, which, because of inter-oxidase ET, attains a rate greater than the CO 'off' rate, thus leading to complete oxidation of cytochrome a before full oxidation of cytochrome a_3^{2+}. Thus the autocatalytic process is not a property of the isolated functional unit but of interactions between subunits. In Fig. 1 we illustrate the time-courses and give the kinetic difference spectra of the two processes in the α-band region and out to 900 nm. The results in the α-band region are in excellent agreement with those of Greenwood and Gibson [4]. In addition, we show that the return of the 830 nm absorption band on oxidation is also more rapid than oxidation of cytochrome a_3^{2+}. This latter result indicates that electrons from the Cu$_A$ centre of one functional unit may also be transferred to other functional units which have been oxidized.

The slopes of logarithmic plots of cytochrome a oxidation increase with time, in keeping with its autocatalytic nature (see insert Fig. 1), reaching a maximum value. This value has been taken as a measure of the inter-functional unit ET rate constant under study. It is not a simple constant, depending on the values of the more fundamental rate constants, on the extent to which species in the reaction pathway are populated and on their relative extinction coefficients. It is nevertheless an easily measured phenomenological parameter. The value of this apparent rate constant, while being reproducible for a given set of experiments with the same enzyme preparation, was quite variable (0.1–0.5 s^{-1}) from preparation to preparation and on the age of a given preparation, on the

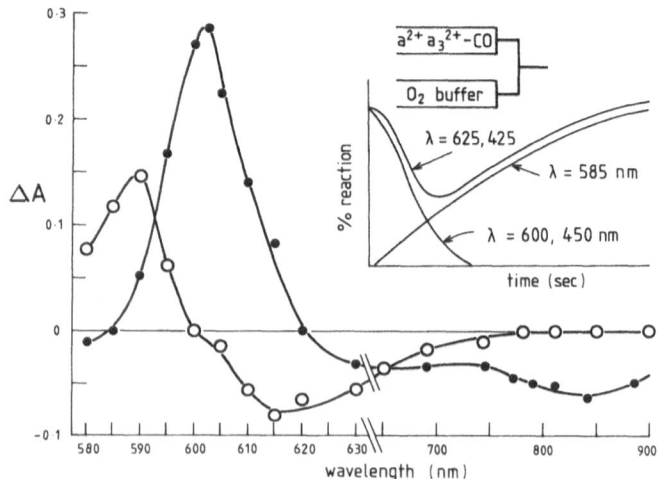

Fig. 1. Kinetic difference spectra of the separated phases of the time-courses of the reaction between 5 μM carbonmonoxy cytochrome c oxidase and 600 μM oxygen. The insert shows the experimental arrangement and illustrates the form of the progress curves at a number of wavelengths during the first 60 s after mixing. The wavelength distribution of the amplitudes of the slow exponential phase (○) corresponds to the difference spectra $a_3^{2+} \cdot$ CO minus a_3^{3+}, while that of the faster phase (●) to $a^{2+} \cdot$ Cu$_A^+$ minus $a^{3+} \cdot$ Cu$_A^{2+}$ $\cdot$ 0.1 M phosphate buffer, pH 7.4 containing 1 % Tween 80. Temp. 20° C. Durrum-Gibson stopped-flow, 2 cm light path.

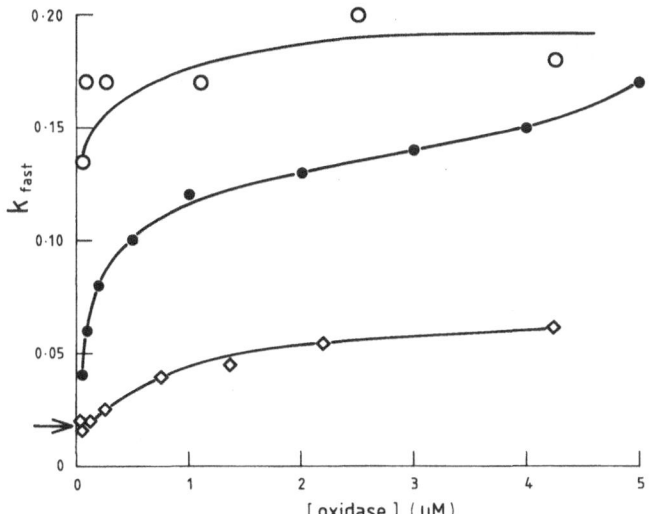

Fig. 2. The dependence of the rate of oxidation of cytochrome *a* on oxidase concentration (functional units). The monitoring wavelengths were either 600 or 450 nm. Bovine dimers ○; Bovine subunit III–less monomers, ●; shark monomers, ◇. Other conditions as Fig. 1. The arrow indicates the CO 'off' rate, in s⁻¹.

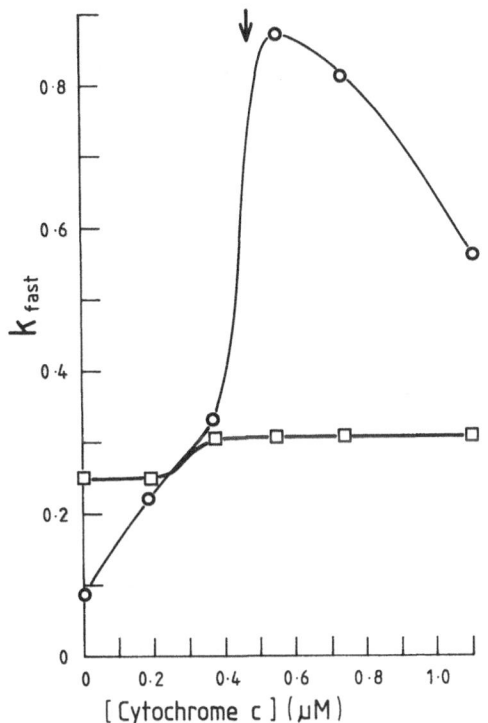

Fig. 3. Effect of cytochrome *c* on the rate of oxidation of cytochrome *a*. ○, 5 mM Hepes, buffer pH 7.3 and oxidase concentration 0.43 μM. The arrow indicates the stoichiometry of 1 cytochrome *c*/oxidase functional unit; □, 5 mM Hepes, buffer pH 7.3 plus 0.2 M-NaCl and oxidase concentration 0.37 μM. Temp. 20 °C. *K* in s⁻¹.

number of times a preparation had been frozen and thawed, and was dependent on solution parameters.

Figure 2 shows the dependence of this rate constant on the oxidase concentration for different aggregation states of the enzyme. If, as proposed, the oxidation of cytochrome *a* proceeds through inter-oxidase ET, then the rate may be expected to show a concentration dependence typical of a second-order reaction, eventually reaching an upper limiting value characteristic of the constant governing ET between functional units within a collisional complex. Such behaviour is indeed observed with the subunit III-depleted mammalian

enzyme and the enzyme isolated from the Hammerhead shark. Both these preparations are monomeric and at very low concentration the rate of cytochrome *a* oxidation approaches that of the CO 'off' rate. This also is expected, for where the inter-oxidase transfer rate falls below 0.02 s⁻¹, because of extreme dilution, each cytochrome *a* is oxidized by the single cytochrome a_3 within the same functional unit and at the rate at which this cytochrome a_3 is oxidized, i.e. the CO dissociation constant. The dimeric enzyme, however, shows different behaviour. In Fig. 2 we see that little or no concentration dependence is observed down to below 10⁻⁷ M enzyme. It is very unlikely that a concentration dependent phase, occurring below 5 × 10⁻⁸ M enzyme, is being missed, as this would require a second order rate constant for dimer/dimer interaction approaching a value of 10⁷ M⁻¹ s⁻¹. For such large complexes this value seems improbably large and thus we interpret the autocatalytic profile, yet lack of concentration dependence, as reflecting ET within the dimer.

Stopped-flow experiments have also been carried out in which oxygenated buffer was mixed with reduced dimeric oxidase which was only partially saturated with CO. In these experiments (not shown) the rate constant for oxidation of cytochrome *a* increased in step with the proportion of the dimers which had a single CO molecule bound (as calculated from the binomial distribution). The auto-catalytic nature of the time-course also became progressively less evident on lowering the CO saturation. These results may be rationalized by noting that upon mixing with oxygen any unliganded functional units would be oxidized within the dead time of the apparatus. Thus the maximum rate enhancement was observed when each cytochrome *a* which remained reduced after mixing (because CO was bound to the cytochrome a_3 of the same functional unit) was within a dimer containing a fully oxidized functional unit. In this circumstance intra-dimer ET would facilitate oxidation of the remaining reduced cytochrome *a* leading to an enhanced rate constant. As each reduced cytochrome *a* had an oxidized cytochrome a_3 available in a neighbouring functional unit, the initial rate would not be limited by the production of oxidized cytochrome a_3 (at the CO dissociation rate) and hence the autocatalytic nature would vanish.

These experiments therefore support the idea that the progress curves for cytochrome *a* oxidation in the fully CO saturated native enzyme are a reflection of the dimeric nature of that enzyme.

A possible mechanism for intra-dimer ET may be via small, bound, contaminating ET agents which co-purify with the enzyme. Examination of the absorbance spectrum of the enzyme, however, indicated that the concentration of the other cytochromes (c/c_1 and *b*) was very low ($\sim 2\%$). Nevertheless, in order to examine the effect of such agents, experiments were carried out in which cytochrome *c*, the *in vivo* electron donor for cytochrome *c* oxidase, was added back to the enzyme and the oxidation of the CO adduct by oxygen monitored. In these experiments the cytochrome *c* was oxidized (as monitored at 550 nm) in an autocatalytic process and in parallel with cytochrome a^{2+}. Fig. 3 shows the dependence of the maximum oxidation rate on the added cytochrome *c* concentration under two ionic strength regimes. At low ionic strength, where separate gel filtration experiments showed that at sub-stoichiometric concentrations the added cytochrome c was fully complexed to the oxidase, the rate increased with cytochrome c up to the value of 1 cytochrome c per functional unit. This indicates that although closely associated with the enzyme, the cytochrome c could facilitate ET between functional units within the dimer, presumably being able to slide over the surface to which it is bound and ferry electrons from centres in neighbouring functional units. In this particular experiment (one of several which were performed) the rate in the absence of added cytochrome c was approximately 0.1 s^{-1}; examination of the figure would suggest that for this to be due to contamination would require the presence of cytochrome c at a concentration equivalent of about 20 % of the oxidase, clearly a requirement not in agreement with the observed spectrum. Above the concentration at which cytochrome c is stoichiometric with oxidase functional units the rate changes dramatically, presumably as some cytochrome c is now free in solution and is able to transfer electrons between functional units in different dimers. This allows a given oxidized cytochrome a_3 to catalyse the oxidation of many cytochrome a centres and hence leads to a large rate enhancement. On addition of further cytochrome c, the apparent rate constant drops as more electrons have to pass through the enzyme (from cytochrome a^{2+} to oxygen) and thus more time is needed. When the experiment is carried out at high salt concentration (Fig. 3), no dramatic effect of increasing cytochrome c from below to above the stoichiometric concentration was observed. This reflects the fact that there is no transition from having all the cytochrome c bound to a situation in which some is free. In addition, the rate constant for ET between cytochrome c and cytochrome a falls from approximately 5×10^7 M^{-1} s^{-1} in Hepes buffer (pH 7.3) to approximately 10^6 M^{-1} s^{-1} in this buffer plus 0.2 M-NaCl. Fig. 3 also shows that in the absence of cytochrome c the rate of oxidation is enhanced by 0.2 M-MaCl. This rate enhancement by ions was most noticeable for phosphate.

Figure 4 illustrates the results of experiments in which the effects of phosphate concentration on the rate of oxidation of cytochrome a in both dimeric and subunit III–less monomeric oxidase was studied. In these experiments the oxidase concentrations was 5 μM (in functional units), a concentration at which both dimers and monomers show autocatalytic oxidation, both the first order and second order inter-unit electron transfer being faster than CO dissociation (see Fig. 2). The final stage in the isolation of the enzyme was carried

out in the absence of phosphate and the enzyme preparation for the experiment were diluted with Hepes buffer (pH 7.3) and allowed to stand for some hours in the cold to allow bound phosphate to dissociate. Under these conditions the maximum rate constant for cytochrome a oxidation is only about twice the CO dissociation constant, and hence the time-course is only slightly autocatalytic. This rate increased markedly on addition of even relatively low concentrations (e.g. 5 mM) of phosphate, reaching a maximum at approximately 25 mM and thereafter decreasing. In the particular experiments shown in Fig. 4 the maximum rate attained by the monomer was greater than that of the dimer; however, as stated earlier the maximum rate can be rather variable and at this stage we do not wish to place undue emphasis on this point. As can be seen, allowing the dimer to 'age' in Hepes for a longer period decreased the maximum rate achieved.

Discussion

It is clear from our results and those of others [3, 4] that ET between functional units can occur in cytochrome c oxidase and that at low enzyme concentration, ET between monomers *within* a dimer plays a major role. This ET pathway may be disrupted by monomerization, as shown by the subunit III–less and the shark enzyme. However, it is clear that other factors may also disrupt or at least impair ET within the dimer. It may be that the exact way the monomers are juxtaposed may vary, leading to the observed variability of this rate (also noticed by Gibson and Greenwood [5]). Changes in the way monomers interact are possibly at the root of the marked effect of phosphate. It has recently been shown that ATP binds to and changes the conformation of cytochrome c oxidase [14]. The reported structural changes are propagated throughout the molecular complex and have marked effects on the interaction between cytochrome c and its oxidase. This indicates that ATP binding triggers conformational changes in the vicinity of the ET site, i.e. cytochrome a. It is feasible that such changes may also affect cytochrome a/cytochrome a interactions. Phosphate, at significantly higher concentrations, may perhaps substitute for ATP. We also note that polylysine (at concentration $\sim$ 5 μM) reduced the oxidation rate of cytochrome a (results not shown). This polycation binds to the enzyme in the region of the cytochrome c-binding site and presumably disrupts the intra-dimer pathway.

Although the evidence presented strongly suggests that ET within a dimer can account for the autocatalytic time course for cytochrome a oxidation, quantitative analysis of models describing such a system reveal considerable gaps in our understanding. Computer simulation on models in which only the haem groups are considered reproduce qualitatively the features of the curves obtained experimentally (as in Fig. 1), but cannot accommodate the large rate increase. Introduction of the copper atoms, however, enlarges the model very significantly. One such model, following the formulation introduced by Malmström and Andréasson [15] is given in Scheme 1. The complexity of this mechanism is daunting but some properties may be discerned. For example, every path from the CO complex to an enzyme in which both cytochrome a centres are oxidized contains at least one CO displacement reaction. No matter what values are assigned to the rate constants for other steps this limits the rate

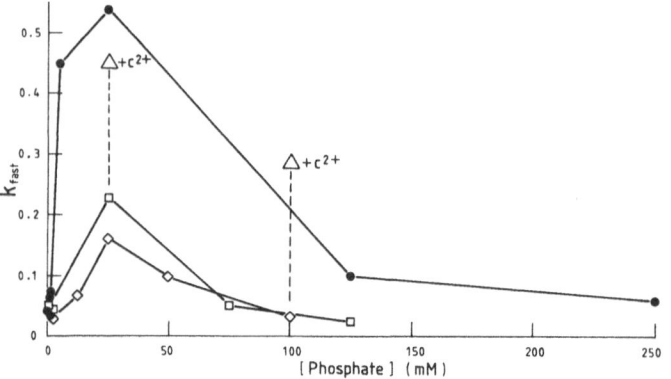

Fig. 4. Effect of phosphate concentration on the oxidation of cytochrome a. ●, subunit III–less enzyme; □, native dimer; ◇, 'aged' (stored 8 h) native dimer; △, addition of cytochrome c at stoichiometric concentration. 5 mM Hepes buffer pH 7.3. Temp. 20 °C. k in s^{-1}.

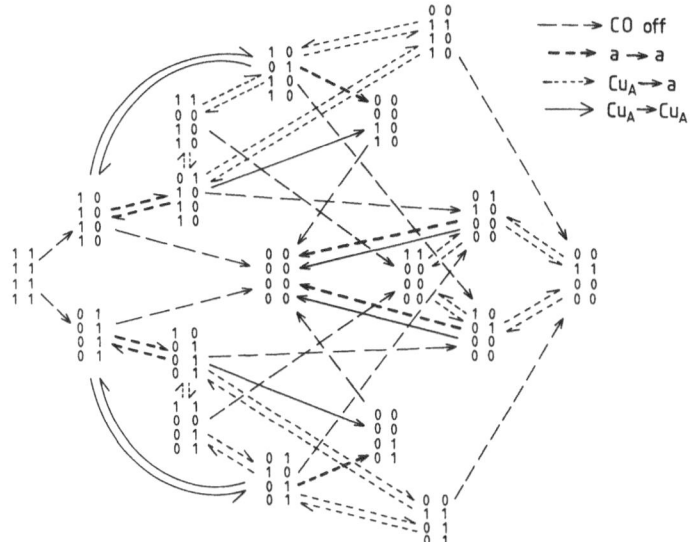

Scheme 1. Model for the oxidation of the carbonmonoxide complex of dimeric cytochrome *c* oxidase by oxygen. The redox state of the four sites (Cyt *a*, Cu_A, Cu_B and Cyt a_3. in descending order in the diagram) are denoted by 1 (reduced) on 0 (oxidized). Electrons are donated to oxygen in pairs. The Scheme is symmetrical around the horizontal axis.

enhancement during cytochrome *a* oxidation to a factor of only two. However, in Scheme 1 we also see that in some paths electrons leaving cytochrome *a* may be replenished from Cu_A; thus the extinction changes we observe may be suppressed at some stages. This in principle could increase apparent rate enhancement during a time-course. Also it is possible that inter-monomer interactions influence the spectral features of individual chromophores. This seems rather unlikely, however, as monomers and dimers have similar, if not identical, spectral properties. One other possibility is that the rate of CO dissociation from cytochrome a_3^{2+} is sensitive to the oxidation state of the other cytochrome a_3 within the same dimer. At present we do not know which, if any, of these possibilities is correct.

Finally, even though we do not as yet have a good quantitative understanding of this system, CO displacement experiments at low protein concentration do provide a rather simple way to distinguish monomer from dimeric oxidase. We have used this approach [6] to examine the aggregation state of cytochrome *c* oxidase reconstituted into phospholipid vesicles with the finding that monomeric oxidase, bovine subunit III–less, behaves as dimers when replaced into vesicle membrane. At present we do not know the behaviour of the monomeric shark enzyme, but our findings leave open the question regarding the ability or otherwise of the monomeric enzyme to pump protons.

Acknowledgements

We thank Professor Paola Vecchini for the analysis of sedimentation velocity experiments and Dr D. Bickar for the gift of purified shark enzyme. This work was partially supported by Stimulation Action Grant STI-086-J-C(CD) from the EEC and grants from the Ministero della Pubblica Instruzione of Italy (to M. Brunori) and the SERC of the U.K. (to M. T. Wilson).

References

1. Kawato, S., Lehner, C., Müller, M. and Cherry, R., *J. Biol. Chem.* **257**, 6470 (1982).
2. Kawato, S., Sigel, E., Carafoli, E. and Cherry, R., *J. Biol. Chem.* **256**, 7518 (1981).
3. Antonini, E., Brunori, M., Greenwood, C. and Malmström, B. G., *Nature* **228**, 936 (1970).
4. Brunori, M., Sarti, P., Malatesta, F., Antonini, G. and Wilson, M. T., in *Cytochrome Systems: Molecular Biology and Bioenergetics* (ed. S. Papa). Plenum Press (in the press).
5. Gibson, Q. and Greenwood, C., *J. Biol. Chem.* **239**, 586 (1964).
6. Antonini, G., Brunori, M., Malatesta, F., Sarti, P. and Wilson, M. T. *J. Biol. Chem.* **262**, 10077 (1987).
7. Yonetani, T., *J. Biol. Chem.* **236**, 1680 (1961).
8. Puettner, I., Carafoli, E. and Malatesta, F., *J. Biol. Chem.* **260**, 3719 (1985).
9. Hinkle, P., Kim, J. J. and Racker, E., *J. Biol. Chem.* **247**, 1338 (1979).
10. Sarti, P., Jones, M. G., Antonini, G., Malatesta, F., Colosimo, A., Wilson, M. T. and Brunori, M., *Proc. Natl. Acad. Sic. USA* **82**, 4876 (1985).
11. Love, B., Chan, S. H. P. and Stotz, E. *J. Biol. Chem.* **245**, 6664 (1970).
12. Wilson, M. T., Lalla-Maharahj, W., Darley-Usmar, V. M., Bonaventura, J., Bonaventura, C. and Brunori, M., *J. Biol. Chem.* **255**, 2722 (1980).
13. Georgevich, G., Darley-Usmar, V. M., Malatesta, F. and Capaldi, R. A., *Biochemistry* **22**, 1317 (1983).
14. Bisson, R., Schiavo, G. and Montecucco, C., *J. Biol. Chem.* **262**, 5992 (1987).
15. Malmström, B. G. and Andréasson, L. K., *J. Inorg. Biochem.* **23**, 233 (1985).

Chemica Scripta 1988, 28A, 63–69

Early Molecular Events in the Reaction of Fully-Reduced Cytochrome Oxidase with Oxygen at Room Temperature

Yutaka Orii

Department of Public Health, Faculty of Medicine, Kyoto University, Kyoto, 606, Japan

Paper presented at the Nobel Conference 'Biophysical Chemistry of Dioxygen Reactions in Respiration and Photosynthesis', Fiskebäckskil, Sweden, 1–4 July, 1987

Abstract

Rapid reactions of fully reduced cytochrome oxidase with oxygen were investigated by changing the initial oxygen concentration from 35 to 700 μM. An absorbance change followed at 445 nm was biphasic at low oxygen concentrations [Orii, Y., *J. Biol. Chem.* **259**, 7178–7181 (1984)], but a plateau appeared in between as the oxygen concentration increased above 280 μM. An apparent rate constant for the slow change was independent of the oxygen concentration and 1000 s^{-1} on the average. The rate constant for the rapid change increased hyperbolically with the oxygen concentration, suggesting a transient formation of a reaction intermediate prior to the primary oxygen compound of cytochrome oxidase. The dissociation constant for the compound was calculated to be 5.1×10^{-5} M at pH 7.4 and 25 °C. A duration of the plateau became longer as the reaction temperature was lowered, and an involvement of a temperature-dependent conformational change with an apparent rate constant of 6×10^{3} s^{-1} was suggested. Scrutiny of the spectral changes recorded by rapid-scanning spectrophotometry has revealed that in the initial rapid change the peroxy intermediate formation is also involved. Based on these observations, the early molecular events in cytochrome oxidase have been delineated.

Introduction

Crystallographic studies on the structures of deoxy- and oxyhemoglobins revealed an intricate mechanism for the origin of the allosteric interactions induced by oxygenation. In deoxyhemoglobin an iron atom of protoheme is displaced from the porphyrin plane by 0.75 A to proximal histidine in F8 [1]. Upon oxygenation the iron moves back to the porphyrin plane pulling the ligating histidine residue and induces small changes in the tertiary structure of a subunit. These changes are further transmitted to the interfaces between the subunits, thus provoking the allosteric interactions [1].

With cytochrome oxidase it is conceivable that oxygenation, or the formation of the primary oxygen compound [2], similarly triggers small structural changes in subunit(s) associated with either heme a, or copper, or both. Such structural changes would be associated intimately with the intramolecular electron transfer and the proton pumping activity of cytochrome oxidase as well [3–12] while it catalyzes reduction of molecular oxygen into water. The author has speculated previously that the oxygenation might induce an alignment of the aromatic residues to 'pave' the pathway for the intramolecular electron transfer (ET) [2], and it is possible that such a structural rearrangement also creates a proton channel or opens a gate of the channel. Anyway, it is quite certain that these changes occur in a very early stage after oxygenation, and the transient kinetic analyses must be

useful and essential in revealing the molecular mechanism of the coupling of water production and proton pumping.

In order to follow the reaction of cytochrome oxidase with oxygen, Gibson and Greenwood introduced a flow–flash technique, the principle of which was to mix the CO compound of the reduced enzyme with oxygen in a stopped-flow apparatus and to initiate the reaction by releasing the reduced enzyme with a pulse of flash [13]. The apparatus was improved in several respects since then [14–17], and recently supplemented by rapid scanning spectroscopy which enabled direct detection of the primary oxygen compound formation at room temperatures for the first time [2]. Aided by such an advance in technology, the author studied the reaction of fully reduced cytochrome oxidase with oxygen in more detail. Thus, the formation of a peroxy intermediate immediately followed the primary oxygen compound formation, and some features related to structural changes which would occur prior to the intramolecular ET were revealed.

Materials and methods

Cytochrome oxidase† was purified from bovine heart muscle as described previously [18], dissolved in 50 mM sodium phosphate buffer (pH 7.4)–0.5 % (w/v) sodium cholate, and stored in liquid nitrogen until used. The molecular activity was the same as reported previously [19]. The concentration (in terms of heme a) was determined spectrophotometrically [18]. The experimental details for the flow–flash measurements were essentially the same as reported previously [2]. Equilibration of buffer solutions with oxygen was achieved by bubbling the solutions for at least 15 min with air containing different concentrations of oxygen that had been processed by a Toray Oxygen Pump (Toray Co. Ltd., model SEP-104). By this procedure up to 0.49 mM oxygen after mixing was provided, and 0.7 mM oxygen was attained by employing 100 % oxygen for equilibration.

The kinetic constants for the biphasic rapid reaction, k_f and k_s, were determined by nonlinear least squares regression for the experimental data obtained with oxygen concen-

---

† Cytochrome oxidase of eukaryotes is a multi-subunit enzyme and contains two heme a molecules and two copper atoms in a monomeric unit of about 180 000 Da [60–62]. In order to differentiate two heme a molecules, subscripts A and B will be used (heme a_A and heme a_B) in the present paper. These were taken after Cu$_A$ and Cu$_B$ [63]. Thus, heme a_B might be taken as being associated both structurally and functionally with traditional cytochrome a_3 [64], and accordingly participates in the reaction with oxygen and other respiratory poisons.

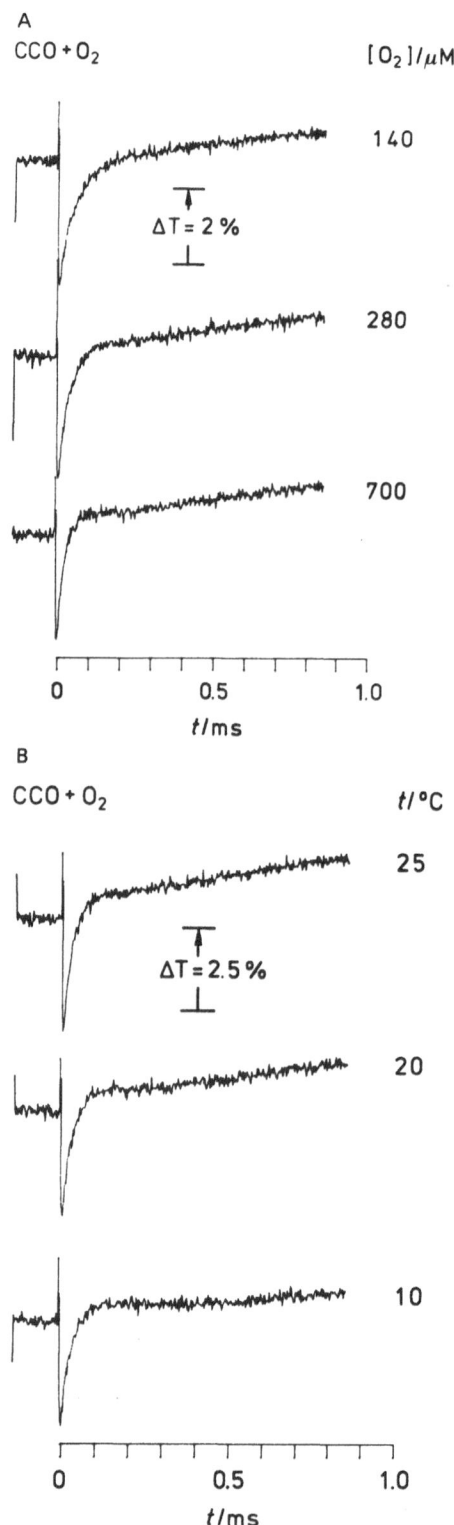

Results

Appearance of a plateau in biphasic changes. Figure 1 A top illustrates a biphasic absorbance change at 445 nm for the reaction of fully reduced cytochrome oxidase with 140 μM dioxygen at pH 7.4 and 25 °C. Previously the initial rapid change was ascribed to the formation of an oxygen compound of the enzyme whereas the slow phase to the intramolecular ET mainly from heme a_A to heme a_B [2]. As the oxygen concentration was increased to 280 μM, a plateau appeared after the initial rapid change and before the slow one, and the plateau became more pronounced at 700 μM oxygen. These results confirm the findings reported previously [2, 25]. Furthermore, the duration of the plateau became longer as the reaction temperature was lowered, lasting about 0.4 ms at 10 °C in the presence of 700 μM oxygen (Fig. 1 B). Although such a phenomenon may occur when the rate constants for the rapid and slow changes have different temperature dependence [2], the temperature effect also suggests a possible involvement of structural changes which provide a proper pathway for the intramolecular electron transfer from heme a_A (and Cu_A) and heme a_B (and Cu_B), as will be discussed later.

Effects of oxygen concentration on the oxygen kinetics. As illustrated in Fig. 2, k_f showed a hyperbolic dependence on the oxygen concentration whereas k_s remained constant at 1000 s^{-1} on the average. The hyperbolic relationship suggests the two-step binding process.

$$CCO + O_2 \overset{K}{\rightleftharpoons} CCO \cdot O_2 \underset{K-2}{\overset{K+2}{\rightleftharpoons}} CCO^* \cdot O_2 \qquad (1)$$

where it is assumed that the rate of equilibration of $CCO \cdot O_2$ is much faster than of $CCO^* \cdot O_2$, and that the observed signal is due to the appearance of the latter [21]. Thus, the experimental data were analysed based on this reaction model, yielding $K = 0.98$ mM, $k_{+2} = 9.9 \times 14^4$ s^{-1},

Fig. 1. Reaction of cytochrome oxidase with dioxygen. (A) Time-courses for reaction of fully-reduced cytochrome oxidase with dioxygen at pH 7.4 and 25 °C were followed at 445 nm. The concentrations of cytochrome oxidase and oxygen after mixing were 0.85 μM and as indicated in the figure, respectively. (B) Time-courses for reaction of fully-reduced cytochrome oxidase (0.85 μM) with dioxygen at temperatures as indicated in the figure were followed at 445 nm. The concentration of oxygen after mixing was 700 μM at 25 °C, and assumed to be the same for other temperatures.

trations at and below 0.35 mM. At and above 0.42 mM oxygen the initial rapid change and the following slow one were separated by a plateau and both phases decayed exponentially, so that pseudo-first-order rate constants for each phase were determined.

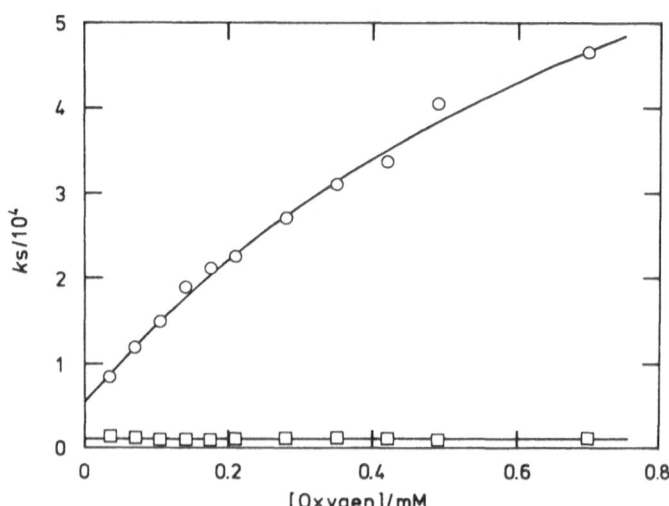

Fig. 2. Dependence of the rate constants for the fast and slow phases of the reaction between cytochrome oxidase and oxygen on its concentration. The apparent first-order rate constants of the fast ($\bigcirc$) and slow ($\square$) changes at every oxygen concentration were determined by the procedure as described in the text. The concentrations of cytochrome oxidase after mixing was 1.0 μM. The reaction was allowed to proceed at pH 7.4 and 20 °C. The solid line for the fast change is the theoretical curve derived from $k = k_{-2} + k_{+2}(1 + K/[O_2])$ with $k_{-2} = 5.4 \times 10^3$ s^{-1}; $k_{+2} = 9.9 \times 10^4$ s^{-1}, and $K = 0.98$ mM.

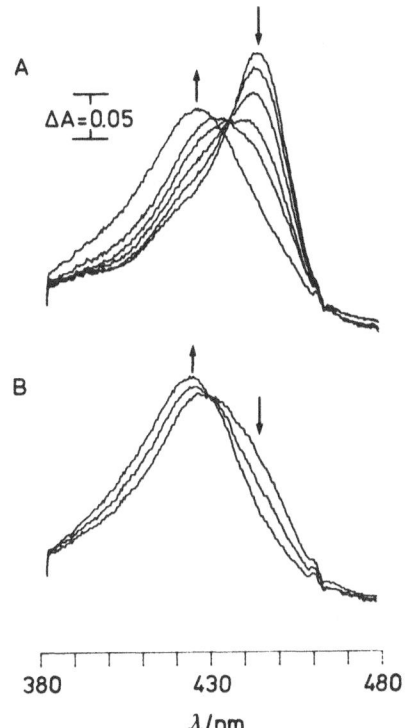

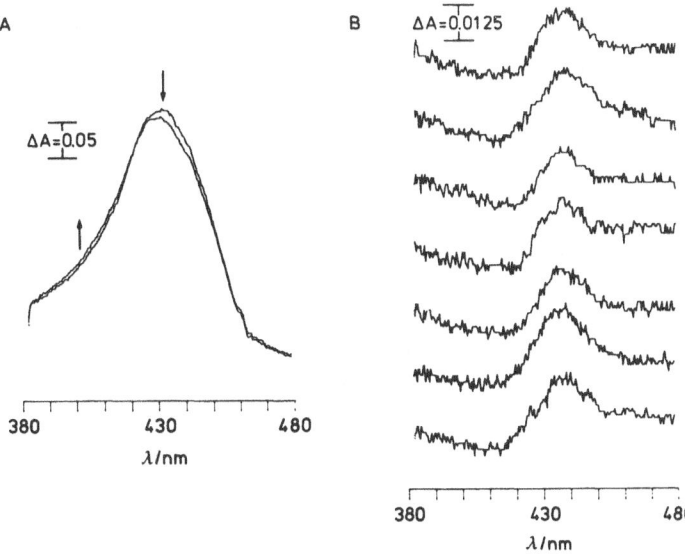

Fig. 4. Absorption spectra of cytochrome oxidase during reaction with oxygen. (A) Spectra recorded at 100 and 250 μs after initiation of the reaction by the flow–flash method at pH 7.4 and 25 °C. The concentrations of cytochrome oxidase and oxygen after mixing were 5.85 and 700 μM, respectively. (B) Time-difference spectra of (100–250 μs). These were recorded independently under the same conditions as in (a).

Fig. 3. Spectral changes of cytochrome oxidase during reaction with dioxygen. (A) spectra recorded after initiation of the reaction by the flow–flash method at pH 7.4 and 25 °C. The reaction times were 5, 25, 50, 100, 200 μs, and 1 ms in the order as indicated by arrows. (B) spectra recorded at reaction times of 500 μs, 1 and 2 ms. The concentrations of cytochrome oxidase and oxygen after mixing were 3.85 and 140 μM, respectively.

and $k_{-2} = 5.4 \times 10^3$ s^{-1}. An apparent dissociation constant of CCO*·O$_2$, K_{diss}, was calculated to be 5.1×10^{-5} M according to equation (2).

$$K_{diss} = K/(1 + k_{+2}/k_{-2}) \qquad (2)$$

This value is comparable to 1.8×10^{-5} M, which has been obtained previously based on the slope (k_{on}) and intercept (k_{off}) of a linear plot of k_f against the oxygen concentration up to 235 μM [2]. The oxygen-independent rate constant again supports the previous proposal that this process would represent the intramolecular ET.

Time-resolved absorption spectra recorded during reaction of fully-reduced cytochrome oxidase with oxygen. Spectral changes of cytochrome oxidase upon reaction of its fully-reduced form with 140 μM oxygen are illustrated in Fig. 3. These spectra were recorded at times indicated in the figure after initiation of the reaction. In agreement with the biphasic absorbance change at 445 nm, the spectra recorded during the initial rapid change, between 5 and 200 μs after initiation of the reaction, apparently shared an isosbestic point at 435 nm. The 5 μs spectrum showed a peak at 443 nm for the reduced enzyme and this shifted to shorter wavelengths with time accompanying a decrease in intensity. The 100 μs spectrum had rather a flat peak extending from 434 to 441 nm, thus contribution of multiple species being suggested. The 200 μs spectrum showed an apparent peak at 430 nm. On the other hand, the spectra recorded during the slow phase later than 0.5 ms showed an isosbestic point at 430 nm. The 1 and 2 ms spectra had an apparent peak at 425 and

424 nm, respectively, but a time-difference spectrum of (2–1 ms) showed a peak at 415 nm and a trough at 444 nm.

The isosbestic point at 435 nm for the spectra in the first group may indicate participation of only two kinds of spectral species, the reduced form and the primary oxygen compound. However, this is not the case because time-difference spectra (50–5 μs) and (100–5 μs), are not similar in shape, implicating an involvement of a third species at least. The former difference spectrum had a peak at 427 nm and a trough at 445 nm (data not shown but see fig. 2B in ref. [2]), and was ascribed mainly to the formation of the primary oxygen compound. Compared with this, the peak height of the latter was relatively smaller than the depth of the 445 nm trough and a small but definite bump was noticed around 410 nm. Such spectral profiles clearly indicate a further conversion of the oxygen compound into succeeding species over a period between 50 and 100 μs as will be described shortly. On the other hand, time-difference spectra of (2–1 ms) and (1–0.5 ms) were similar in shape, both having a peak at 415 nm and a trough at 444 nm. Thus, it was concluded that after 0.5 ms the oxidation of the heme a moiety was the main event.

An interesting feature of spectral changes which give a clue to understand the molecular event between 50 and 100 μs as described above emerges from comparison of the spectra obtained at 100 and 250 μs after initiation of the reaction in the presence of 700 μM oxygen (Fig. 4A). Each spectrum represents the states of the enzyme at the beginning and end of the plateau period in the time-course recorded at 445 nm, respectively. Both were very similar in shape but there was a small but a definite difference between them, which was shown more clearly by a time-difference spectrum (100–250 μs). Then, seven independent measurements were made yielding seven time difference spectra as illustrated in Fig. 4B. Although each spectrum was skewed somehow differently reflecting a subtle drift of instrumental conditions, the spectra were almost the same in shape. This result indicates that the reproducibility and reliability of the measurement are satisfactory to elicit such very minute spectral changes. Averaging of the spectra helped to identify the peak at 436 nm. The position of the trough was scattered between 418 and 413 nm but found mostly at 413 nm. It is noticeable that the difference spectrum quite similar in shape to this, with a peak at 434 nm and a trough at 412 nm, has been obtained [29] by adding hydrogen peroxide to Intermediate III, the active oxidized form of cytochrome oxidase. Thus, the difference spectrum obtained here indicates a relative abundance of the peroxy form at 100 μs compared with an abundance of Intermediate III at 250 μs. This interpretation inherently implies that at 100 μs after initiation of the reaction there already occurs the peroxy form, which would have been produced by two-electron transfer to the bound oxygen from heme a_B and Cu_B in the primary oxygen compound. The extent of conversion of the peroxy form to intermediate III during the plateau period, however, was about 20 % of the total change which dominated after 0.5 ms. Thus, this change accounts for a minor fraction of the molecular events during this period.

Time courses of absorbance changes at various wavelengths during the oxygen reaction. As complementary data to the spectral changes shown in Fig. 4, the absorbance changes upon reaction of fully-reduced cytochrome oxidase with 700 μM O_2 were followed at various wavelengths (Fig. 5). In the right column are presented the time-courses recorded at the wavelengths either for the peak or the trough in the time-difference spectrum of 2 ms–5 μs. A trace at 445 nm was a mirror image of that at 420 nm.

Observations of interest are summarized as follows:

(1) Up to 100 μs the trace at 435 nm does not change except for an abrupt shift in absorbance at time 0 induced by a flash. This result coincides well with the appearance of an isosbestic point at 435 nm for the spectra recorded during the initial rapid change (Fig. 3A).

(2) At 436 nm an initial rapid change is almost completed within 100 μs, immediately followed by a gradual decrease in absorbance. The latter change includes a decrease in absorbance at 436 nm for the peak of the time-difference spectrum illustrated in Fig. 4B.

(3) From 434 to 430 nm the absorbance reached a maximal level at 100 μs and then decreased afterwards. An absorbance decrease at 432 nm, however, ceased at 300 μs and remained unchanged at least for a period of 600 μs. From the absorbance decrease a pseudo first-order rate constant was calculated to be 6×10^3 s^{-1}, significance of which will be discussed later. The steady state level between 300 and 900 μs would correspond to the appearance of an isosbestic point at 430 nm for the spectra recorded during the slow phase of the reaction (Fig. 3B). Since the location of an isosbestic point can be determined more accurately by the present method, 432 nm will be assigned to this isosbestic point hereafter.

(4) At 425 nm the absorbance level was constant between 100 and 200 μs. This result indicates that both spectra depicted in Fig. 4A have the isosbestic point at 425 nm.

Discussion

The hyperbolic dependence of k_f on the oxygen concentration (Fig. 2) does not fit a simple bimolecular reaction model as

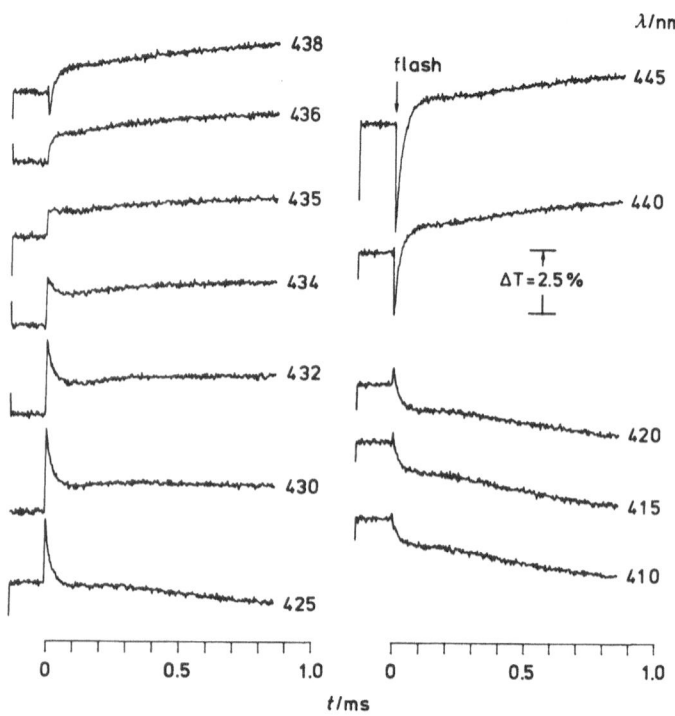

Fig. 5. Time-courses for the reaction of cytochrome oxidase with dioxygen followed at various wavelengths. The reaction was carried out at pH 7.4 and 25 °C. The concentrations of cytochrome oxidase and oxygen after mixing were 0.98 and 700 μM, respectively.

proposed previously [2] but can be explained more satisfactorily by the two-step reaction model as represented by equation (1). The two-step reaction model for the binding of a gaseous ligand to cytochrome oxidase itself is not unique because the same model has been proposed to elucidate the reaction of CO with this enzyme [22] as well as *Vitreoscilla* cytochrome *o* [23]. The initial step assumes the establishment of a rapid equilibrium of oxygen between the external medium and the space in a heme pocket of the enzyme. The oxygen in the heme pocket, which has been shown to accommodate one molecule of oxygen or CO [24, 25], is not yet bound to heme *a*, and the binding occurs in the second step accompanying a spectral change. The equilibrium constant for oxygen was estimated to be 0.98 mM at pH 7.4 and 20 °C. Previously the second-order rate constant of 1.1×10^8 M^{-1} s^{-1} for the presumed bimolecular reaction of fully-reduced oxidase and oxygen has been obtained from the linear relationship between k_f and the oxygen concentration, which was practically valid below 230 μM oxygen [2]. However, with the adoption of the two-step reaction model this second-order rate constant loses its original kinetic significance. The dissociation constant of the primary oxygen compound as defined by equation (2) is 5.1×10^{-5} M, which is close to 1.8×10^{-5} M obtained based on the bimolecular reaction model [2]. The present dissociation constant is one order of magnitude smaller than 4.8×10^{-4} M which was given by Chance *et al.* to Compound A at −91 °C [26] and an apparent K_m of 2.85×10^{-4} M determined by Hill and Greenwood at room temperature [20], but is still higher than the K_m values for oxygen determined in the steady state kinetics of the cytochrome oxidase reaction. Using soluble cytochrome oxidase Petersen *et al.* determined the K_m to be 0.95 μM [27]. This value was confirmed again by a later study [28], in which K_m of 1.1 μM for oxygen was also determined with liposomal cytochrome oxidase in the presence of uncouplers. Albeit some accounts have been given to explain the discrepancy between the dissociation constant and K_m [20, 26], our explanation is as follows. In a simple Michaelis–Menten equation it is assumed that the decomposition of ES is rate-limiting, and an apparent dissociation constant K_d ($= k_{-1}/k_{+1}$) is always smaller than K_m which is equal to $(k_{-1}+k_{+2})/k_{+1}$. Contrary to this, if we assume that, in the reaction model represented by equation (1), the rate-limiting step is not at the immediate decay of $CCO^* \cdot O_2$ but lies beyond that, the opposite relation that K_d is larger than K_m is held (see Appendix). Although it has been implicitly taken that the K_m for oxygen in the cytochrome oxidase reaction reflects the affinity of the reaction system to oxygen and thus approximately equates with the dissociation constant of the oxygen compound, this is not the case. The intrinsic dissociation constant of the oxygen compound itself is even larger than that of the CO compound of cytochrome oxidase, 2.0×10^{-7} M at pH 7.4 and 25 °C (Orii, to be published), and it would be kinetic maneuvers that endow the cytochrome oxidase reaction system with the high affinity for oxygen.

A rapid scanning technique combined with the flow–flash method has enabled spectral scrutiny of the early events in the reaction of cytochrome oxidase with oxygen and indicated that the peroxy intermediate is formed very rapidly following the formation of the primary oxygen compound. Although the rate constant of the conversion of the oxygen compound to the peroxy form was not determined directly, it must be smaller than k_{+2} of 9.9×10^4 s^{-1} because otherwise the primary oxygen compound would not have been detected spectrally. By assuming the occupancy of the primary oxygen compound at 50 μs to be 30% based on the spectral comparison and with a use of k_{+2} as well as k_{-2} given, k_{+3} was estimated to be about 3×10^4 s^{-1}. In this context, it should be pointed out that equation (1) is the minimum requirement to explain the experimental results but does not necessarily describe the molecular events completely. In fact, in a time range in which k_f values have been derived the peroxy form, or Intermediate I [29], is formed already. Thus, $CCO^* \cdot O_2$ in equation (1) is well taken as collectively representing the oxygen compound and the peroxy form, at least.

So far the intramolecular electron transfer from heme a_A to heme a_B in cytochrome oxidase has been studied extensively by initiating the anaerobic reduction of the oxidized enzyme with reductants such as sodium dithionite [30–32], reduced phenazine methosulfate [33], hexaaquochromium [34], hexaamineruthenium [35, 36], and ferrocytochrome c [37–45]. The estimated rate constants ranged from 0.02 to 7 s^{-1} for either the resting or the 'oxygenated' enzyme. Apparently even the maximal rate constant is too small to account for the oxidase reaction [46]. On the contrary, the rate constant of the intramolecular ET initiated by pulsing the reduced enzyme with oxygen is much higher than the above values ranging from 700 to 1500 s^{-1} [2, 13, 14, 16, 20], and this step well constitutes the reaction cycle of the enzyme. In order to explain such a remarkable discrepancy in rate constant, the author is tempted to propose that the binding of oxygen to heme a_B triggers a specific kind of structural change in the oxidase molecule providing an electron transfer pathway of physiological significance [2]. In other words, only when the reaction is initiated by the binding of oxygen to reduced heme

a_B the intramolecular ET would be gated properly. The rate constant for the structural changes was estimated as follows. The duration of the plateau in the absorbance change at 445 nm became longer as the reaction temperature was lowered (Fig. 1 B). This phenomenon can be correlated with the previous finding that the structural changes occur less readily at low temperatures [18, 30, 47, 48]. Therefore, an absorbance change which occurs specifically during the plateau period at other wavelengths would be taken as representing the process of the structural changes. The trace at 432 nm in Fig. 5 seemingly fulfills this requirement. During the plateau period at 445 nm which lasted for 100–150 μs the 432 nm trace decayed rapidly giving a rate constant of 6000 s^{-1} and stayed at a constant level thereafter. It is to be noted that this rate constant is similar to 6400 s^{-1} reported for the structural changes from the R- to T-state of human deoxyhemoglobin at 20 °C [49, 50]. Hill and Greenwood observed that an absorbance change at 830 nm [20] for reaction of the fully-reduced enzyme with oxygen was biphasic and obtained apparent rate constants of 7000 s^{-1} and 800 s^{-1}. These authors ascribed the former to a presumed ET from Cu_A to the oxygen-binding centre(s), but this value is much higher than the values of 700–1500 s^{-1} reported for this intramolecular process as described above. Based on the same reasoning, it is less likely that during this plateau period a third electron is supplied to the peroxy intermediate as suggested by Brunori and Gibson [51] because the rate constant of 5000 s^{-1} is assigned for this process at 2 °C. However, whether the intramolecular ET from cuprous Cu_A is able to proceed with a much higher rate constant than that from reduced heme a_A or not should be determined by further studies.

The reaction scheme which has been derived from the present study is summarized in Fig. 6. This also incorporates the result of a preliminary experiment which suggests that protons utilized for the water formation are consumed with a rate constant of 1000 s^{-1} at room temperature (Orii, to be published). As an alternative, however, an oxene compound $Fe_B^{3+} \cdot O$ may be formed after the peroxy intermediate releasing one molecular of water as is the case in the peroxidase reactions [52]. Recently, based on optical and EPR studies Witt *et al.* proposed the formation of a ferryl Fe_B intermediate during turnover of cytochrome oxidase [53]. Optically this intermediate is characterized by a spectrum with peaks at 580 and 537 nm, which has been observed originally by Orii and King [54, 55] as recognized by Wikstrom [56]. But these measurements were done taking several tens of seconds differing from the present study which required the time resolution of ms or even less. Thus, further studies are required for identification of the ferryl heme among the intermediates of the oxygen reduction. Time-resolved resonance Raman studies would be most powerful for that purpose [57–59].

Appendix

We consider the following reaction scheme as the simplest case.

$$E + S \underset{}{\overset{K_1}{\rightleftharpoons}} E \cdot S \underset{}{\overset{K_2}{\rightleftharpoons}} E^* \cdot S \underset{}{\overset{K_3}{\rightleftharpoons}} E^{**} \cdot S \underset{}{\overset{k_p}{\rightleftharpoons}} E + P \tag{3}$$

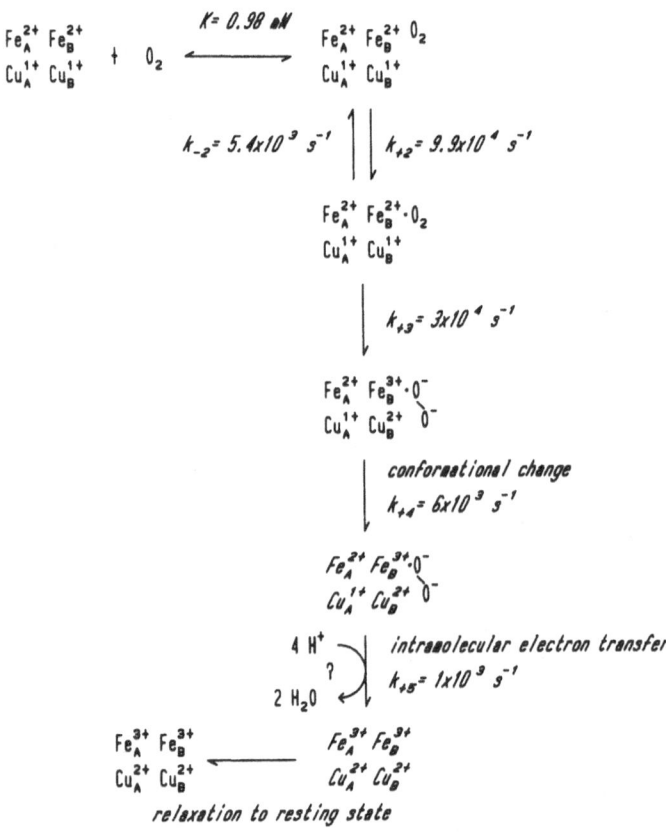

Fig. 6. A scheme for reaction of fully-reduced cytochrome oxidase with dioxygen. Intermediate species which are involved in the intramolecular ET of physiological significance with specific conformations are represented by italic characters. Subscripts for the rate constants are simply for labeling.

Suppose that the dissociation constant of $E^* \cdot S$ can be determined by transient kinetics as in the case of equation (1) to be

$$K_{\text{diss}} = K_1 \cdot K_2 / (1 + K_2) \tag{4}$$

with application of a rapid-equilibrium analysis. On the other hand, suppose that under a steady-state condition the decay of $E^{**} \cdot S$ is rate-limiting. Then,

$$K_{\text{m}} = K_1 \cdot K_2 \cdot K_3 / (1 + K_3 + K_2 \cdot K_3) \tag{5}$$

Accordingly,

$$K_{\text{m}} / K_{\text{diss}} = (K_3 + K_2 \cdot K_3) / (1 + K_3 + K_2 \cdot K_3) < 1 \tag{6}$$

Acknowledgements

This work was supported in part by a research grant of the Fujiwara Foundation of Kyoto University, and a Grant-in-Aid for Special Research on Molecular Mechanism of Bioelectric Response (61107002) and a Grant-in-Aid for Scientific Research on Priority Areas of 'Bioenergetics' to Y.O. from the Ministry of Education, Science and Culture of Japan.

I am grateful to Professor Seiyo Sano for stimulating discussions and encouragement throughout this work.

References

1. Perutz, M. F., *Nature* 228, 726–739 (1970).
2. Orii, Y., *J. Biol. Chem.* 259, 7187–7190 (1984).
3. Wikström, M. F. K., *Nature* 266, 271–273 (1977).
4. Alexandre, A., Reynafarje, B. and Lehninger, A. L., *Proc. Natl. Acad. Sci. USA* 75, 5296–5300 (1978).
5. Sigel, E. and Carafoli, E., *Eur. J. Biochem.* 89, 119–123 (1978).
6. Krab, K. and Wikström, M., *Biochim. Biophys. Acta* 548, 1–15 (1979).
7. Azzone, G. F., Pozzan, T. and Di Virgilio, F., *J. Biol. Chem.* 254, 10206–10212 (1979).
8. Al-Shawi, M. and Brand, M. D., *Biochem. J.* 200, 539–546 (1981).
9. Wikström, M., *Nature* 308, 558–560 (1984).
10. Wikström, M. and Casey, R., *FEBS Lett.* 183, 293–298 (1985).
11. Mitchell, P., Mitchell, R., Moody, A. J., West, I. C., Baum, H. and Wrigglesworth, J. M., *FEBS Lett.* 188, 1–7 (1985).
12. Miki, T. and Orii, Y., *J. Biol. Chem.* 261, 3915–3918 (1986).
13. Gibson, Q. H. and Greenwood, C., *Biochem. J.* 86, 541–554 (1963).
14. Greenwood, C. and Gibson, Q. H., *J. Biol. Chem.* 242, 1782–1787 (1967).
15. Erecinska, M. and Chance, B., *Arch. Biochem. Biophys.* 151, 304–315 (1972).
16. Ludwig, B. and Gibson, Q. H., *J. Biol. Chem.* 256, 10092–10098 (1981).
17. Hill, B. C. and Greenwood, C., *Biochem. J.* 215, 659–667 (1983).
18. Orii, Y., Manabe, M. and Yoneda, M., *J. Biochem. (Tokyo)* 81, 505–517 (1977).
19. Orii, Y., *J. Biol. Chem.* 257, 9246–9248 (1982).
20. Hill, B. C. and Greenwood, C., *Biochem. J.* 218, 913–921 (1984).
21. Gutfreund, H., *Enzymes: Physical Principles*, pp. 205–206. John Wiley & Sons Ltd., New York and London (1972).
22. Orii, Y., *Proc. Biophys. Soc. Japan*, p. 128 (1980).
23. Webster, D. A. and Orii, Y., *J. Biol. Chem.* 260, 15526–15529 (1985).
24. Sharrock, M. and Yonetani, T., *Biochim. Biophys. Acta* 462, 718–730 (1977).
25. Orii, Y., *J. Inorg. Biochem.* 23, 263–271 (1985).
26. Chance, B., Saronio, C., Leigh, J. S., Ingledew, W. J. and King, T. E., *Biochem. J.* 171, 787–797 (1978).
27. Petersen, L. C., Nicholls, P. and Degn, H., *Biochim. Biophys. Acta* 452, 59–65 (1976).
28. Hansen, F. B., Miller, M. and Nicholls, P., *Biochim. Biophys. Acta* 502, 385–399 (1978).
29. Orii, Y., in *Oxygenases and Oxygen Metabolism* (ed. M. Nozaki, S. Yamamoto, Y. Ishimura, M. J. Coon, L. Ernster and R. Estabrook), pp. 137–149. Academic Press, New York (1982).
30. Orii, Y. and Miki, T., in *Oxidases and Related Redox Systems* (ed. T. E. King, H. Mason and M. Morrison), pp. 1181–1200. Pergamon Press, Oxford (1982).
31. Halaka, F. G., Babcock, G. T. and Dye, J. L., *J. Biol. Chem.* 256, 1084–1087 (1981).
32. Jones, G. D., Jones, M. G., Wilson, M. T., Brunori, M., Colosimo, A. and Sarti, P., *Biochem. J.* 209, 175–182 (1983).
33. Halaka, F. G., Barnes, Z. K., Babcock, G. T. and Dye, J. L., *Biochemistry* 23, 2005–2011 (1984).
34. Greenwood, C., Brittain, T., Brunori, M. and Wilson, M. T., *Biochem. J.* 165, 413–416 (1977).
35. Scott, R. A. and Gray, H. B., *J. Am. Chem. Soc.* 102, 3219–3224 (1980).
36. Reichardt, J. K. V. and Gibson, Q. H., *J. Biol. Chem.* 257, 9268–9270 (1982).
37. Gibson, Q. H., Greenwood, C., Wharton, D. C. and Palmer, G., *J. Biol. Chem.* 240, 888–894 (1965).
38. Antonini, E., Brunori, M., Greenwood, C. and Malmstrom, B. G., *Nature* 228, 936–937 (1970).
39. Andreasson, L.-E., Malmström, B. G., Strömberg, C. and Vänngard, T., *FEBS Lett.* 28, 297–301 (1972).
40. Wilson, M. T., Greenwood, G., Brunori, M. and Antonini, E., *Biochem. J.* 147, 145–153 (1975).
41. Andreasson, L.-E., *Eur. J. Biochem.* 53, 591–597 (1975).
42. Greenwood, C., Brittain, T., Wilson, M. and Brunori, M., *Biochem. J.* 157, 591–598 (1976).
43. Petersen, L. C. and Cox, R. P., *Biochim. Biophys. Acta* 590, 128–137 (1980).
44. Veerman, E. C. I., Wilms, J., Casteleijn, G. and Van Gelder, B. F., *Biochim. Biophys. Acta* 590, 117–127 (1980).
45. Antalis, T. M. and Palmer, G., *J. Biol. Chem.* 257, 6194–6202 (1982).
46. Nicholls, P. and Chance, B., in *Molecular Mechanism of Oxygen Activation* (ed. O. Hayaishi), pp. 479–534. Academic Press, New York (1974).
47. Yoshida, S., Orii, Y., Kawato, S. and Ikegami, A., *J. Biochem. (Tokyo)* 86, 1443–1450 (1979).
48. Kawato, S., Ikegami, A., Yoshida, S. and Orii, Y., *Biochemistry* 19, 1598–1603 (1980).

49. Sawicki, C. A. and Gibson, Q. H., *J. Biol. Chem.* **251**, 1533–1542 (1976).
50. Sawicki, C. A. and Gibson, Q. H., *J. Biol. Chem.* **252**, 5783–5788 (1977).
51. Brunori, M. and Gibson, Q. H., *EMBO J.* **2**, 2025–2026 (1983).
52. Dunford, H. B. and Stillman, J. S., *Coord. Chem. Rev.* **19**, 187–251 (1976).
53. Witt, S. N., Blair, D. F. and Chan, S. I., *J. Biol. Chem.* **261**, 8104–8107 (1986).
54. Orii, Y. and King, T. E., *FEBS Lett.* **21**, 199–202 (1972).
55. Orii, Y. and King, T. E., *J. Biol. Chem.* **251**, 7487–7493 (1976).
56. Wikström, M., *Proc. Natl. Acad. Sci. USA* **78**, 4051–4054 (1981).
57. Babcock, G. T., Jean, J. M., Johnston, L. N., Palmer, G. and Woodruff, W. H., *J. Am. Chem. Soc.* **106**, 8305–8306 (1984).
58. Babcock, G. T., Jean, J. M., Johnston, L. N., Woodruff, W. H. and Palmer, G., *J. Inorg. Biochem.* **23**, 243–251 (1985).
59. Ogura, T., Yoshikawa, S. and Kitagawa, T., *Biochim. Biophys. Acta* **832**, 220–223 (1985).
60. Robinson, N. C. and Capaldi, R. A., *Biochemistry* **16**, 375–381 (1977).
61. Saraste, M., Penttila, T. and Wikstrom, M., *Eur. J. Biochem.* **115**, 261–268 (1981).
62. Orii, Y. and Miki, T., in *Frontiers in Biochemical and Biophysical Studies of Proteins and Membranes* (ed. T.-Y. Liu, S. Sakakibara, A. N. Schechter, K. Yagi, H. Yajima and K. T. Yasunobu), pp. 279–287. Elsevier Science Publishing Co., Inc., New York (1983).
63. Wikström, M., Krab, K. and Saraste, M., *Cytochrome Oxidase A Synthesis*. Academic Press, New York (1981).
64. Keilin, D., *The History of Cell Respiration and Cytochrome*. Cambridge University Press, Cambridge (1966).

Chemica Scripta 1988, **28A**, 71–74

Mechanism of Cell Respiration

Properties of individual reaction steps in the catalysis of dioxygen reduction by cytochrome oxidase

Mårten Wikström

Department of Medical Chemistry, University of Helsinki, Siltavuorenpenger 10A, SF-00170 Helsinki, Finland

Paper presented at the Nobel Conference 'Biophysical Chemistry of Dioxygen Reactions in Respiration and Photosynthesis', Fiskebäckskil, Sweden, 1–4 July, 1987

Abstract

The pH-dependence of two of the four catalytic one-electron steps of dioxygen reduction at the binuclear centre of cytochrome oxidase was studied in further detail. Reduction of the 'peroxy' intermediate (P) to the 'ferryl' iron state (F) is associated with uptake of $2H^+$ over the pH range studied. The further reduction of F to the ferric/cupric state of the oxidized enzyme (O) is linked to uptake of $2H^+$ at lower pH, but becomes pH-independent at higher pH. This phenomenon is explained as due to protonation of metal-bound hydroxyls with a pK about 8.5.

The midpoint redox potentials of the four catalytic one-electron steps are assessed in some detail, and the influence of the protonmotive force analysed. The results suggest how the catalytic centre has achieved rapid kinetics of reduction of dioxygen to water, and how this reaction might be controlled by the mitochondrial energy state.

Introduction

Cytochrome oxidase (EC 1.9.3.1) catalyses cell respiration in eukaryotes and in many aerobic prokaryotes. The enzyme appears structurally much more complicated in higher organisms than in bacteria. However, only two subunits (I and II) carry the four redox centres, namely the haems of cytochromes a and a_3, Cu_A and Cu_B (but see Fee *et al.*, this volume, for a possibly yet simpler case). The primary structures of these subunits, and the physicochemical properties of the redox centres are highly conserved throughout evolution [1, 2].

Holm *et al.* [3] recently showed that the two haems and Cu_B might form a compact structure within subunit I, utilizing only three predicted transmembranous helices of this subuit. This structure includes the binuclear haem a_3/Cu_B centre, which catalyses the reduction of dioxygen to water. The model depends on the thesis that cytochrome oxidases of aa_3 type, whether from higher animals or bacteria, share a common 'catalytic core' [4].

Results and discussion

Catalytic mechanism of O_2 reduction

Figure 1 shows the proposal by Wikström [4–7] for the catalytic mechanism of dioxygen reduction. It is based on observations in connection with partial reversal of the catalytic cycle in mitochondria at a high electrochemical proton gradient. In such conditions the ferric/cupric binuclear centre (state O) may be converted into state F, and further into state P. Both P (Compound C [8, 9] and F [10, 11] have also been observed in the forward reaction of half-reduced and fully reduced enzyme with O_2, respectively, at

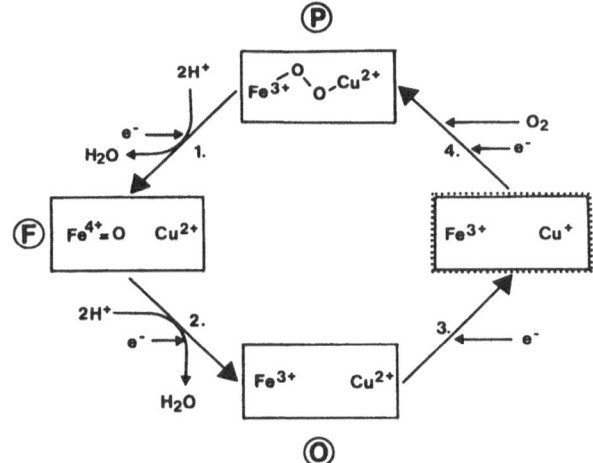

Fig. 1. Proposed catalytic mechanism of reduction of dioxygen at the binuclear centre of cytochrome oxidase.

very low temperatures. A compound with spectral properties similar or indentical with P is also formed when the oxidized enzyme (state O) is reacted with hydrogen peroxide ([12–15]; cf. Wever *et al.* and Orii *et al.*, this volume), though then with low occupancy (see ref. [7]).

The proposed oxidation states of iron and copper in F and P ([5–7]; Fig. 1) are based on the observation that the transition from O (in which the oxidation states of iron and copper are known) to F, and from F to P, are both one-electron oxidation steps [5] with ferricytochrome c as the electron acceptor [5–7]. This chemical evidence is consistent with the limited physical data available. However, final proof must await physical confirmation, e.g. from Mössbauer and X-ray absorption spectroscopy of these states.

The protonation states of F and P in Fig. 1 are based on observed pH-dependences of the transitions ([6, 7], cf. below).

pH-dependence

Based on the strong pH-dependence of formation of F and P from O it was suggested that both are associated with release of two hydrogen ions ([6], Fig. 1). As reported earlier ([6], data to be published elsewhere), this pH-dependence is specific for the matrix space of the mitochondrion.

We have now performed more detailed pH titrations than before to focus better on the F/O transition (Fig. 2). The data have been plotted more appropriately (cf. ref. [6]), as log (F)/(O) and log (P)/(F) *vs.* pH. The results reveal an apparent pK

of about 7.2 for the F/O transition (cf. ref. [7]), whereas the P/F transition still shows a constant 2H⁺ dependence throughout the titrated range.

This strongly suggests that whilst there is uptake of 2H⁺ in the (forward) one-electron reduction of P to F, this is so for the one-electron reduction of F to O only below an apparent

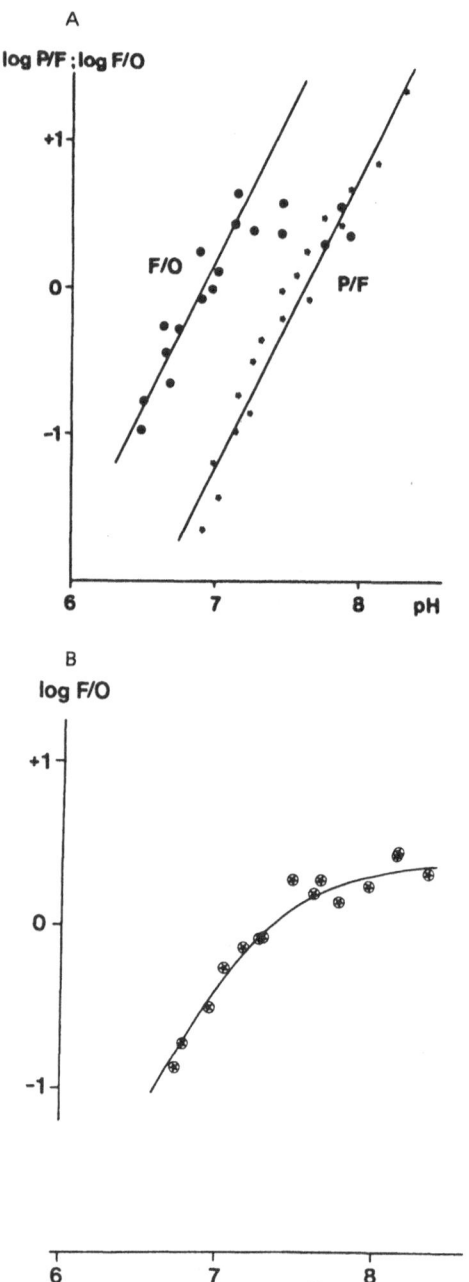

Fig. 2. pH-dependence of the P/F and F/O equilibria. (A) Rat liver mitochondria (0.77 μM of cytochrome aa_3) were suspended in 0.2 M sucrose, 20 mM-KCl, 0.1 mM EGTA, 20 mM Hepes-Tris medium, and supplemented with 2μM rotenone, 5 μM myxothiazol, 4 mM potassium ferricyanide and 30 ng/ml of nigericin. Temperature 25 °C. The pH of the medium was preadjusted to different values, and the final pH determined immediately after each test. The reaction was started by addition of 3 mM of ATP. Measurements were made at 585–630 nm and at 607–630 nm for each pH. Concentrations of F and P were determined from the ATP-induced absorbance changes after deconvolution of spectral overlap, and using extinction coefficients of 6 and 12 mM⁻¹ cm⁻¹, respectively, for F at the former and P at the latter wavelength pair. The concentration of oxidized enzyme (O) was assumed to equal total enzyme minus (F + P). (B) Data from another experiment as in (A). The curve is fitted to the data under the assumption that one-electron reduction of F yields an alkaline form of O without net proton uptake, and that two groups on this form may be protonated, both with a pK of 7.2 (see the text).

pH of 7.2. Above that value this reaction becomes independent of pH. As shown by the simulated curve in Fig. 2B, this behaviour is consistent with the existence of two protonable groups of the O state, with similar or identical pK values near 7.2.

The apparent pK of 7.2 relates to the set pH in the mitochondrial suspension. Due to the inclusion of nigericin in the reaction medium, there is an electroneutral exchange of potassium and hydrogen ions across the membrane [16] in these experiments. Based on the external and internal [17] concentrations of K⁺ it can be calculated that matrix pH is approx. 6.8 when medium pH is 7.2 in these conditions. However, the relevant proton activity is that localised to the binuclear centre in the membrane, which is protonically connected to the matrix phase (see Fig. 3). This local proton activity, and the true pK, may be estimated as described below.

Conversion of P to F

We argue that the 'peroxy' structure (Fig. 1) is reasonable for the intermediate P. The bound peroxide probably bridges between haem iron and copper in the fully deprotonated state [4, 5]. This is consistent with the results of Orii *et al.* (this volume) showing that the earliest phases of the reaction between reduced enzyme and O_2 are independent of pH.

We suggest that the pH-dependences of the reduction of P to F, and of F to O, may *a priori* be attributed to protonation of bound oxygen. It is recognized, however, that the reactions could instead be linked to protonation of protein side-chains (Bohr effects). At present we cannot exclude this possibility. Primarily, however, it may be more productive to consider a direct protonation of the bound ligand before more complicated alternatives. This is also in line with the finding by Shaw *et al.* [18] that when reduced enzyme reacts with ¹⁸O_2 in a single turnover, all oxygen isotope is recovered as water.

Chan *et al.* [19] suggested that one-electron reduction of P would primarily yield a ferrous-cupric peroxy intermediate with special EPR-characteristics from the copper, and with uptake of one proton. The reported properties of this species

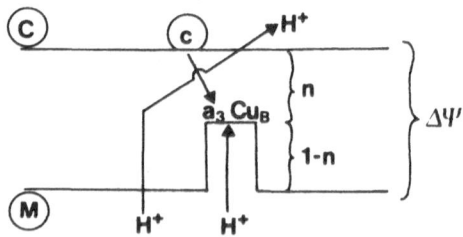

① $E_h^b = E_h^c + n\Delta\Psi + \Delta\bar{\mu}_{H^+}$

② $pH^b = pH^m + \dfrac{(1-n)\Delta\Psi}{60}$

Fig. 3. Scheme of cytochrome oxidase in the membrane. Electron transfer from cytochrome c to the binuclear centre is shown to be associated with translocation of 1 H⁺/e⁻ (proton pump). The binuclear centre is depicted inside the membrane at a site connected protonically to the matrix (M) phase. The drop in electrostatic potential from the cytoplasmic (C) phase to this site is a fraction n of the membrane potential. Equations (1) and (2) describe the local redox potential (E_h^b) and pH (pH^b) at the binuclear centre, as functions of redox potential of cytochrome c (E_h^c), membrane potential ($\Delta\Psi$), protonmotive force ($\Delta\bar{\mu}_{H^+}$), pH of the matrix (pH^m) and n.

were such that it would not be observed in the reversed electron transfer experiments at room temperature.

Then rapid intramolecular two-electron transfer from iron to bound peroxide follows in the forward reaction, with scission of the O–O bond and formation of F (Fig. 1). To account for the overall uptake of $2H^+$ in conversion of P to F ([6, 7], cf. above), we postulate that this step is associated with uptake of another proton.

Conversion of F into O

This is a strongly pH-dependent reaction only below the observed pK (Fig. 2).

One possible explanation for this is that one-electron reduction of F (Fig. 1) primarily leads to

$$Fe^{3+}-OH \ HO-Cu^{2+} \qquad (i)$$

i.e. an alkaline form of the O state that is stable above the pK. Water might already be bound to Cu_B in state F [6], or then the reaction may be associated with uptake of both H^+ and OH^-.

At a lower pH both metal-bound hydroxyls in (i) may be protonated to water with similar pK values. This would account for the observed uptake of $2H^+$ below an apparently single pK (Fig. 2), yielding the acid form of state O shown in Fig. 1.

Assessment of local properties of the binuclear centre

The structures of the intermediates P and F obviously cannot be decided precisely with the present information. Nevertheless, the available data helps to understand some of the basic chemical features of the catalytic process.

At first demonstrated in [6], this approach also yields estimates of the thermodynamic properties of the individual steps. To obtain this, certain assumptions must be made, which were not discussed in detail previously.

In redox titrations of the P/F and F/O equilibria in energized mitochondria, the redox potential of cytochrome c (E_h^c) is defined by the potential of the ferri-/ferrocyanide couple [5]. The corresponding potential at the binuclear centre (E_h^b) is offset from this by two terms, namely the difference in electrostatic potential between the outside medium and the binuclear site, and the electrochemical proton gradient across the membrane [Fig. 3, eqn. (1)]. The latter is due to the coupling of electron transfer between cytochrome c and the binuclear site to translocation of $1H^+$ [4, 20]. Both terms can be considered to be constants during a typical redox titration so that the local potential is a linear function of the potential of cytochrome c. This is supported by the linear relationships observed [5].

In contrast, the midpoint redox potentials (E_m) of the redox couples P/F and F/O depend quite differently on the protonmotive force, being functions of local 'pH' at the binuclear centre (pH^b). Since the latter is in protonic equilibrium with the aqueous M-phase [6], the difference in electrochemical proton potential between the binuclear centre and the M-phase is zero. Consequently, a difference in proton activity between these sites balances the difference in electrostatic potential. This is described approximately by equation (2) (Fig. 3), which relates the local (time-averaged) 'pH' with pH in the M-phase and the appropriate fraction of total membrane potential.

Local (true) pK of state O

In the pH titrations described by Fig. 2, carried out in the present of nigericin, the protonmotive force may be estimated to be about 180 mV. At pH = 7.2, equalling the observed pK, matrix pH may be estimated to be 6.8 (cf. above). In such conditions the membrane potential is 204 mV. Assuming that the binuclear centre is roughly halfway in the membrane with respect to the electrostatic potential profile (i.e. $n = 0.5$ in Fig. 3), the local pH may be calculated to be about 8.5. A pK of 8.5 may be reasonable for protonation of the iron- and copper-bound hydroxyls in (i).

Assessment of midpoint redox potentials of the catalytic steps

Apparent $E_{m,7}$ values of 375 and 465 mV were found in redox potential titrations of the P/F and F/O couples [5]. These relate to the redox potential and pH in the extramitochondrial C-phase, and must hence be corrected as outlined above (Fig. 3), to get more realistic thermodynamic properties of the catalytic reaction steps. Table I shows the appropriate calculations, and gives the resulting $E_{m,7}$ values. Although these are estimates their accuracy is sufficient for drawing some general conclusions.

The midpoint potentials (adjusted to pH = 7) for the four catalytic one-electron steps differ relatively little from one another, and from the $E_{m,7}$ of the O_2/H_2O couple [s.d. about 0.3 V, which is mainly due to reaction (3)]. This means that the midpoint potentials of the metals of the binuclear site, measured in anaerobic potentiometric titrations [22, 23], are not very relevant for the physiological situation. The only exception may be reaction (3). But it is possible that dioxygen binds already to the ferric/cuprous intermediate (Fig. 1),

Table I. *Calculated midpoint redox potentials (pH = 7) of the four redox couples in the catalytic cycle*
Reaction numbers correspond to those in Fig. 1. Assumptions: Extramitochondrial pH = 7, pH^m = 7.8, protonmotive force = 180 mV, membrane potenitial = 132 mV, $n = 0.5$ (cf. Fig. 3). The observed $E_{m,7}$ values based on the potential of cytochrome c are 375 and 465 mV for reactions (1) and (2), respectively [5]. The local E_h and pH values were calculated as described in the text and in Fig. 3. The E_m values obtained for reactions (1) and (2) (at a local pH of 8.9) were corrected to pH = 7 on the basis of the pH-dependences shown on Fig. 2. The $E_{m,7}$ for reaction (4) is calculated from the values for reactions (1)–(3), and the $E_{m,7}$ for the overall reduction of O_2 to water (800 mV at $pO_2 = 0.2$ atm; ref. [6]).

	Reaction			
	(1) (P/F)	(2) (F/O)	(3)	(4)
$E_{m,7}$ (obs) (mV)	465	375	—	—
Protonmotive force (mV)	180	180	—	—
$0.5 \times$ membrane potential (mV)	66	66	—	—
Sum (E_m at local pH) (mV)	711	621	380	1080
Correction to pH = 7 (mV)	228	180	0	0
$E_{m,7}$ (mV)	939	801	308	1080
Reference	*	*	[22, 23]	*

which could enhance the $E_{m,7}$ of reaction (3) at the expense of the $E_{m,7}$ of reaction (4).

In contrast, the four midpoint potentials of one-electron reduction of dioxygen in solution (pH = 7) differ considerably from one another (s.D. about 1.1 V; cf. refs [4, 7]). Hence the chemistry of catalysis and the binding of oxygen intermediates to the binuclear centre have achieved considerable energetic 'smoothening' of the overall reaction. Such an effect tends to equalise the probabilities of the different catalytic intermediates in the steady state, with obvious kinetic advantages.

It also follows from our findings that the kinetics of electron transfer to the binuclear site should be very different (accelerated) in aerobic *vs,* anaerobic conditions. The $E_{m,7}$ values for anaerobic reduction of iron and copper of the binuclear centre are in the 200–400 mV range [4, 22, 23], i.e. much lower than the $E_{m,7}$ values for the appropriate acceptor couples in the presence of dioxygen (Table I). This may account for the old dilemma of very slow (anaerobic) reduction of cytochrome a_3 in comparison with the aerobic turnover number [24].

Control by the protonmotive force

The protonmotive force tends to reverse all four catalytic reaction steps due to its effect on the local E_h [Fig. 3, equation (1)]. However, due to the effect of the membrane potential on the local 'pH' [Fig. 3, equation (2)], the reversal of the pH-dependent reactions (1) and (2) is favoured over the pH-independent reactions (3) and (4). This may be seen from Table I, which also lists the approximate E_m values for the four reactions at the high local 'pH' of energized mitochondria. This asymmetric effect may explain why the catalytic cycle is easily reversed only partially.

Conclusions

We have shown here and elsewhere how essential chemical and thermodynamical properties of the catalytic mechanism of cell respiration are revealed from room-temperature studies of the reversal of the reaction in mitochondria.

The strongest evidence for a 'ferryl' iron intermediate (F) still comes from the demonstration of its formation by one-electron oxidation of the oxidized ferric/cupric binuclear centre (state O). Although this is supported by other more recent chemical data [10, 11], the proposed structure is yet to be confirmed by direct physical methods. This also applies for the 'peroxy' (P) intermediate, even if the proposed structure appears reasonable.

Our experimental approach has indicated the steps of proton uptake in reduction of dioxygen to water. This is an important aspect of the catalytic mechanism, but also has energy conservation, thermodynamic and control implications.

Two of the four catalytic reaction steps were found to be highly pH-dependent due to their linkage, in the forward direction, to proton uptake from the mitochondrial matrix. The directionality of proton uptake renders the activity of the

binuclear centre energy conserving since the electrons are accepted from the opposite side of the membrane. This mode of generation of electrochemical proton gradient should not be confused with the function of the proton pump, which is linked to electron transfer between cytochrome *c* and the binuclear centre (Fig. 3). The two generators of protonmotive force are serially coupled, and contribute equally to the energy-conserving function of cytochrome oxidase [4].

Finally, it is expected that the mitochondrial energy state exerts particular control over reactions (1) and (2) due to their strong pH-dependence and the dependence of local pH on protonmotive force [6]. This may well contribute to the component of respiratory control ascribed to cytochrome oxidase.

Acknowledgements

This work has been supported by grants from the Sigrid Juselius Foundation, the Finnish Academy (MRC), and the University of Helsinki. I am grateful to Hilkka Vuorenmaa and Anneli Sundström for expert technical assistance and help with preparation of the manuscript. Helpful comments by G. T. Babcock, Tuomas Haltia and Anne Puustinen are gratefully acknowledged.

References

1. Wikström, M., Saraste, M. and Penttilä, T., in *The Enzymes of Biological Membranes* (ed. A. N. Martonosi), vol, 4, pp. 111–148. Plenum Publ. Corp. (1985).
2. Raitio, M., Jalli, T. and Saraste, M., *EMBO J.* **6**, 2825–2833 (1987).
3. Holm, L., Saraste, M. and Wikström, M., *EMBO J.* **6**, 2819–2823 (1987).
4. Wikström, M., Krab, K. and Saraste, M., *Cytochrome Oxidase – A Synthesis.* Academic Press, London and New York (1981).
5. Wikström, M., *Proc. Natl. Acad. Sci. USA* **78**, 4051–4054 (1981).
6. Wikström, M., *Chemica Scripta.* **27B**, 53 (1987).
7. Wikström, M., in *Cytochrome Systems – Molecular Biology and Bio-energetics* (ed. S. Papa *et al.* Elsevier (in the press).
8. Chance, B., Saronio, C. and Leigh, J. S., Jr., *J. Biol. Chem.* **250**, 9226–9237 (1975).
9. Clore, G. M., Andreasson, L.-E., Karlsson, B., Aasa, R. and Malmström, B. G., *Biochem. J.* **185**, 155–167 (1980).
10. Witt, S. N., Blair, D. F. and Chan, S. I., *J. Biol. Chem.* **261**, 8104–8107 (1900).
11. Witt, S. N. and Chan, S. I., *J. Biol. Chem.* **262**, 1446–1448 (1987).
12. Orii., *J. Biol. Chem.* **257**, 9246–9248 (1982).
13. Bickar, D., Bonaventura, J. and Bonaventura, C., *Biochemistry* **21**, 2661–2666 (1982).
14. Wrigglesworth, J. M., *Biochem. J.* **217**, 715–719 (1984).
15. Kumar, C., Naqui, A. and Chance, B., *J. Biol. Chem.* **259**, 11668–11671 (1984).
16. Mitchell, P., *Chemiosmotic Coupling and Energy Transduction*, Glynn Research Ltd., Bodmin, U.K. (1968).
17. Rossi, E. and Azzone, G. F., *Eur. J. Biochem.* **7**, 418–426 (1969).
18. Shaw, R. W., Rife, J. E., O'Leary, M. H. and Beinert, H., *J. Biol. Chem.* **256**, 1105 (1981).
19. Blair, D. F., Witt, S. N. and Chan, S. I., *J. Am. Chem. Soc.* **107**, 7389–7399 (1985).
20. Wikström, M., *Nature* **266**, 271–273 (1977).
21. Wikström, M., Krab, K. and Saraste, M., *Annu. Rev. Biochem.* **50**, 623–655 (1981).
22. Lindsay, J. G., Owen, C. S. and Wilson, D. F., *Arch. Biochem. Biophys.* **169**, 492–505 (1975).
23. Wilson, D. F. and Nelson, D., *Biochim. Biophys. Acta* **680**, 233–241 (1982).
24. Lemberg, M. R., *Physiol. Rev.* **49**, 48–121 (1969).

Chemica Scripta 1988, **28A**, 75–78

The Cytochrome *caa₃* from *Thermus thermophilus*

James A. Fee, Barbara H. Zimmermann and Carmen I. Nitsche

Isotope and Structural Chemistry Group, Los Alamos National Laboratory, Los Alamos, New Mexico 87545, USA

and Frank Rusnak and Eckard Münck

The Gray Freshwater Biological Institute, The University of Minnesota, Navarre, Minnesota 55392, USA

Paper presented by James A. Fee at the Nobel Conference 'Biophysical Chemistry of Dioxygen Reactions in Respiration and Photosynthesis', Fiskebäckskil, Sweden, 1–4 July, 1987

Abstract

Over the past few years, it has been shown that cytochromes aa_3 are widely distributed among bacteria. The distribution and properties of the enzyme isolated from a variety of bacteria have recently been reviewed [Ludwig, B., *FEMS Microbiol. Rev.* **46**, 41 (1987)], and we have provided a comprehensive review of the cytochrome caa_3 complexes isolated from *Thermus thermophilus* and bacillus isolate PS3 [Fee, J. A., Kuila, D., Mather, M. W. and Yoshida, T., *Biochem. Biophys. Acta* **853**, 153 (1986)]. In this article, we will provide a concise summary of past studies with the cytochrome caa_3 from *Thermus*, briefly describe our preliminary Mössbauer characterization of its a_3 site, and present some new information on its reaction with peroxides.

The subject enzyme is strongly hydrophobic and is associated with the plasma membrane. It has the following properties:

(i) Contains 2 gram-atom of Cu and 3 of Fe, the latter are present in 1 heme C and 2 heme A.

(ii) Exhibits optical spectra virtually indistinguishable from a 1:1 mixture of mammalian cytochromes c and aa_3.

(iii) The aa_3 portion has been examined by a wide range of spectroscopic probes and found to be similar to the mammalian enzyme.

(iv) Catalyzes the oxidation of reduced cytochromes c by O_2 with formation of H_2O and active pumping of protons across the plasma membrane.

(v) Exhibits complicated redox behavior reminiscent of the eukaryotic enzyme.

The following observations suggest that the redox components of *Thermus* aa_3 are associated solely with the 55 kDa polypeptide (A protein):

(i) Inability to resolve the cytochrome caa_3 into more than two subunits.

(ii) Purification of the intact cytochrome c and the demonstration that it contains no heme A or Cu.

(iii) Purification of a cytochrome aa_3 possessing cytochrome c oxidase activity and having only the 55 kDa polypeptide.

(iv) Observation of a baa_3 complex, lacking the 33 kDa peptide, but retaining the 55 kDa A protein.

The present data thus suggest that all the complex behavior associated with the biological function of cytochrome aa_3 may be associated with a single polypeptide. This is possibly the most important new insight which has emerged from the recent examination of the bacterial enzymes.

The ease with which the bacterial enzyme can be enriched with ^{57}Fe allows the possibility for Mössbauer investigations. At present, this effort has revealed that:

(i) The fully reduced protein contains two low-spin Fe^{2+} hemes, assigned to cytochromes c and a, and one high-spin Fe^{2+} heme, assigned to a_3.

(ii) The oxidized cyano complex contains two low-spin Fe^{3+} hemes, assigned to cytochromes c and a, and one low-spin Fe^{3+} heme *ferro*magnetically coupled to Cu_B^{2+}.

(iii) The fully oxidized enzyme contains two low-spin Fe^{3+} hemes, assigned to cytochromes c and a, and one Fe^{3+} heme which exhibits physically complicated and preparation-dependent behavior.

Here we report further on the Mössbauer properties of cytochrome a_3. The essential new finding is that a_3 in the 'as-isolated' enzyme can assume a number of forms. In the original report [1], we described one preparation of enzyme (preparation I) in which the a_3 had Mössbauer features typical of high-spin Fe^{3+} heme and which was diamagnetic. In preparation II, a_3 had the properties of a high-spin Fe^{3+} heme weakly coupled to Cu_B^{2+} [1]. More recently, working with a new preparation of enzyme and highly concentrated samples ($\geq$ 1 mM a_3) we have been able to characterize a previously undetected heterogeneity in a_3. The detailed analysis is to be published elsewhere [2]; the conclusions are given in Table I.

While the oxidized enzyme displays such heterogeneity, it is noteworthy that the reduced enzyme and the oxidized cyano complex do not appear to be heterogeneous nor to depend on the preparation. It thus seems that the oxidized enzyme may exist in several energetically similar but structurally different forms; such information may be of

Table I. *Distribution of cytochrome* a_3 *species (in percentage of cytochrome* a_3*) at pH 5.7, 6.5, 7.8 and 9.3*

		pH			
		5.7	6.5	7.8	9.3
I	High spin	21	20	45	46
II	High spin	26	68	15	10
III	High spin	30	0	0	0
IV	Low spin	14	0	18	18

immediate relevance to the enzymes from other sources. For example, it has been known for many years from chemical behavior that the oxidized mammalian protein exists in unique states, e.g. resting and pulsed states (cf. ref. [3] for review) – which exhibit distinct reactivity and/or spectral features. It is reasonable to suggest that the *Thermus* enzyme may also possess such behavior although this has not been carefully investigated.

For some time it has been known that oxidized, mammalian cytochrome oxidase reacts with H_2O_2 to form a new compound involving cytochrome a_3 [4–7]. We have carried out limited studies of this reaction with the *Thermus* enzyme. Figure 1 shows the difference spectra (cytochrome c_1aa_3 + ROOH minus oxidized cytochrome c_1aa_3) resulting when enzyme was incubated with hydrogen peroxide (panels A and D), ethyl hydroperoxide (panels B and E), or t-butyl hydroperoxide (panels C and F). The difference spectra are

characterized by a peak at 433 nm and trough at 413 nm in the Soret, and a peak centered at ~ 580 nm and trough at ~ 650 nm in the visible region. Although the rates of formation decrease dramatically as the steric bulk of ROOH is increased, the resulting spectrum is nominally identical, suggesting that the same compound is formed from all three peroxides. As expected, the decay rate of the complex is independent of the peroxide from which it was formed. In our conditions the enzyme returns to the oxidized form with $t_{\frac{1}{2}} \sim 400$ min; this is probably due to the presence of endogenous reducing equivalents in the solution.

The optical spectrum of oxidized cytochrome c_1aa_3 may be regenerated from the spectrum of peroxo-complex by addition of reducing equivalents or CO. We have attempted to titrate the peroxo-complex with various reductants. These experiments have not been successful because of the requirement for stringent anaerobicity and because of the slow reaction of the reductants with the peroxo-complex. Nevertheless, it is clear that addition of reducing agents under anaerobic conditions regenerate the oxidized enzyme. In the case of CO, the peroxo-complex was formed, separated from excess peroxide by passage over a short gel filtration column, and placed under Ar in the anaerobic cuvette; the solid line in Fig. 2 was then obtained. The contents of the cuvette were flushed anaerobically several times with CO, and the spectrum taken as the dotted line in Fig. 2. The regeneration of oxidized enzyme occurs within the time taken to exchange the gas in the cuvette from Ar to CO (< 5 min).

We have carried out Mössbauer studies of peroxo-complex formed by incubating the oxidized enzyme with a 9-fold excess of H_2O_2. Comparing the spectra of peroxo-complex with the corresponding spectra of the oxidized enzyme at

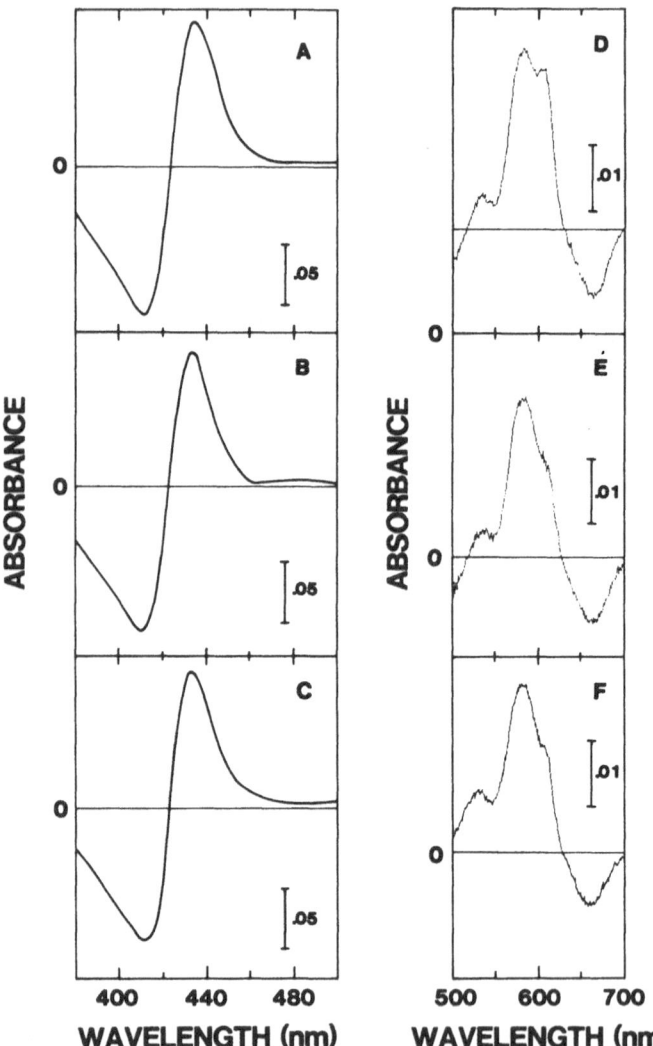

Fig. 1. Spectral characteristics of the complexes formed by oxidized cytochrome c_1aa_3 with three different peroxides. The difference spectra, cytochrome c_1aa_3 + ROOH minus cytochrome c_1aa_3, were recorded after incubating with the appropriate peroxide. (A) 0.84 mM hydrogen peroxide, 9.18 μM cytochrome c_1aa_3; (B) 0.62 mM ethyl hydroperoxide, 10.0 μM cytochrome c_1aa_3; (C) 76 mM t-butylhydroperoxide, 9.60 μM cytochrome c_1 aa_3; (D) 0.62 mM hydrogen peroxide, 7.0 μM cytochrome c_1aa_3; (E) 0.62 mM ethyl hydroperoxide, 7.1 μM cytochrome c_1aa_3; (F) 63. mM t-butylhydroperoxide, 7.1 μM cytochrome c_1aa_3. Experimental conditions were: 10 mM TrisCl, pH 7.8, 0.1 mM EDTA, and 0.1% lauryl maltoside at 25 °C. The cuvette pathlength was 0.4 cm for spectra (A)–(C) and 1 cm for panels (D)–(E).

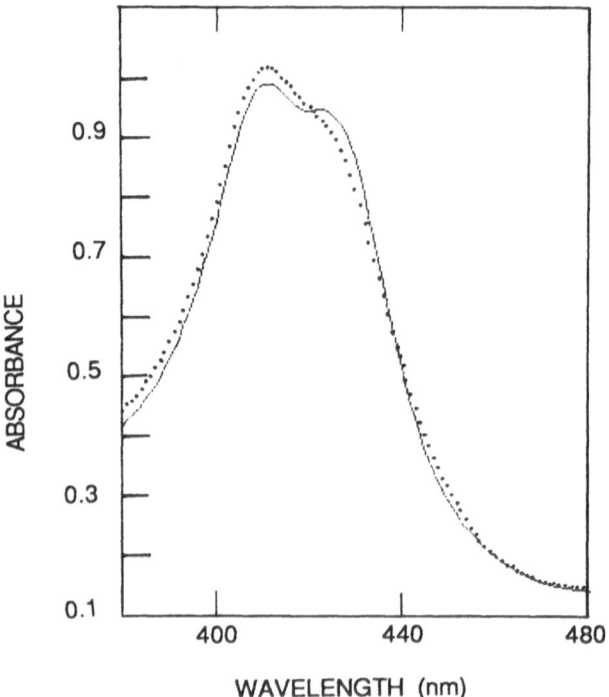

Fig. 2. The reaction of cytochrome c_1aa_3 peroxo-complex with CO. Peroxo-complex was formed, separated from excess peroxide, and placed under Ar in an anaerobic cuvette as described in the text. Solid line, peroxide-free peroxo-complex under Ar; dotted line, ~ 5 min after addition of CO. Experimental conditions were: 10 mM MES pH 6.5, 0.1 mM EDTA, and 0.1% lauryl maltoside at 25 °C, 4.22 μM cytochrome c_1aa_3. Path length was 1 cm.

different temperatures and applied magnetic fields have convinced us that the spectra of cytochromes c_1 and a are unaffected by the addition of peroxide. Mössbauer spectra of the peroxo-complex of cytochrome c_1aa_3 recorded at 190 K are shown in Fig. 3. We have plotted over the data in Fig. 3a an experimental spectrum of cytochromes c_1 and a obtained by analyzing the oxidized enzyme. In order to illustrate more clearly the effect of peroxide on cytochrome a_3, we have removed the spectral contribution (67% of total ⁵⁷Fe) of cytochromes c_1 and a, as well as some additional (3–4% of total ⁵⁷Fe) low-spin ferric component; the difference in shown in Fig. 3b, and is attributed to cytochrome a_3. The spectrum in Fig. 3b consists of two components, shown separately above the data. The minor component, A, accounts for ∼ 30% of cytochrome a_3 and has a quadrupole splitting and isomer shift similar to resting cytochrome a_3. This species probably represents unreacted enzyme. The more prominent species, component B, has $\Delta E_q = 1.7$ mm/s and $\delta = 0.06$ mm/s (relative to iron metal at 300 K), and accounts for ∼ 70% of cytochrome a_3. These parameters, in particular δ, unambiguously identify this species as a low-spin ($S = 1$) Fe(IV)-porphyrin.

Mössbauer spectra of the peroxo-complex taken at 4.2 K (not shown) reveal that only about 35%, rather than 70%, of

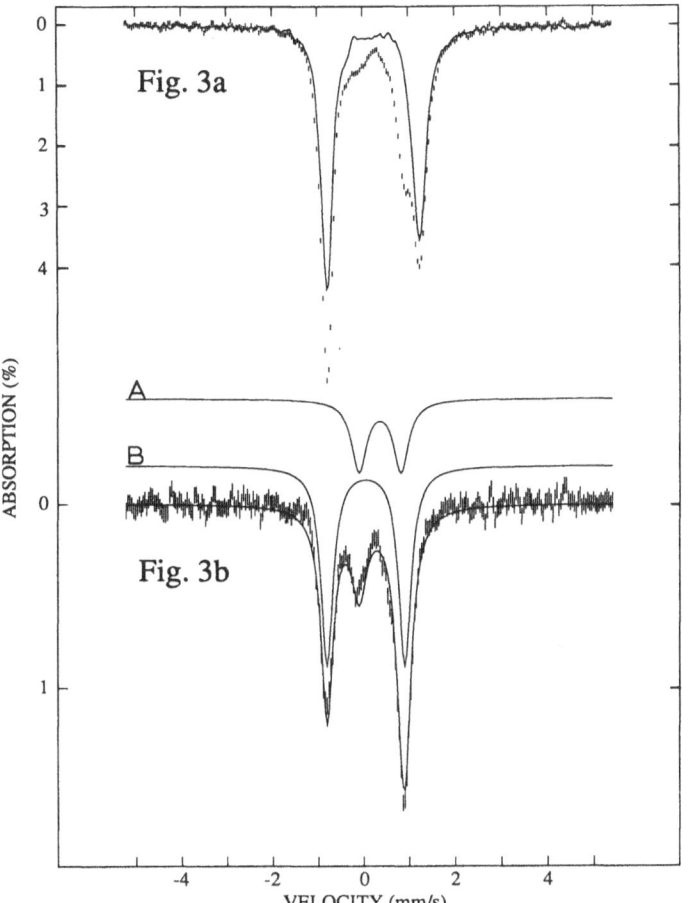

Fig. 3. Mössbauer spectra of the peroxo-complex of cytochrome c_1aa_3 (*a*) and cytochrome a_3 (*b*) are shown at 190 K in the absence of an applied field. Drawn over the data in (*a*) is an experimental spectrum of cytochrome c_1 and a plotted to represent 67% of the ⁵⁷Fe in the sample. The spectrum of cytochrome a_3 is shown in (*b*), obtained as described in the text. The solid lines above the data are simulations of component A and B, obtained by a least-squares-fit to the spectrum. The weighted sum (see text) is shown plotted over the data.

cytochrome a_3 contributes to the quadrupole doublet of component B. What has happened to the other 35%? We can offer two suggestions. First, doublet B observed at 190 K may consist of a mixture of two distinct Fe(IV) species; half being in a non-Kramers state [such as Fe(IV) uncoupled from copper], the other half in a Kramers state [such as Fe(IV)–unpaired electron, where the unpaired electron resides either on the porphyrin or on the copper]. If the Kramers fraction would exhibit at 4.2 K weak magnetic hyperfine splittings (< 3 mm/s), only the non-Kramers component would persist as doublet B. Alternatively, doublet B may consist of two Fe(IV) components exhibiting the same ΔE_q and δ at 190 K, but distinct ΔE_q values at 4.2 K. The rather complex 4.2 K spectra are compatible with both assumptions. Examination of the EPR spectra has revealed no signal attributable to the peroxo-complex; this does not, however, necessarily imply that the peroxo-complex is in the non-Kramers state.

The presence of a ferryl ion at the a_3 site in the peroxo-complex of *Thermus* enzyme is not surprising, since this valence has been proposed to exist in intermediates occurring during turnover of bovine cytochrome aa_3 [8]. The reaction of cytochrome aa_3 with peroxide to form a ferryl complex one or two oxidizing equivalents above oxidized enzyme is reminiscent of the reaction with peroxide of another class of heme enzymes: the peroxidases and catalases. Horseradish peroxidase participates in the following reactions [9]:

$$P^\cdot Fe^{3+} + ROOH \rightarrow P^{\cdot+}Fe(IV)(I)$$
$$P^{\cdot+}Fe(IV)(I) + AH_2 \rightarrow PFe(IV)(II) + AH^\cdot$$
$$PFe(IV)(II) + AH_2 \rightarrow PFe^{3+} + H_2O + ROH$$

where AH_2 is an eletron donor, and P represents the porphyrin ring of the heme. Compounds I and II have distinctive optical absorption spectra. Relative to the spectrum of oxidized enzyme, Compound (I) has a greatly attentuated Soret spectrum, whereas the Soret peak of Compound II is red-shifted and slightly enhanced [9]. Inasmuch as heme A can be expected to behave like heme B in these reactions, the spectral changes associated with the complex are more similar to compound II than to compound I: bleaching of the Soret absorbance is not observed upon formation of the complex, instead, the spectrum becomes red-shifted; this would imply a one electron-oxidized cytochrome a_3. A strong argument against this idea is that the oxidation of cytochrome a_3 to Fe(IV) would modify the spin coupling between the iron and the copper at the oxygen-binding site, presumably leaving an EPR visible Cu_B^{2+} as described earlier by Hansson *et al.* [10], or an EPR active $Fe^{4+}–Cu^{2+}$ pair. As noted above, we have found no evidence for an appropriate EPR signal. Unfortunately, the Mössbauer spectra do not help us here. The presence of cytochromes a and c_1 obscure the details which distinguish the Fe(IV) porphyrins of compounds I and II [11–13].

Carbon monoxide is a two-electron reductant requiring a two-electron acceptor. CO has been implicated as the reducing agent in the formation of the mixed-valence state of bovine cytochrome aa_3 [14], which consists of cytochrome a_B^{2+} and Cu_B^{1+}, while the other two sites remain oxidized. In light of the absence of an EPR signal from Cu_B^{2+}, and the ease with which CO acts as a reductant, we favour the idea that the *Thermus* peroxo-complex is actually two equivalents more oxidized than resting enzyme and suggest two tentative structures:

$$Fe^{4+}-O^{2-} = -Cu^{3+} \tag{1}$$

and

$$Fe^{4+}-O^{2-} = -Cu^{2+}-R^{\cdot} \tag{2}$$

where $R^{\cdot}$ is a radical. We know of no precedent for structure (1) in biological systems; however, the formation of model compounds of trivalent copper complexes with amino acid ligands is well documented [15]. Support for structure (2) is found in the observation that an EPR silent complex is formed in the reaction of galactose oxidase with ferricyanide, and it appears to consist of Cu^{2+} magnetically coupled to a radical species [16]. Furthermore, in the reaction of cytochrome *c* peroxidase, the second reducing equivalent is abstracted from an amino acid residue near the active site, and thus generates ES, an intermediate isoelectronic with Compound I with structure $Fe(IV)-R^{\cdot}$ [17].

References

1. Kent, T. A., Münck, E., Dunham, W. R., Filter, W. F., Findling, K. L., Yoshida, T. and Fee, J. A., *J. Biol. Chem.* **257**, 12489 (1982).
2. Rusnak, F. M., Münck, E., Nitsche, C. I., Zimmermann, B. H. and Fee, J. A., *J. Biol. Chem.*, (in press).
3. Baker, G. M., Noguchi, M. and Palmer, G., *J. Biol. Chem.* **262**, 595 (1987).
4. Bickar, D., Bonaventura, J. and Bonaventura C., *Biochem.* **21**, 2661 (1982).
5. Orii, Y., *J. Biol. Chem.* **257**, 9246 (1982).
6. Naqui, A. and Chance, B., *Ann. Rev. Biochem.* **55**, 137 (1986).
7. Wrigglesworth, J., *Biochem. J.* **217**, 715 (1984).
8. Witt, S. N., Blair, D. F. and Chan, S. I., *J. Biol. Chem.* **261**, 8104 (1986).
9. Schonbaum, G. and Chance, B., in *The Enzymes*, 3rd edn. (ed. P. D. Boyer), vol. XII, Academic Press, New York (1976).
10. Hansson, Ö., Karlsson, B., Aasa, R., Vänngård, T. and Malmström, B. G., *EMBO J.* **1**, 1295 (1982).
11. Schulz, C. E., Devaney, P. W., Winkler, H., Debrunner, P. G., Doan, N., Chiang, R., Rutter, R. and Hager, L. P., *FEBS Lett.* **103**, 102 (1979).
12. Harami, T., Maeda, Y., Morita, Y., Trautwein, A. and Gonser, U., *J. Chem Phys.* **67**, 1164 (1977).
13. Rutter, R., Valentine, M., Hemdrich, M. P., Hager, L. P. and Debrunner, P. G., *Biochemistry* **22**, 4769 (1983).
14. Brzezinski, P. and Malmström, and B. G., *FEBS Lett.* **187**, 111 (1985).
15. Kurtz, J. L., Burce, G. L. and Margerum, D. W., *Inorg. Chem.* **17**, 2454 (1978).
16. Winkler, M. E. and Bereman, R. D., *J. Am. Chem. Soc.* **102**, 6244 (1980).
17. Hewson, W. D. and Hager, L. P., in *The Porphyrins* (ed. D. Dolphin), vol. VII, 294 (1979).

Chemica Scripta 1988, **28A**, 79–84

A Di-haem Cytochrome *c* Peroxidase (*Pseudomonas aeruginosa*): its Activation and Catalytic Cycle

Colin Greenwood and Nicholas Foote

Schools of Biological Sciences, University of East Anglia, Norwich NRH 7TJ, Norfolk, United Kingdom

and Paul M. A. Gadsby and Andrew J. Thomson

Schools of Chemical Sciences, University of East Anglia, Norwich NR4 7TJ, Norfolk, United Kingdom

Paper presented by Colin Greenwood at the Nobel Conference 'Biophysical Chemistry of Dioxygen Reactions in Respiration and Photosynthesis', Fiskebäckskil, Sweden, 1–4 July, 1987

Abstract

The di-haem cytochrome *c* peroxidase isolated from *Pseudomonas aeruginosa* has been examined by electron paramagnetic resonance (EPR) and magnetic circular dichroism (MCD) spectroscopy. Potentiation of the enzyme for catalysis involves reduction of the high-potential haem which causes the low-potential ferric haem to shed a ligand and become high-spin ferric. The half-reduced enzyme reacts rapidly with peroxoide to form compound I containing one ferric and one ferryl haem.

The MCD signature of the ferryl haem in *Pseudomonas* cytochrome *c* peroxidase has been compared to the ferryl state in a number of other proteins including horseradish peroxidase, yeast cytochrome *c* peroxidase and the acid form of the peroxide compound of myoglobin. Single electron reduction of compound I results in the formation of compound II, the EPR spectrum of which shows a peak at $g = 3.45$ and a sharp feature close to $g = 2.0$. Cooling this species from 10 to 5 K results in a 2.5-fold intensification of the $g = 3.45$ signal and it is suggested that this arises from a magnetically coupled pair of low-spin ferric haems.

Introduction

Cytochrome c_{551} peroxidase (EC $1 \cdot 11 \cdot 1 \cdot 5$) isolated from *Pseudomonas aeruginosa* (Ps CCP) catalyses the oxidation of ferrocytochrome c_{551} by peroxide in a reaction essentially identical with that catalysed by the peroxidases isolated from yeast (YCCP) and horseradish root tissue (HRP). The bacterial enzyme differs from other peroxidases in having two haem *c* moieties covalently attached to a single polypeptide chain [1]. A number of studies have clearly demonstrated the inequivalence of the haem centres in structure [2, 3], reactivity [4, 5] and, strikingly, in terms of their oxidation–reduction potential which Ellfolk *et al.* [6] have determined at about $+320$ mV and -330 mV respectively. The latter value is close to the redox potentials of other peroxidases, suggesting strongly that the low-potential (LP) haem centre is probably the site of H_2O_2 binding. Both the single haem peroxidases HRP and YCCP react rapidly with added peroxide and operate through an accessible and relatively stable Fe(IV), ferryl oxidation state which is apparently an essential characteristic of the group as a whole. In order to store the two equivalents of oxidizing capacity available from peroxide the two single haem proteins also use non-metal based centres. In HRP an electron is removed from the porphyrin ring [7] whilst in YCCP a stable radical is produced on an amino-acid residue [8]. The presence of two haem *c* groups per unit molecular weight in Ps CCP probably obviates the need for radical species in the mechanism, cf. HRP and YCCP. In marked contrast to the single haem proteins Ps CCP, as isolated in the fully ferric state, is inactive towards reaction with peroxide. Addition of peroxide to this species causes small, slowly established changes, compared to HRP and YCCP where rapid spectral changes corresponding to the formation of compound I follow peroxide addition to the ferric enzyme. Reduction of the high-potential (HP) haem centre by physiological donors like cytochrome c_{551} Fe^{II} and azurin Cu^I to generate a half-reduced (mixed valence) enzyme is the trigger which activates Ps CCP and potentiates peroxide binding at the LP centre. Magnetic circular dichroism (MCD) spectroscopy, especially in the near infra-red spectral region where band positions can be correlated to haem axial ligation, has revealed the dramatic effects that occur following activation of the enzyme [9]. In the resting oxidized enzyme neither haem is accessible to added ligands. The LP, putative peroxide-binding haem centre is low spin, 6-coordinate with two axial histidine ligands. The other, HP haem has a histidine, methionine ligand set and is partially high spin but, we presume, buried in the protein. On reduction this latter haem becomes fully low spin perhaps with a shortened iron methionine bond length. This acts as the trigger to cause further conformational changes the culmination of which is that the LP haem group loses one of its histidine ligands becoming high spin and accessible to other ligands as well as reactive towards peroxide.

In the present paper we report an examination of the ferryl state mainly using MCD spectroscopy to establish the nature of this species in various protein environments and compare the features which Ps CCP has in common with other peroxidases.

Experimental

Enzymes and chemicals

Horseradish peroxidase was further purified from freeze-dried HRP (Sigma type VI) using FPLC ion-exchange chromatography. The enzyme, dissolved in 20 mM-Na acetate pH 4.4 was loaded onto a mono-S cation-exchange column (Pharmacia) equilibrated with the same buffer. The tightly bound enzyme was eluted with a linear gradient 10 mM/ml of NaCl in 20 mM-Na acetate pH 4.4 and the main protein peak was collected. This procedure increased the spectroscopic purity ratio A_{403}/A_{280} from 3.1 to 3.51. The salt was exchanged for water on a Sephadex G25 (Sigma) column.

Cytochrome c_{551} peroxidase was isolated from *Pseudomonas aeruginosa* (NCTC 6750) as described previously [10, 11]. The purity ratio A_{407}/A_{208} was 4.65. Tween 80 (0.01 %) was added to stabilize samples that required degassing.

Yeast cytochrome c peroxidase, given by Dr G. Sievers was further purified by FPLC. Horse heart myoglobin (Sigma type III) was treated immediately before use with sub-stoichiometric potassium ferricyanide, which was then removed by gel filtration.

Preparation of intermediates

HRP compound II was prepared by adding hydrogen peroxide to HRP; impurities in the sample caused the slow decay of the green compound I so formed to a stable, red compound II state. YCCP compound ES was formed by adding 1.1 M equiv. of peroxide to the ferric enzyme. Myoglobin at the Fe(IV) level was prepared by adding 2–4 M equiv. of H_2O_2 to the ferric protein at pH 8–9. In order to obtain the short-lived acid form of this species the stable alkaline form in 5 mM TAPS buffer at pH 8.5 containing 50 % etanediol, was rapidly mixed with 0.1 M acetic acid, also containing 50 % ethanediol, and frozen in liquid nitrogen within 5 s. The final pH was measured as 3.5. Compound I of Ps CCP was prepared by rapidly mixing and freezing half-reduced Ps CCP and excess H_2O_2 in 25 mM-Mes pD5.6 containing 50 % ethanediol.

In order to collect MCD spectra in the infra-red region D_2O was used in place of H_2O and adjustments made with NaOD (Aldrich Chemical Co.). Reported pD values are the measured pH reading $+0.4$ [12].

The samples were introduced into cells of 1.0–1.2 mm pathlength and placed in an Oxford Instruments SM4 split-coil magnet, capable of generating fields up to 5 T. MCD spectra in the region 250–1050 nm were measured with a Jasco J500D spectropolarimeter, and in the region 800–2000 nm with a dichrograph constructed in these laboratories and described previously [13]. EPR spectra were recorded with a Bruker ER 200D SRC spectrometer. Temperature regulation of EPR samples was by an Oxford Instruments ESR – 900 continuous flow cryostat, with a DTC-2 temperature controller.

Results

Potentiation for peroxide binding and peroxidatic activity

The activation process which occurs when the HP (histidine: methionine) haem is reduced is well documented and some of the physical evidence has already been presented [9]. Figure 1 compares the low temperature (4.2 K) MCD spectra recorded at 5 T between 800 and 2000 nm for oxidized and the half-reduced forms of *Pseudomonas* cytochrome c_{551} peroxidase. The 1850 nm porphyrin to ferric iron-charge transfer band which we assign to the HP haem is lost on reduction; meanwhile the intensity of the signal from the other LP (bis-histidine) haem centre drops dramatically. This we take to be a manifestation of haem–haem interaction related to the activation process. A similar phenomenon is also evident in EPR; on half reduction the signal at g_z 3.3 which we assign to the haem with histidine:methionine

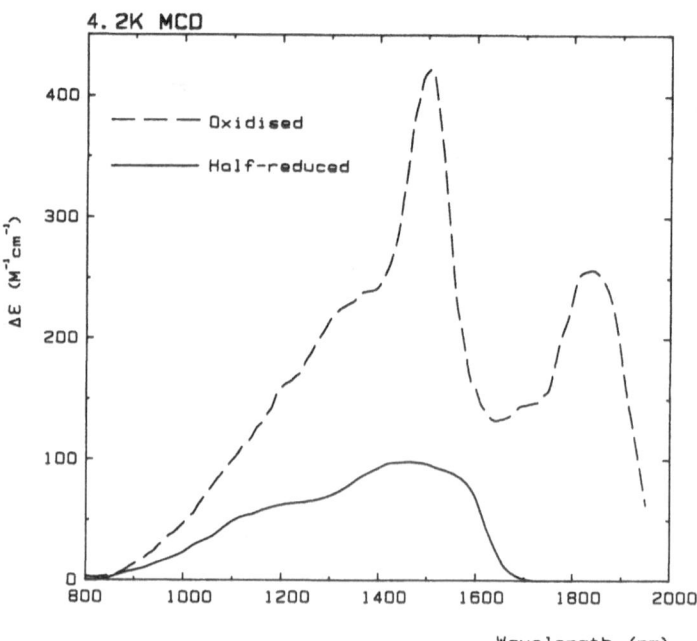

Fig. 1. Near infrared MCD spectra of oxidized and half-reduced cytochrome c_{551} peroxidase at 4.2 K. The enzyme concentration was approximately 200 μM, in 25 mM-Mes pD 5.6 containing 50 % (v/v) glycerol. The half-reduced sample was prepared anaerobically by the addition of 400 μM-NADH and 4 μM phenazine methosulphate. Magnetic field strength, 5 T.

ligands disappears and the g_z value of the bis-histidine haem centre shifts from 3.0 in the oxidized enzyme to 2.85.

It is a feature of this enzyme that only low-spin signals are generally observed at low temperature even when room temperature spectra show high-spin characteristics. Presumably the cooling process results in binding of a sixth ligand.

Figure 2 shows the absorption spectrum recorded at room temperature. In addition to the sharp β- and split α-bands of the ferrous HP (His:Met) haem there is also a band at about 630 nm corresponding to the high-spin ferric state. However, examination of the oxidized spectrum reveals a similar high-

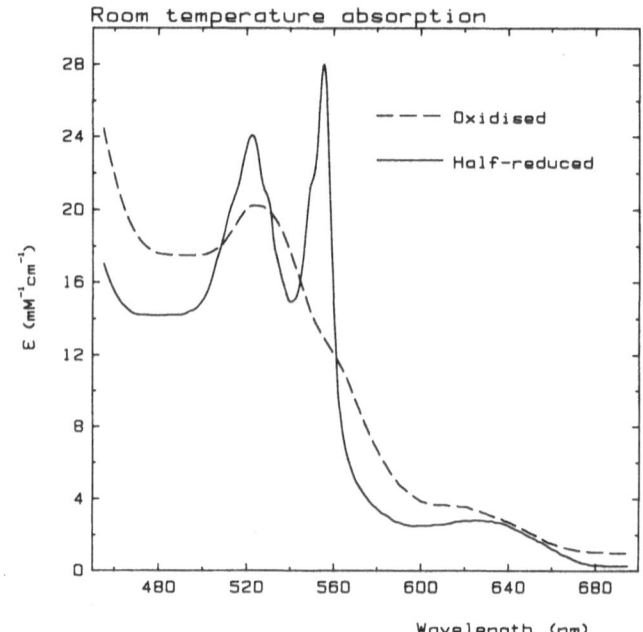

Fig. 2. Room temperature absorption spectra of oxidized and half-reduced cytochrome c_{551} peroxidase. 590 μM Ps CCP, in 50 mM-Mes pD5.6, was reacted with 4 mM-Na ascorbate in a 1 mm pathlength cuvette.

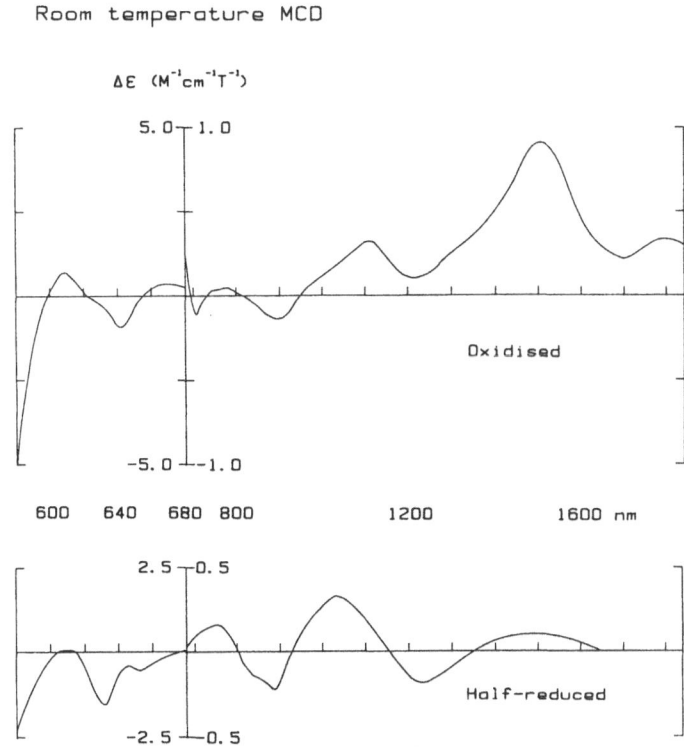

Fig. 3. Near infrared MCD spectra of oxidized and half-reduced cytochrome c_{551} peroxidase at room temperature. 410 μM Ps CCP, in 50 mM-Mes pD 5.6, was reacted anaerobically with 800 μM-NADH and 8 μM phenazine methosulphate. Pathlength, 5 mm; magnetic field strength, 6 T.

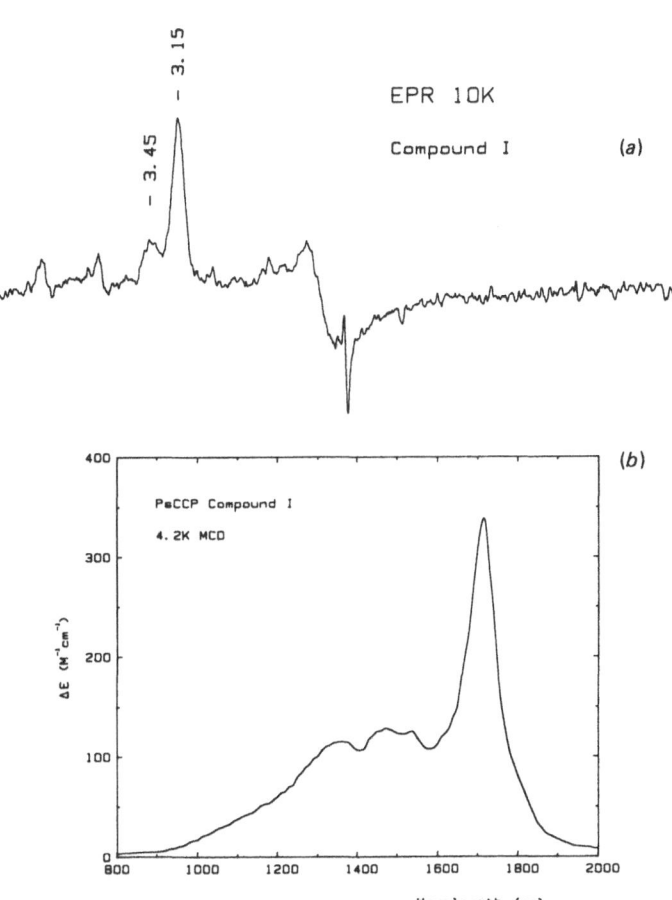

Fig. 4. Low-temperature spectra of cytochrome c_{551} peroxidase compound I. (*a*) EPR Ps CCP, half-reduced with ascorbate, was mixed with excess hydrogen peroxide and frozen within a few seconds. Final concentrations: Ps CCP, 135 μM; Na ascorbate, 1 mM; H_2O_2, 150 μM; 25 mM-Mes pH 6.0 containing 50% (v/v) ethanediol. Conditions: temperature, 10 K; microwave frequency, 9.40 GHz; power 2.02 mW; gain, 10^6. (*b*) Near infrared MCD Excess ascorbate was removed by gel filtration from half-reduced Ps CCP, which was then mixed with excess hydrogen peroxide and rapidly frozen. Final concentrations: Ps CCP, 160 μM; H_2O_2, 320 μM; 25 mM-Mes pD 5.6 containing 50% (v/v) ethanediol. Temperature, 4.2 K; magnetic field strength, 5 T.

spin band together with a typical low-spin ferric haem spectrum. The near IR MCD spectra recorded at room temperature and presented in Fig. 3 show for oxidized enzyme that the band at 1500 nm of the LP haem is of normal intensity, whilst the peak at 1800 nm due to the low-spin ferric form of the HP haem (His:Met) has an intensity corresponding only to about 50% of one haem. We postulated previously [11] that the other half of the HP haem is represented by the high-spin charge transfer bands in the range 600–1100 nm. Reduction of the HP haem renders it virtually MCD silent, the characteristic peak at 1800 nm being lost. A series of bands in the 700–1300 nm region belonging to the high-spin ferric peroxidatic site appears while the peak at 1500 nm due to the LP low-spin haem coordinated by two histidine groups is lost.

We believe that the best explanation for these observations and the activation process is that the high-spin bands in oxidized and half-reduced enzyme arise from different haem groups. Thus, in the oxidized enzyme, at room temperature, the HP haem is in a spin equilibrium, 50% low spin and 50% high spin, whereas the putative peroxidatic site is fully low spin. Reduction of the HP haem converts it to low-spin ferrous, causing the LP site to shed a histidine ligand to become 5-coordinate and high spin. Thus the half-reduced enzyme contains the required peroxide-binding site in company with a second low-spin ferrous haem centre which can potentially act as a store for one of the two oxidizing equivalents of peroxide.

Compound I

Figure 4(*a*, *b*) shows the EPR and MCD spectra of the compound I of Ps CCP prepared by mixing the half-reduced

enzyme with an excess of hydrogen peroxide and freezing rapidly. The EPR spectrum (*a*) is dominated by a low-spin ferric signals with g_z of 3.15. The small peak at $g = 3.45$ most probably represents a small amount of contaminating compound II (see later) formed prior to freezing. Figure 4*b* shows the low-temperature MCD spectrum, with a peak at 1720 nm which is consistent with the presence of a single, low-spin ferric haem having a histidine:methionine ligand set. It should be noted, however, that both the EPR g_z value of 3.15 and the peak location in the MCD spectrum are shifted relative to those of the oxidized enzyme. We take this as further evidence for a changed conformation of the HP haem related to the redox processes. It thus appears that the HP haem oxidizes to the ferric state on compound I formation. However, the experiments leave unanswered questions about the nature of the LP haem in compound I. One obvious possibility is that the haem is in the EPR silent ferryl form identified in mono-haem peroxidases. The evidence for the structure of ferryl haem in other peroxidases has come from a number of techniques. EXAFS identifies a short Fe–O bond of 1.6 Å in HRP [14, 15] which testifies to the multiple character of the bond. Resonance Raman spectroscopy shows a resonantly enhanced vibration of Fe =

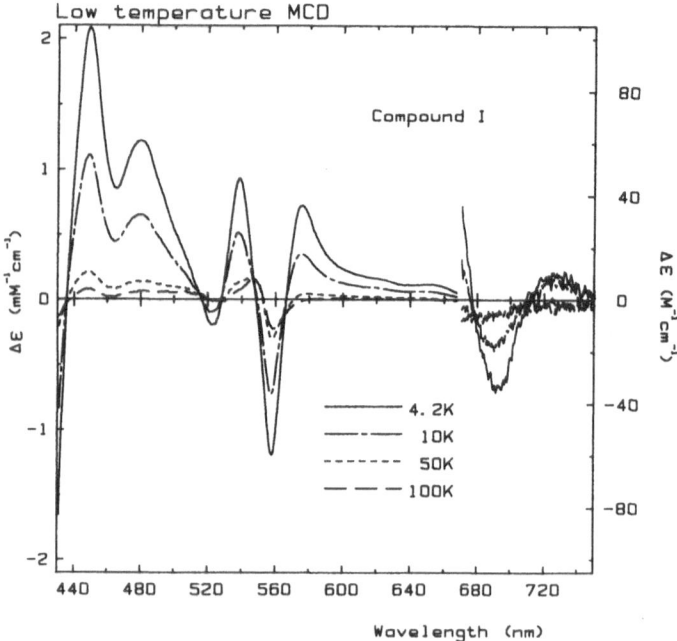

Fig. 5. Visible region MCD. spectra of cytochrome c_{551} peroxidase compound I at various temperatures. Half-reduced Ps CCP, with ascorbate removed by gel filtration, was mixed with excess hydrogen peroxide and rapidly frozen. Final concentrations: Ps CCP, 160 μM; H_2O_2, 1.6 mM; 25 mM-Mes pD 5.6 containing 50 % (v/v) ethanediol. Magnetic field strength, 5 T.

O at 790 cm^{-1} [16] which is again indicative of a multiple bonded oxygen. Mossbauer spectroscopy shows it to be a d^4, $S = 1$ magnetic state. We have recorded the low temperature absorption spectrum for Ps CCP compound I and this shows peaks at 525 and 550 nm as would be expected from a ferryl haem such as compound ES of YCCP, but they are superimposed on the spectrum of the ferric (HP) haem and the evidence is inconclusive. Therefore, we have turned to the MCD spectrum to assist in the identification of the ferryl group.

Figure 5 shows the mcd spectrum of compound I over the range 440–750 nm and recorded at a number of low temperatures. Because the MCD of low-spin ferric haems is very intense the results presented in Fig. 5 are dominated by the features characteristic of one low-spin ferric haem. Confirmation that this centre is, as expected, the histidine:methionine ligated haem, comes from the negative band at 695 nm which is the MCD counterpart of the methionine–ferric iron charge transfer band often referred to as the '695 nm' absorption band. The problem is thus one of observing the ferryl state in an MCD spectrum dominated by the low-spin ferric, HP haem. The solution to this difficulty lies in the differential response of the ferric and ferryl haems to temperature. Inspection of the data reveals that the spectrum does not reduce in intensity evenly throughout the wavelength range, in particular at 540–560 nm where a sharp derivative shaped feature persists at higher temperatures. The magnetic properties of different species in MCD spectra can be obtained by a method, introduced by Thomson and Johnson [17], which studies the field and temperature dependences of the MCD intensity at wavelengths were one or other dominates.

Figure 6 compares the MCD intensity at constant magnetic field (5 T) as an inverse function of absolute temperature (I/T) for the case of a spin $= \frac{1}{2}$ system arising from a low-spin

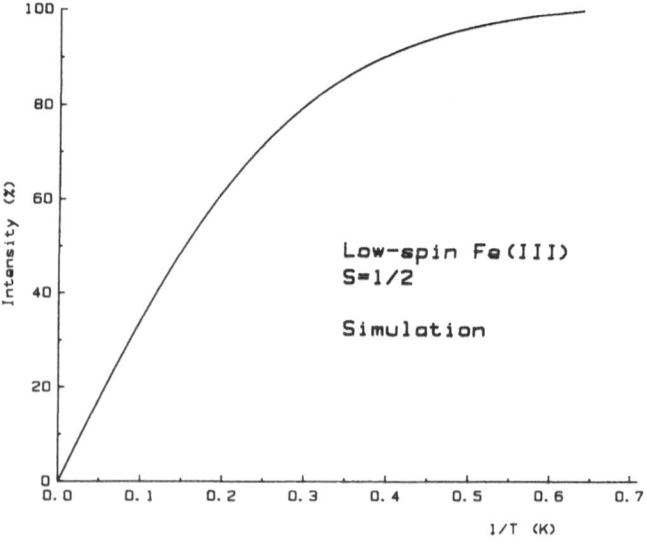

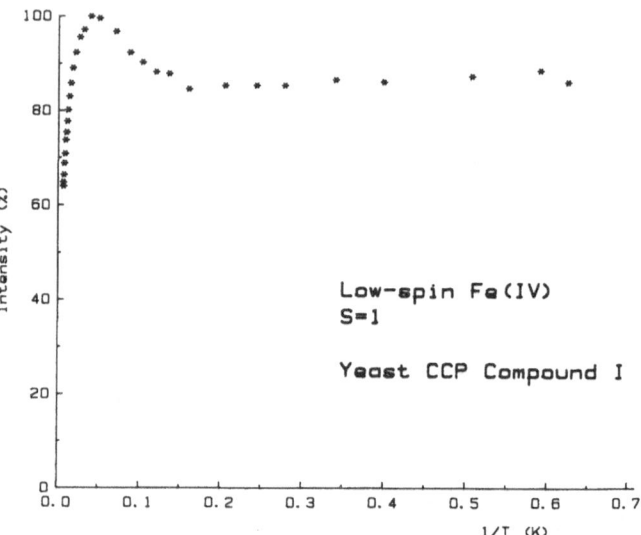

Fig. 6. MCD signal intensity as a function of temperature for two different haem states. Upper: computer simulation for low-spin Fe(III) haem. Lower: experimental data for yeast CCP Compound I, measured as the peak (555 nm)-to-trough (564 nm) separation at 5 T between 1.6 and 100 K.

ferric haem centre and that of the spin = 1, ferryl species found in yeast cytochrome c peroxidase compound ES. The form of the latter curve corresponds to a $S = 1$ ground state with a positive axial zero field splitting of at least 20 cm^{-1} the analysis of which has been discussed elsewhere [18]. The important point to note is that at 10 K the MCD intensity of a ferryl haem is about the same as at 1.6 K whereas for the $S = \frac{1}{2}$ system (low spin ferric) the intensity is strongly dependent on temperature, at 100 K being only 3% of its value at 1.6 K.

Therefore, warming a mixture of $S = \frac{1}{2}$ and $S = 1$ species from 1.6 to 100 K will resolve the features characteristic of the ferryl state since the ferric signals will almost disappear. Figure 7 presents the high-temperature (100 K) MCD spectrum of Ps CCP compound I and compares it with the high-temperature spectra of YCCP compound I, HRP compound II and the acid form of the peroxide compound of horse myoglobin all of which have haems established to be in the ferryl state. Inspections of the figure shows very clearly that, in all these proteins, the form of the MCD spectrum is similar. We take this to be strong confirmatory evidence for the presence of a ferryl haem in compound I of Ps CCP. Thus

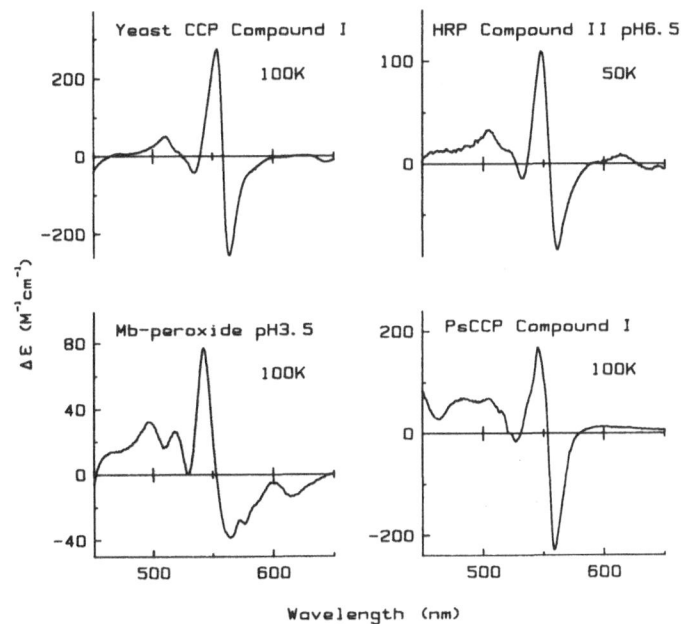

Fig. 7. Visible region MCD spectra of proteins containing ferryl haem. Spectra were recorded at 5 T and at the indicated temperatures. Details of sample preparation are in Experimental section and legend to Fig. 5.

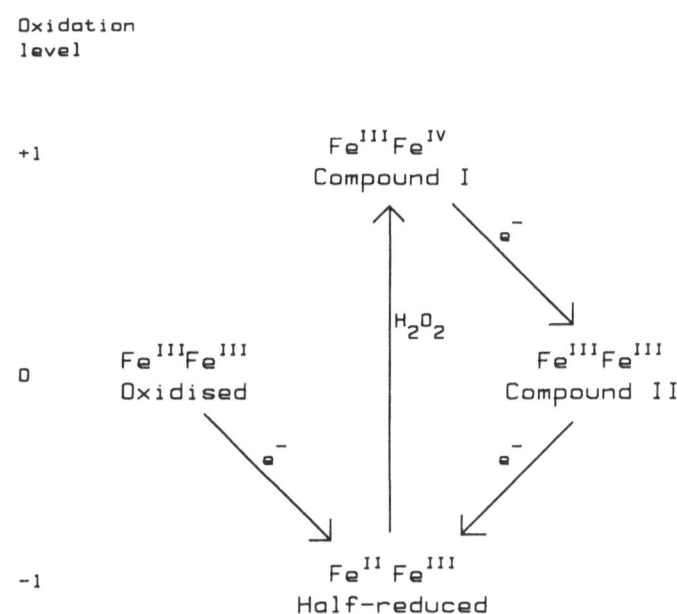

Fig. 8. A reaction scheme for cytochrome c_{551} peroxidase.

we suggest that in compound I the HP haem is ferric and the LP haem is ferryl.

Table I summarizes the structures of both haem sites in three derivatives of the enzyme namely, fully oxidised, half reduced and compound I observed at low temperatures (4.2 K for MCD and 10 K for EPR). The table shows the extent of haem–haem interactions. The character of the LP haem clearly reflects the redox state of its HP companion and we can identify the mechanistically important spin state changes at both haems in oxidized and half-reduced enzyme. It is also clear from the table that, even when constrained by low temperature to adopt a low-spin 6-coordinate state, the LP haem has different magnetic and magneto optical

Table I. *EPR* g_z *values (10 K), near infrared MCD band maxima (4.2 K) and axial ligand assignments for various derivatives of cytochrome* c_{551} *peroxidase*

Oxidised		
EPR g_z	3.0	3.35
MCD	1500 nm (sharp)	1850 nm
Redox potential	−330 mv	+320 mv
Ligands	His–Fe^{III}–His	His–Fe^{III}–Met
Half-reduced		
EPR g_z	2.85	–
MCD	1450 nm (broad)	–
Ligands	His–Fe^{III}–His	His–Fe^{II}–Met
Compound I		
EPR g_z	–	3.15
MCD	–	1720 nm
Ligands	His–Fe^{IV} = 0	His–Fe^{III}–Met

properties in the half-reduced compared to the oxidized enzyme. A possible molecular explanation may be that there has been a change in orientation of the axial ligands, certainly it has been suggested [19, 20] that histidine orientation markedly influences *g* values and the relationship that exists between *g* value and position of the near IR MCD bands in a range of proteins and model compounds has been investigated [21]. Comparison of the properties of the HP ferric haem in the oxidized enzyme and compound I clearly shows that this centre responds to the redox status of the LP site with both the g_z value and the energy of the porphyrin to ferric iron charge transfer bands being affected. One possible explanation is an increase in negative charge at the axial histidine ligand caused either by deprotonation or by a strengthening of hydrogen bonding to other residues.

The catalytic cycle

The scheme in Fig. 8 summarizes the events of the catalytic cycle which starts with the reaction between the half-reduced enzyme and hydrogen peroxide generating compound I two oxidizing equivalents higher. Single electron donation to compound I results in the formation of compound II which is at the same oxidation level as the resting oxidized enzyme but must be very different in terms of reactivity since the reduction of the latter is too slow for it to be a competent intermediate in the cycle.

The simplest mechanism that explains all the data so far is a two-state model. The oxidized enzyme is unreactive towards peroxide because the LP peroxidatic centre has a six coordinate low-spin configuration. Reduction of the HP haem promotes a conformational change throughout the molecule shifting a spin equilibrium at the HP site over to low spin. This causes the LP site to lose one of its histidine ligands and become high spin, peroxide receptive when the enzyme assumes an active conformation.

The EPR signature of compound II is quite unexpected and provides evidence of a direct interaction between the haems. Compound II is expected to possess two low-spin ferric haems, since it is one reducing equivalent below

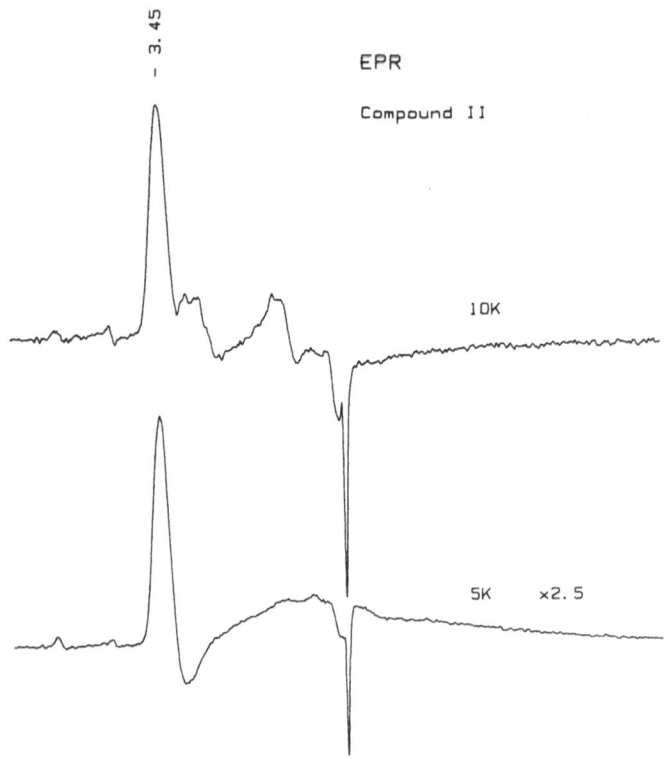

Fig. 9. EPR. spectra assigned to compound II of cytochrome c_{551} peroxidase. The specifies was formed during the decay of compound I, produced by reacting half-reduced Ps CCP with 2 molar equivalents of H_2O_2. Conditions: microwave frequency, 9.42 GHz; power, 2.01 mW; gain, 5×10^5 (upper) and 2×10^5 (lower); temperature as indicated.

compound I which contains one ferric and one ferryl haem. However, the EPR spectrum does not reveal two distinct low-spin ferric haems. Instead compound II EPR exhibits a peak at $g = 3.45$, first reported by Aasa *et al* [3], and a sharp feature close to $g = 2.0$. We find that on cooling the sample from 10 to 5 K the peak at $g = 3.45$ intensifies by a factor of about 2.5 and reveals a derivative line-shape. These EPR characteristics are quite unlike either a single or a pair of isolated low-spin ferric haem. We are, therefore, led to the conclusion that this signal arises from a magnetically coupled pair of low-spin ferric haems. Variable frequency EPR spectroscopy is necessary to verify this conclusion.

These results point to a striking difference between the low temperature form of the oxidized, resting state of the enzyme and compound II. Both contain two low-spin ferric haems which in the oxidized case are magnetically isolated and, in the latter, interact magnetically. We take this as another piece of evidence for rather extensive conformational changes between the resting oxidized and the active states of the enzyme and for a high degree of co-operativity between the haems, triggered by redox and spin-state changes.

Acknowledgements

Professor A. J. Thomson and Professor C. Greenwood thank the SERC and the Royal Society for grants in support of this work.

We wish to thank Dr G Sievers for a sample of yeast cytochrome *c* peroxidase.

References

1. Ellfolk, N. and Soininen, R., *Acta Chem. Scand.* **25**, 1535 (1971).
2. Ronnberg, M., Osterlund, K. and Ellfolk, N., *Biochim. Biophys. Acta* **626**, 23 (1980).
3. Aasa, R., Ellfolk, N., Ronnberg, M. and Vanngard, T., *Biochim. Biophys. Acta* **670**, 170 (1981).
4. Ronnberg, M. and Ellfolk, N., *Biochem. Biophys. Acta* **581**, 325 (1979).
5. Ronnberg, M., Araiso, T., Ellfolk, N. and Dunford, H. B., *Arch. Biochem. Biophys.* **207**, 197 (1981).
6. Ellfolk, N., Ronnberg, M., Aasa, R., Andreasson, L. E. and Vänngård, T., *Biochim. Biophys. Acta* **743**, 23 (1983).
7. Dolphin, D., Forman, A., Borg, D. C., Fajer, J. and Felton, R. A., *Proc. Nat. Acad. Sci. USA* **68**, 614 (1971).
8. Yonetani, T., Schleyer, H. and Ehrenberg, A., *J. Biol. Chem.* **241**, 3240 (1966).
9. Foote, N., Peterson, J., Gadsby, P., Greenwood, C. and Thomson, A. J. *Biochem. J.* **230**, 227 (1985).
10. Foote, N., Thompson, A. C., Barber, D. and Greenwood, C., *Biochem. J.* **209**, 701 (1983).
11. Foote, N., Peterson, J., Gadsby, P., Greenwood,C. and Thomson, A. J., *Biochem. J.* **223**, 369 (1984).
12. Perrin, D. D. and Dempsey, B., in *Buffers for pH and Metal Ion Control*, p. 81. John Wiley & Son, New York (1974).
13. Eglinton, D. G., Johnson, M. K., Thomson, A. J., Gooding, P. E. and Greenwood, C., *Biochem. J.* **191**, 319 (1980).
14. Penner-Hahn, J. E., Eble, K. S., McMurray, T. J., Renner, A. L., Balch, J. T., Groves, J. H., Dawson, J. H. and Hodgson, K. O., *J. Amer. Chem. Soc.* **108**, 7819 (1986).
15. Chance, B., Powers, L., Ching, Y., Poulos, T., Schonbaum, G. K., Yamazaki, I. and Paul, K. G., *Arch. Biochem. Biophys.* **235**, 596 (1984).
16. Terner, J., Sitter, A. J. and Reczek, C. M., *Biochem. Biophys. Acta* **828**, 73 (1985).
17. Thomson, A. J. and Johnson, M. K. *Biochem. J.* **191** 411 (1980).
18. Thomson, A. J., Greenwood, C., Gadsby, P. M. A. and Foote, N., *Cytochrome Systems: Molecular Biology and Bioenergetics.* (ed. S. Papa *et al.*) Plenum Publishing Corp. (1987).
19. Carter, K. R., Tsai, A. and Palmer, G., *FEBS Lett.* **132**, 243 (1981).
20. Salerno, J. C. and Leigh, J. S., *J. Am. Chem. Soc.* **106**, 2156 (1984).
21. Gadsby, P. M. A. and Thomson, A. J., *FEBS Lett.* **197** 253 (1986).

Photosynthesis

Chemica Scripta 1988, **28A**, 87–91

Spectroscopic Studies of Manganese Involvement in Photosynthetic Oxygen Evolution

Kenneth Sauer*, R. D. Guiles, Ann E. McDermott and James L. Cole

Laboratory of Chemical Biodynamics, Lawrence Berkeley Laboratory and Department of Chemistry, University of California, Berkeley, CA 94720, USA

and Vittal K. Yachandra, Jean-Luc Zimmermann and Melvin P. Klein

Laboratory of Chemical Biodynamics, Lawrence Berkeley Laboratory, University of California, Berkeley, CA 94720, USA

and S. L. Dexheimer and R. David Britt

Laboratory of Chemical Biodynamics, Lawrence Berkeley Laboratory and Department of Physics, University of California, Berkeley, CA 94720, USA

Paper presented by Kenneth Sauer at the Nobel Conference 'Biophysical Chemistry of Dioxygen Reactions in Respiration and Photosynthesis', Fiskebäckskil, Sweden, 1–4 July, 1987

Abstract

The oxidation of water to molecular oxygen accompanying photosynthesis by green plants and cyanobacteria is mediated by a membrane-bound enzyme containing manganese. Electron paramagnetic resonance and X-ray absorption spectroscopic studies show that the manganese is directly involved in storing oxidizing equivalents during the four electron transfer steps that lead to O_2 release. The complex involves four Mn atoms, at least two of which are joined by a μ-oxo bridge; other ligands are furnished by N or O atoms from amino-acid side-chains. The location of the site of interaction is suggested to be at the C-termini of the D1 and D2 polypeptides of the Photosystem II reaction center. To date we have not found evidence of the stage at which water is incorporated into the complex or when the chemical rearrangements occur that lead to O_2 formation.

Introduction

The oxidation of water to molecular oxygen in chloroplasts of higher plants and algae and in cyanobacteria reflects a major advance in the evolution of photosynthetic organisms. By utilizing water as the source of electrons for electron transport (ET) leading to carbon fixation from CO_2, it accomplished the transition from less abundant, albeit more powerful, reductants such as H_2, H_2S and other sulfur-containing compounds. At the same time the accumulation of the product of water oxidation, O_2, led to the conversion of the earth's atmosphere from a reducing to an oxidizing one. Not only did this alter the chemical nature of the environment at the surface of the earth, but it also enabled the development of the higher forms of life that we know today. Oxygenic photosynthesis provided not only the oxygen needed for respiration but also the reduced forms of organic carbon that lie at the base of the food chain.

Our knowledge of the mechanistic nature of the steps that lead to photosynthetic oxygen evolution and of the apparatus

responsible for carrying out the water oxidation reactions is still sketchy. It is becoming clear, however, that a single solution to this problem has been used both for eukaryotic higher plants and algae and for the more primitive pro-karyotic cyanobacteria. In both cases the water-splitting enzyme consists of a complex containing four manganese atoms which is associated with a photosynthetic reaction center (Photosystem II) that is derived from reaction centers of non-oxygen-evolving photosynthetic bacteria. These bacteria, many of which are intolerant of oxygen in their environment, presumably predate the organisms that are capable of oxygen evolution.

Comparison of the sequences of nucleotides in genes coding for proteins involved in the reaction center activity has indicated those regions that are significantly conserved between non-oxygenic purple bacteria and PS II of plants, algae and cyanobacteria [1]. These intrinsic membrane proteins include two, designated D1 and D2, that are responsible for forming the framework of the reaction center components – the chlorophyll primary electron donors, the pheophytin intermediate electron carriers and the iron-quinone complex – that stabilizes the separated electron against recombination with the hole remaining on the primary donor species following light activation. Because the structure of the reaction centers of purple bacteria have been determined by X-ray crystallography [2], we have a model that can be utilized as a basis for conjecture concerning the corresponding structure of the reaction center of PS II. When the amino-acid sequences of D1 and D2 polypeptides from plants or algae [3, 4] are compared with those of the L and M subunits, respectively, from reaction centers of purple bacteria [1, 4], strong homologies are seen especially in the region of binding of the reaction center co-factors. For each poly-peptide, 5 *trans*-membrane helical segments have been assigned [5]. These assignments, together with an orientation having the N-termini at the exterior (stromal) interface and the C-termini at the interior (lumenal) interface, have been supported by studies using antibodies to particular segments of the D1 polypeptide [6]. The primary donor cholorophylls are located near the lumenal interface at positions identifiable by conserved histidines near the corresponding ends of the membrane spanning D helices. The iron–quinone electron acceptor complex occurs near the stromal interface, and conserved histidines involved as Fe ligands are seen at the corresponding stromal ends of the D and E helices.

Water oxidation in oxygen evolving organisms is known to occur at or near the lumenal interface of the thylakoid membranes. This is consistent with the location of the site of the bound manganese associated with this process [7]. Furthermore, three extrinsic proteins (especially one of 33 kDa) that are thought to play a role in stabilizing manganese binding or in catalyzing the water-splitting reactions, are also located at the lumenal interface [8]. The C-termini of the D1 and D2 polypeptides, which are highly conserved in all O_2 evolving organisms but which are largely missing in the L and M subunits of the purple bacteria, are rich in amino acids such as histidine, glutamate and aspartate that are reasonable candidates for ligands to manganese. Furthermore, the D1 protein, which has been extensively studied because of its involvement in the site of interaction of quinone-site inhibitors on the acceptor side of PS II, has also been implicated not only in ET reactions on the donor side of PS II but also in binding of Mn and the facilitation of water oxidation [9, 10].

At present it is not clear what is the minimum number of polypeptides that need to be present to enable O_2 evolution to occur. In addition to the D1, D2 and possibly the extrinsic 33 kDa polypeptides, two larger chlorophyll-binding poly-peptides of 47–50 and 40–43 kDa and two smaller poly-peptides of 5 and 9 kDa associated with cytochrome *b*-559 are always present in highly purified preparations that exhibit significant O_2 evolution activity [11]. What role, if any, these proteins play in electron transport related to water oxidation is not known.

Mechanism of water oxidation

Based on the pioneering work of Joliot, Kok and co-workers [12, 13], it is apparent that the four-electron oxidation of two water molecules to produce O_2 occurs in four sequential steps, each involving a single turnover of the reaction center of PS II. This period-4 character can be seen using single-turnover light flashes by monitoring either the release of O_2 or through the measurement of spectroscopic changes associated with ET components involved in connecting the reaction center to the site of water oxidation. In addition to the removal of four electrons, the mechanism must also accomplish the splitting of O–H bonds, the formation of an O–O bond and the final oxidation and release of O_2. From both a mechanistic and energetic point of view it is evident that water must be bound into a molecular complex, and this is generally supposed to involve the participation of Mn. In addition to binding water and providing a locus for the chemical rearrangements, changes in the oxidation state of Mn provide a mechanism for the accumulation of oxidizing equivalents generated by successive PS II-light-reaction turnovers.

Spectroscopy of Mn in photosystem II

We have used a combination of electron paramagnetic resonance (EPR) and X-ray absorption spectroscopy to investigate the reactions involved in the oxidation of water. The first such evidence came from an analysis of the decay kinetics of Signal II [14], an EPR signal that is characteristic of an organic free radical, long thought to originate from plastoquinone but recently shown more likely to result from a tyrosine radical [15]. Upon repetitive flashes of dark-adapted samples, the amplitudes and sub-millisecond relaxation kinetics of this signal were found to undergo oscillations of period 4 that were correlated with the pattern of O_2 release [14, 16].

The involvement of Mn as a redox component in this cycle was demonstrated through the observation of a complex EPR multiline signal detectable at low temperature (< 10 K) following either single-flash turnovers or continuous illumination under conditions that allow the water splitting complex to advance only to the S_2 state [17, 18]. Amplitude oscillations with a period of four flashes occur also for this signal. The set of at least 19 hyperfine components of this line led to its modeling as a binuclear or tetranuclear complex with complex splittings resulting from interactions with the nuclear spins on the Mn atoms [19, 20]. Because of the many possibilities for combinations of Mn oxidation states that should result in detectable EPR signals, it has not been possible to make this assignment on the basis of EPR evidence alone. Nevertheless, the presence of this signal provides a valuable indicator of at least one state of the water-splitting complex.

X-ray absorption spectroscopy of Mn provides information that is more directly interpretable in terms of oxidation level and coordination environment [21, 22]. Comparison of the absorption edge energies for two states, such as S_1 and S_2, provides a direct measure of the oxidation state difference between them only if the coordination environment is similar in the two states [23–25]. Furthermore, whatever heterogeneity exists among the Mn atom environments will result in a superposition of spectra that is very difficult to resolve. Despite these reservations some important conclusions can be reached on the basis of Mn X-ray absorption measurements.

(1) The oxidation state of Mn appears to increase upon advancement of S_1 to S_2. This conclusion is based on an absorption edge energy shift from 6551.4 to 6552.4 eV (Fig. 1), with little if any change in the higher energy EXAFS region that is particularly sensitive to the coordination environment [25]. Assuming that four Mn are present per reaction center, the magnitude of the edge shift is what one expects from the loss of 1 or, at most, 2 electrons from the complex.

(2) The water-splitting complex in the S_1 and S_2 states contains Mn atoms predominantly in the Mn(III) and Mn(IV) oxidation states. This conclusion is consistent with the absolute edge energies, which correspond to values observed for a variety of model compounds in those oxidation states (Fig. 2); however, the nature of the coordinating ligand atoms can have a significant influence on these energies, and our present knowledge from EXAFS studies allows us only to restrict the candidates to O or N atoms. More direct evidence comes from detailed analysis of high quality

Spinach S_0, S_1 and S_2 Mn K-edges

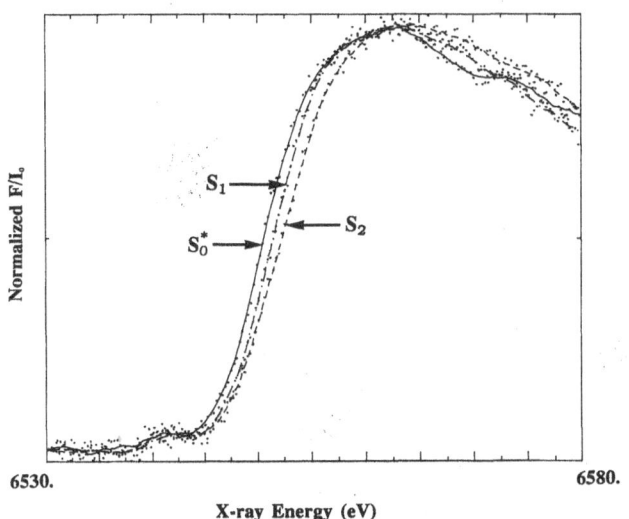

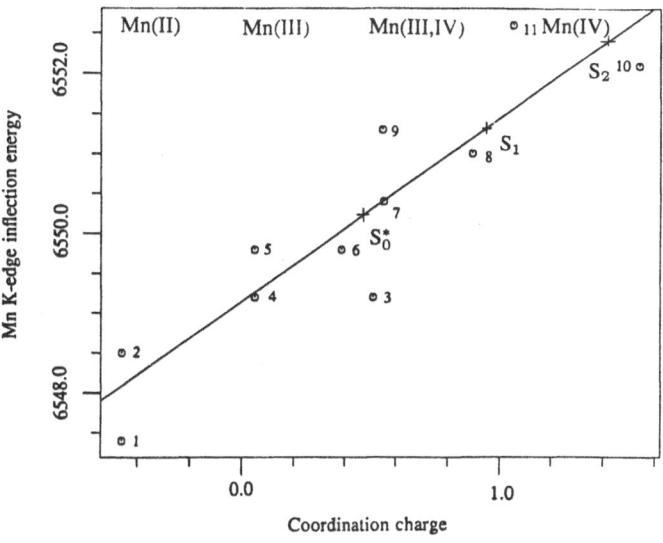

Fig. 1. X-ray absorption edge spectra of O_2-evolving PS II preparations from spinach. Spectra are plotted for a sample containing 40 μM hydroxylamine which was illuminated at 190 K (S_0^*), a dark adapted sample without hydroxylamine (S_1) and a similar sample illuminated at 190 K (S_2). The Mn K-edge inflection energies for S_0^*, S_1 and S_2 are 6550.2, 6551.4 and 6552.4 eV, respectively. Details of the experiment and sample preparation are given in [32].

X-ray spectra in the pre-edge region, where features around 6541 eV that result from 1s to 3d transitions are discernible.

(3) A comparison of the pre-edge region of the Mn K-edge spectrum of PS II preparations from spinach or *Synechococcus* with the pre-edges of a series of inorganic manganese complexes (Fig. 3) shows certain differences between S_1 and S_2 which can be associated with Mn(III) and Mn(IV) respectively. Clearly the S_1 state in preparations from spinach and *Synechococcus* more closely resembles Mn(III), while the S_2 state more closely resembles Mn(IV). The pre-edge features reflect orbitally forbidden $1s \rightarrow 3d$ transitions which become weakly allowed, probably through the mechanism of vibronic coupling [26]. Certain differences in the structure of the pre-edges between Mn(III) and Mn(IV) and between S_1 and S_2 are immediately apparent. (*a*) The pre-edges of Mn(IV) complexes are generally higher in amplitude than the corresponding Mn(III) pre-edge amplitudes. Also for the Mn(IV) complexes the amplitude of the peak of the $1s \rightarrow 3d$ transition is clearly more pronounced relative to the trough in the intervening region at higher energy just below the continuum threshold. (*b*) Close examination of the pre-edge region reveals underlying structure, indicating the presence of at least two overlapping peaks. The separation between peak maxima of overlapping components in the spectra can be more accurately determined by the separation of minima in the second derivative (not shown). Mn(IV) complexes show smaller splittings between peak maxima than do Mn(III) complexes. This smaller splitting in the Mn(IV) complexes may be partially responsible for the more pronounced peak and its higher amplitude. The higher amplitude of the pre-edge may also reflect the superposition of an additional transition due to the extra vacant d-level present in the Mn(IV) complexes.

We believe that these differences in the pre-edge splitting reflect vibronically induced distortions of high-spin Mn(III) complexes [27] which do not occur in high-spin Mn(IV) complexes [28–30]. Spectroscopic and structural (crystallographic) evidence for vibronic distortion of at least two

Mn K-edge Inflection Energies and Coordination Charges of Manganese Model Compounds

Compound	Coordination Charge	Mn K-edge energy (eV)
1. Mn(II)(acac)$_2$2H$_2$O	-0.460	6547.4
2. Mn^{+2}(aqueous)	-0.460	6548.5
3. Mn(III)salenCl H$_2$O	$+0.511$	6549.2
4. Mn(III)$_2$(μ-O)(μ-O$_2$CCH$_3$)$_2$(HBpz$_3$)$_2$	$+0.051$	6549.2
5. Mn(III)$_2$(μ-O)(μ-O$_2$CCH$_3$)$_2$(tmtacn)$_2$	$+0.051$	6549.8
6. Mn(III)Mn(IV)(μ-O)$_2$(bipy)$_4$(ClO$_4$)$_3$·H$_2$O	$+0.386$	6549.9
7. Mn(III)Mn(IV)(μ-O)$_2$(μ-O$_2$CCH$_3$)(tacn)$_2$	$+0.551$	6550.5
8. Mn(IV)$_2$(μ-O)$_2$(phen)$_4$ClO$_4$·H$_2$O	$+0.892$	6551.0
9. Mn(III)μ_3-O(O$_2$CH$_3$)$_7$HO$_2$CCH$_3$	$+0.546$	6551.3
10.α-MnO$_2$	$+1.540$	6552.1
11.Mn(IV)$_4$(μ-O)$_6$(tacn)$_4$(ClO$_4$)$_4$	$+1.051$	6552.6

where acac=acetylacetonate, bipy=2,2'-bipyridine, HBpz$_3$=hydrotris(1-pyrazolylborate), phen=1,10-phenanthroline, salen=ethane-1,2-diylbis(salicylideneiminate) tacn=1,4,7-triazacyclononane, tmtacn=N,N',N''-trimethyl-1,4,7-triazacyclononane

Fig. 2. X-ray absorption Mn K-edge energies and coordination charge for a variety of model Mn compounds in various oxidation states. A straight line corresponding to the best fit to the points is drawn on the plot of the data, and points corresponding to the edge energies for the S_0^*, S_1 and S_2 (or S_3) are located on that line. Mn K-edge inflection energy regions corresponding to various valences are indicated at the top of the plot.

Jahn–Teller ions [Mn(III) and Cu(II)] is now abundant [27]. We have found that the magnitude of the splittings in the pre-edge of Mn(IV) complexes is comparable to the magnitude of d–d transition energies assigned in Mn(IV) complexes [31]. The difference between the magnitude of the observed splittings in Mn(III) and Mn(IV) is comparable to reported Jahn–Teller stabilization energies for Mn(III) [27, 31]. Also, the spread in the first coordination sphere bond lengths in this crystallograpically characterized series of complexes is significantly greater among all of the Mn(III) complexes ($\geqslant 0.3$ Å) relative to the Mn(IV) complexes. This supports the assertion that vibronic distortion due to the Jahn–Teller effect is the cause of a greater splitting among the d-orbitals. Thus, this comparison with the corresponding features in model compounds indicates that the light-induced changes in the pre-edge peak between S_1 and S_2 correspond to a decrease in Mn(III) and an increase in Mn(IV) content in S_2.

(4) The state S_3 is not markedly different from S_2 in terms of either Mn oxidation state or coordination environment [32]. This is a conclusion based on two different methods of producing samples that, based on measurements of the EPR multiline signal amplitude, consist of 50–66% S_3 [33]. The lack of a significant change of the EXAFS of manganese in the $S_2 \rightarrow S_3$ transition argues strongly against recent proposals [34] that a significant structural rearrangement from a cubane-like structure to an adamantane-like tetranuclear manganese structure occurs at this transition. In fact, the absence of features in simulations of the EXAFS in the

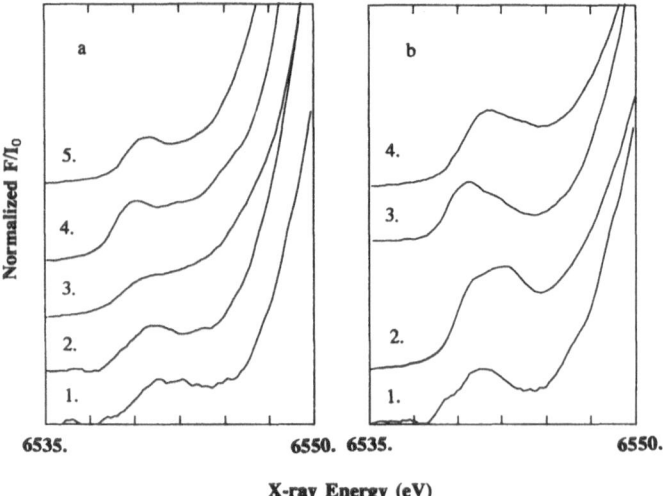

Fig. 3. The pre-edge region of the Mn K-edge spectrum of O_2-evolving PS II preparations from spinach and *Synechococcus*, poised in the S_1 and S_2 states, as described in refs. [32,38]. To emphasize the similarity of the pre-edge structure of the S_1 state to pre-edges of Mn(III) complexes, a series of Mn(III) complexes of different nuclearity has been plotted for comparison. Similarly for the S_2 state a number of Mn(IV) complexes have been plotted. (*a*) From bottom to top, the pre-edge spectra are: (1) a spinach PS II preparation poised in the S_1 state, (2). a *Synechococcus* preparation poised in the S_1 state; (3) the trinuclear Mn(III) complex, $Mn_3(\mu_3$-O$)$ (OAc)$_7$ HOAc; (4) the binuclear Mn(III) complex, $Mn_2(\mu$-O$)$ (OAc)$_2$ (HBpz$_3$)$_2$, and (5) the Mn(III) mononuclear complex, Mn salen Cl. (*b*) The pre-edge spectra of: (1) a spinach PS II preparation poised in the S_2 state; (2) the tetranuclear Mn(IV) complex [Mn_4O_6(tacn)$_4$](ClO$_4$)$_4$; (3) a binuclear Mn(IV) complex ([$Mn_2(\mu$-O$)_2$ (phen)$_4$](ClO$_4$)$_4$CH$_3$CN) and (4) the Mn(IV) mononuclear complex, Mn(salicylate)$_2$ bipyridine. We gratefully acknowledge Drs George Christou (Indiana University, Bloomington), Karl Wieghardt (Ruhr-Universität Bochum) and William H. Armstrong (University of California, Berkeley) for providing the manganese complexes used in this study.

region around 3.2 Å excludes adamantane-like structures as relevant models for the manganese cluster in the S_3-state. We base this statement on a detailed comparison of the EXAFS of a tetranuclear adamantane-like manganese complex provided by Dr Karl Wieghardt [35].

(5) States more reduced than S_1 exhibit correspondingly lower values of the Mn absorption edge energy. Addition of hydroxylamine is known to set back the release of O_2 upon flash illumination by two steps [36]. The X-ray absorption spectrum of a PS II sample treated with NH_2OH and kept in the dark is indistinguishable from that of an untreated (S_1) sample [37]. Upon illumination to produce a single turnover of the reaction center of the NH_2OH-treated sample, the X-ray absorption edge energy decreases to 6550.2 eV (Fig. 1). This state, which we designate S_0^*, thus appears to contain Mn in a more reduced state than that in S_1; however, there are differences in the EXAFS region that are indicative of some accompanying change in the coordination environment. A true S_0 state, produced by a series of three reaction-center turnovers, has so far not been achieved by us in samples suitable for X-ray spectroscopy.

(6) The coordination environment of Mn consists of two or more shells of low Z elements (probably O and/or N atoms) in the range from 1.75 to 2.0 Å and an additional shell at 2.70 Å attributable to Mn [22–25]. This pattern is consistent with the short Mn–O and Mn–Mn distances associated with μ-oxo-bridged binuclear Mn complexes [21,32]. Although the evidence from relative scattering amplitudes is less reliable, the spectra are more consistent with a μ-oxo-bridged

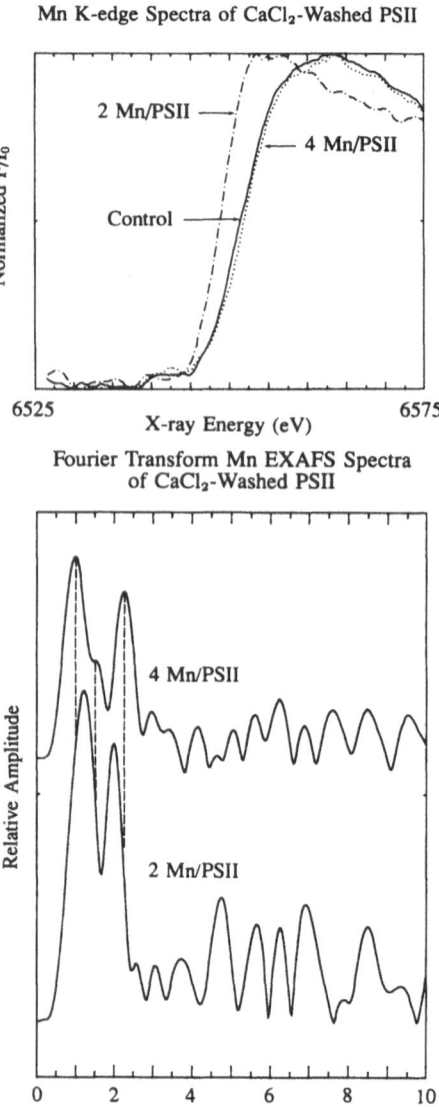

Fig. 4. Effect of $CaCl_2$ washing on the X-ray spectra of oxygen-evolving PS II preparations. Top: Mn K-edge spectra of control (—); $CaCl_2$-washed sample depleted of the 16, 24 and 33 kDa polypetides (....) and $CaCl_2$-washed sample depleted of the extrinsic polypeptides and 2 Mn (----). The edge inflection energies are 6551.3, 6551.8 and 6548.9 eV, respectively. Bottom: Fourier transforms of the Mn EXAFS of the $CaCl_2$-washed sample containing 4 Mn/PS II reaction center and of the $CaCl_2$-washed sample containing 2 Mn/PS II. Details of the procedures are given in ref. [40].

binuclear Mn structure than with a symmetric cubane structure of the type suggested by Brudvig and Crabtree [34]. A feature identified in a preliminary report as possible evidence of scattering by a transition metal at $R > 3$ Å [32] has not been reproduced with our current detection method.

(7) The Mn environment in the water-splitting complex of cyanobacteria is very similar to that seen in thylakoid membranes of PS II-enriched particles from chloroplasts of higher plants [38].

(8) The binding-site ligands for Mn in the water-splitting complex appear to be furnished by intrinsic membrane proteins. Removal using 0.8 M-$CaCl_2$ of the extrinsic 16, 24 and 33 kDa polypeptides, which abolishes O_2 evolution at moderate Cl^- concentrations, does not release significant amounts of bound Mn [39]. Nor is there any effect of this treatment on the Mn X-ray edge and EXAFS spectra compared with that of a dark-adapted control (Fig. 4) [40].

Further treatment of the $CaCl_2$-washed sample by incubation in a low Cl^- buffer for 18 h at 4 °C results in the loss of about half of the Mn and a dramatic shift to lower energy (6548.9 eV) of the X-ray absorption edge. Obvious candidates for intrinsic proteins that provide ligands to coordinate the multi-nuclear Mn complex are the D1 and D2 polypeptides, as discussed above.

Conclusions

Spectroscopic studies confirm the direct involvement of manganese in the oxidation reactions leading to photosynthetic oxygen evolution. Based on X-ray absorption edge energies, the oxidation state increases observed from an analogue S_0 state to S_1 and from S_1 to S_2 are apparently not repeated for the S_2 to S_3 transition. No evidence has been detected that indicates any significant rearrangement of ligands during the S_1 to S_2 to S_3 transitions, such as might be expected to accompany either the incorporation of water or its chemical rearrangement on the way to O_2 production. S_2 is a paramagnetic state that reflects a manganese cluster of predominantly Mn(III) and Mn(IV). It is clear that a dimeric Mn_2(III,IV) core is essential; however, tetrameric Mn_4 (III,III,III,IV) or tetrameric Mn_4(III,IV,IV,IV) are also viable models. The evidence supports a change from Mn(III) to Mn(IV) in the S_1 to S_2 transition. The ligands in the first coordination shell surrounding Mn are light elements, predominantly O and N. The amino acids that are reasonable candidates to provide such ligands (Asp, Glu, His, Tyr, Arg) are found in abundance in strongly conserved regions near the C-termini of the D1 and D2 reaction center proteins of PS II.

Acknowledgements

The research described in this report was supported by grants from the National Science Foundation (PCM84-16676), from the Competitive Research Grants Program of the US Department of Agriculture (85CRGR1-1847) and by the Director, Office of Energy Research, Office of Basic Energy Sciences, Division of Biological Energy Conversion and Conservation of the US Department of Energy under contract DE-AC03-76SF00098. Synchrotron radiation facilities were provided by the Stanford Synchrotron Radiation Laboratory. J.-L. Z. acknowledges support from the Commissariat à l'Energie Atomique (France).

References

1. Hearst, J. E., in *Encyclopedia of Plant Physiology*, vol. 19: *Photosynthesis III* (ed. L. A. Staehelin and C. J. Arntzen), pp. 382–389, Springer-Verlag, Berlin (1986).
2. Deisenhofer, J. Epp, O., Miki, K., Huber, R. and Michel, H., *J. Mol. Biol.* 180, 385–398 (1984); *Nature* 318, 618–624 (1985).
3. Rochaix, J.-D., Dron, M. Rahire, M. and Malnoë, P., *Plant Mol. Biol.* 3, 363–370 (1984).
4. Michel, H., Weyer, K. A., Gruenberg, H., Dunger, I., Oesterhelt, D. and Lottspeich, F., *EMBO J.* 5, 1149–1158 (1986).
5. Trebst, A., *Z. Naturforsch.* 41C, 240–245 (1986).
6. Sayre, R. T., Andersson, B. and Bogorad, L., *Cell* 47, 601–608 (1986).
7. Sauer, K., *Acc. Chem. Res.* 13, 249–256 (1980).
8. Andersson, B., in *Encyclopedia of Plant Physiology*, vol. 19: *Photosynthesis III* (ed. L. A. Staehelin and C. J. Arntzen), pp. 447–456. Springer-Verlag, Berlin (1986).
9. Metz, J. G., Pakrasi, H. B., Seibert, M. and Arntzen, C. J., *FEBS Lett.* 205, 269–274 (1986).
10. Ikeuchi, M. and Inoue, Y., *FEBS Lett.* 210, 71–76 (1987).
11. Yamada, Y., Tang, X.-S., Itoh, S. and Satoh, K., *Biochim. Biophys. Acta* 891, 129–137. (1987).
12. Joliot, P., Barbieri, G. and Chabaud, R., *Photochem. Photobiol.* 10, 309–329 (1969).
13. Kok, B., Forbush, B. and McGloin, M., *Photochem. Photobiol.* 11, 457–475 (1970).
14. Babcock, G. T., Blankenship, R. E. and Sauer, K., *FEBS Lett.* 61, 286–289 (1976).
15. Barry, B. A. and Babcock, G. T., (this volume); *Proc. Natl. Acad. Sci. USA* 84, 7099–7103 (1987).
16. Cole, J. and Sauer, K., *Biochim. Biophys. Acta* 891, 40–48 (1987).
17. Dismukes, G. C. and Siderer, Y., *FEBS Lett.* 121, 78–80 (1980); *Proc. Natl. Acad. Sci. USA* 78, 274–278 (1981).
18. Brudvig, G. W., Casey, J. L. and Sauer, K., *Biochim. Biophys. Acta* 723, 366–371 (1983).
19. Hannson, Ö, and Andréasson, L.-E., *Biochim. Biophys. Acta* 679, 261–268 (1982).
20. Dismukes, G. C., Ferris, K. and Watnick, P., *Photobiochem. Photobiophys.* 3, 243–256 (1982).
21. Kirby, J. A., Robertson, A. S., Smith, J. P., Thompson, A. C., Cooper, S. R. and Klein, M. P., *J. Amer. Chem. Soc.* 103, 5529–5537 (1981).
22. Kirby, J. A., Goodin, D. B., Wydrzynski, T., Robertson, A. S. and Klein, M. P., *J. Amer. Chem. Soc.* 103, 5537–5542 (1981).
23. Goodin, D. B., Yachandra, V. K., Britt, R. D., Sauer, K. and Klein, M. P., *Biochim. Biophys. Acta* 767, 209–216 (1984).
24. Yachandra, V. K., Guiles, R. D., McDermott, A., Britt, R. D., Dexheimer, S. L., Sauer, K. and Klein, M. P., *Biochim. Biophys. Acta* 850, 324–332 (1986).
25. Yachandra, V. K., Guiles, R. D., McDermott, A. E., Cole, J. L., Britt, R. D., Dexheimer, S. L., Sauer, K. and Klein, M. P., *Biochemistry* 26, 5974–5981 (1987).
26. Teo, B. K., *EXAFS: Basic Principles and Data Analysis*, pp. 8–20. Springer-Verlag, Berlin (1986).
27. Bersuker, I. B., *Coord. Chem. Rev.* 14, 357–412 (1975).
28. Pavacik, P. S., Huffman, J. C. and Christou, G., *J. Chem. Soc. Chem. Commun.* 43–44 (1986).
29. Kessissoglou, D. P., Butler, W. M. and Pecoraro, V. L., *J. Chem. Soc. Chem. Commun.* 1253–1255 (1986).
30. Hartman, J. A. R., Foxman, B. M. and Cooper, S. R. *J. Chem. Soc. Chem. Commun.* 583–584 (1982).
31. Lever, A. B. P., *Inorganic Electronic Spectroscopy*, 2nd edn, pp. 190, 430. Elsevier, Amsterdam (1984).
32. Guiles, R. D., Yachandra, V. K., McDermott, A. E., Britt, R. D., Dexheimer, S. L., Sauer, K. and Klein, M. P., in *Progress in Photosynthesis Research* (ed. J. Biggins), vol. I, pp. 561–564, Martinus Nijhoff, Dordrecht (1987).
33. Guiles, R. D., Zimmermann, J.-L. *et al*, (unpublished results).
34. Brudvig, G. W. and Crabtree, R. H., *Proc. Natl. Acad. Sci. USA* 83, 4586–4588 (1986).
35. Wieghardt, K., Bossek, U. and Gebert, W., *Angew. Chem. Int. Ed. Engl.* 22, 328–329 (1983).
36. Bouges-Bocquet, B., *Biochim. Biophys. Acta* 292, 772–785 (1973).
37. Guiles, R. D. *et al*, (unpublished results).
38. McDermott, A., Yachandra, V. K., Guiles, R. D., Britt, R. D., Dexheimer, S. L., Sauer, K. and Klein, M. P., in *Progress in Photosynthesis Research* (ed. J. Biggins), vol. I, pp. 565–568. Martinus Nijhoff, Dordrecht (1987).
39. Ono, T.-A. and Inoue, Y., *FEBS Lett.* 164, 255–260 (1983).
40. Cole, J. L., Yachandra, V. K., McDermott, A. E., Guiles, R. D., Britt, R. D., Dexheimer, S. L., Sauer, K. and Klein, M. P. *Biochemistry* 26, 5967–5973 (1987).

Chemica Scripta 1988, **28A**, 93–98

Ligand-Substitution Reactions of the O₂-Evolving Center of Photosystem II

Warren F. Beck and Gary W. Brudvig*

Department of Chemistry, Yale University, New Haven, CT 06511, USA

Paper presented by Gary W. Brudvig at the Nobel Conference 'Biophysical Chemistry of Dioxygen Reactions in Respiration and Photosynthesis', Fiskebäckskil, Sweden, 1–4 July, 1987

Abstract

Ligand-substitution reactions occur in the O₂-evolving center of photosystem II at two distinct binding sites: a Type 1 binding site located on the Mn tetramer complex, which is assigned to the substrate-binding site, and a Type 2 binding site, not necessarily located on the Mn complex, at which Cl⁻ acts as a co-factor for the H₂O-oxidation reaction. We have detected reactions at both of these binding sites by monitoring the two low-temperature electron paramagnetic resonance (EPR) signals exhibited by the Mn complex in the S_2 state, the $g = 4.1$ and multiline EPR signals. NH₃ coordinates to the Mn complex in the S_2 state at the Type 1 binding site through a nucleophilic addition reaction and alters the ^{55}Mn nuclear hyperfine coupling in the S_2-state multiline EPR signal. This reaction is likely to be similar in mechanism to that involved in the substrate H₂O-binding reaction; however, the coordination of H₂O to the Mn complex is probably triggered by the formation of the S_3 state, while the binding of NH₃, a stronger base, occurs upon formation of the S_2 state. Reactions at the Type 2, or Cl⁻-binding site, in contrast, occur in both the S_1 state, present in dark-adapted samples, and in the S_2 state. The nature of the ligand bound to the Type 2 binding site is proposed to influence an equilibrium between the two conformations of the Mn complex that exhibit either the $g = 4.1$ EPR signal or the $g = 1.98$ multiline EPR signal in the S_2 state. Inhibitors such as F⁻ or small primary amines shift the equilibrium in favor of the conformation of the Mn complex that exhibits the $g = 4.1$ EPR signal; Cl⁻ and Br⁻, which activate the O₂-evolving center, favor the conformation that exhibits the multiline EPR signal in the S_2 state. Depletion of Cl⁻ from the O₂-evolving center causes the $g = 4.1$ EPR signal to be observed in the S_2 state in lieu of the multiline EPR signal. Since loss of Cl⁻ prohibits the stable formation of the S_3 state [Ono *et al.*, *Biochim. Biophys. Acta* **851**, 193 (1986)], apparently by trapping the Mn complex in the S_2-state conformation that exhibits the $g = 4.1$ EPR signal, we propose that Cl⁻ serves a structural role in the O₂-evolving center in its maintenance of a catalytically-active conformation of the Mn complex.

Introduction

The catalysis of the four-electron oxidation of H₂O to O₂ is performed by the O₂-evolving center of photosystem II (PS II), which contains a polynuclear Mn complex [1]. The O₂-evolving center advances through five oxidation states S_i, $i = 0$–4, in each catalytic cycle, with release of O₂ occurring as the S_4 state is reduced to the S_0 state. Electron paramagnetic resonance (EPR) [2,3] and X-ray absorption edge [4] experiments have shown that oxidation of the Mn complex occurs during the S-state cycle. The S_2 state exhibits either a $g = 4.1$ EPR signal or a $g = 1.98$ multiline EPR signal at low temperature [5,7]. These EPR signals have been exploited as probes for the structure and function of the Mn complex [1–3, 5–8]. Analysis of the exchange interactions present in the Mn complex in the S_2 state have led to the conclusion that the Mn complex consists of four exchange-coupled Mn ions, which are proposed to be arranged in a Mn_4O_4 cubane-like configuration [5, 9].

The function of the Mn tetramer complex in the O₂-evolving center very likely involves not only the accumulation of the four oxidizing equivalents required for the H₂O-oxidation reaction but also the binding of two substrate H₂O molecules. The H₂^{17}O-exchange experiments of Hansson *et al.* [8] have shown directly that ^{17}O atoms, originating from the solvent, are coordinated to the Mn complex in the S_2 state, effecting a small broadening of the hyperfine lines of the S_2-state multiline EPR signal via a nuclear hyperfine interaction. This result was interpreted as showing that the two substrate H₂O molecules are coordinated to the Mn complex in the S_2 state [8]. However, it is also possible that the Mn complex contains structural μ-oxo bridges between the Mn ions [2,9] that may easily exchange with solvent H₂O, as has been observed in μ-oxo-bridged Mn(III) dimer complexes [10]. The results of H₂^{18}O-exchange experiments by Radmer and Ollinger [11], who used mass spectrometry to detect O₂ products after a series of saturating light flashes, indicate that the Mn complex does not form a non-exchangeable co-ordination compound with substrate H₂O at least until after formation of the S_3 state.

In order to examine the coordination chemistry involved in the catalytic cycle of the Mn complex, we have characterized its interaction with small molecules that might behave as substrate analogs, such as primary amines and hydroxyl-amines. Through the use of the S_2-state $g = 4.1$ and multiline EPR signals as probes for the structure of the Mn complex, we have detected ligand-substitution reactions at two distinct sites in the O₂-evolving center [6]. Reactions at the Type 1 and Type 2 binding sites can be readily distinguished: binding of NH₃ to the Type 1 binding site alters the ^{55}Mn nuclear hyperfine coupling in the S_2-state multiline EPR signal, implying a direct coordination to the Mn complex [12], whereas coordination of NH₃ to the Type 2 binding site increases the proportion of the S_2 state that exhibits the $g = 4.1$ EPR signal rather than the multiline EPR signal without causing a significant change in the line shape of either EPR signal [6].

Inhibition of O₂-evolution activity also occurs at two types of binding sites. Sandusky and Yocum [13] demonstrated that primary amines inhibit the O₂-evolving center by competing with Cl⁻ for one type of binding site, while NH₃ was unique in its ability to bind at a different binding site independently of the Cl⁻ concentration. Prompted by these findings, we found that the binding of NH₃ to the Type 2 binding site depends inversely on the Cl⁻ concentration [14]. In addition, the rate of reaction of hydroxylamine and its *N*-methyl-substituted analogs with the O₂-evolving center in the S_1 state was also inversely proportional to the Cl⁻ concentration [14].

* To whom correspondence should be addressed.

It is apparent that hydroxylamine and analogous molecules react with the Mn site by an electron-transfer reaction after binding to the Type 2 binding site [14], rather than by coordinating in lieu of substrate H_2O to the Mn complex, as had been previously proposed. In contrast, the coordination of NH_3 to the Mn complex at the Type 1 binding site occurs even at high Cl^- concentrations [6, 15]. Primary amines larger than NH_3 are unable to coordinate to the Mn complex owing to a steric selectivity of the Type 1 binding site [6, 16]. The two EPR-detectable ligand-binding sites are, then, assignable to the two binding sites at which inhibition of O_2 evolution occurs. The NH_3-specific, Cl^--independent Type 1 site on the Mn complex can be assigned to the substrate H_2O-binding site, and the Type 2 site can be assigned to the site at which Cl^- assumes its role as a co-factor for the H_2O-oxidation reaction.

Different mechanisms appear to be involved in ligand-substitution reactions at the Type 1 and Type 2 sites of the O_2-evolving center [1, 14]. While the Mn complex appears to be inert to the coordination of NH_3 at the Type 1 binding site until the S_2 state is present [12], binding of NH_3 or hydroxylamine at the Type 2 binding site apparently occurs in dark, S_1 state samples [6, 14]. Thus, the accessibility of the Type 1 binding site to ligands depends on the formation of a more electron deficient Mn complex; the binding of NH_3 to the Mn complex may then be described as a nucleophilic addition reaction triggered by the formation of the S_2 state [9, 16]. This finding suggests that coordination of substrate H_2O, a weaker base than NH_3, may require the formation of the S_3 state [9]. The coordination of ligands to the Type 2 binding site is evidently not obligately linked to the oxidation state of the Mn complex, since NH_3 can apparently bind in either the S_1 or the S_2 state [6]. However, the presence of NH_3 in the Type 2 binding site in the place of Cl^- affects the relative yield of the two types of S_2-state EPR signals, indicating that the structure of the Mn complex has been affected by the exchange of ligands. In addition, the exchange of Cl^- for hydroxylamine at the Type 2 binding site is followed by a two-electron reduction of the Mn complex to the S_{-1} state; the facility of this reaction suggests that the Type 2 binding site is in close proximity to the Mn complex [14].

In this paper, we examine in more detail the nature of ligand-substitution reactions at the Type 2 site of the O_2-evolving center in the S_1 state. We compare the effects of displacement of Cl^- from the Type 2 site by small primary amines or F^- and the effects of depletion of Cl^- from the O_2-evolving center, by studying the stable S_2-state $g = 4.1$ EPR signal present after illumination of spinach PS II membrane samples at 210 K. The results indicate that ligand-substitution reactions at the Type 2 binding site affect a conformational equilibrium between two forms of the Mn complex that exhibit either the S_2-state $g = 4.1$ or multiline EPR signals.

Experimental section

PS II membranes were isolated in darkness from market spinach leaves according to the procedure of Berthold et al., [17] as modified by Beck et al., [18] and were stored at 77 K in a buffer solution at pH 6.0 containing 20 mM-2-(*N*-morpholino)ethanesulfonic acid (MES), 15 mM-NaCl, and

either 30% (v/v) ethylene glycol or 0.4 M sucrose, which was present as a cryoprotectant. For EPR experiments, PS II membranes were equilibrated in sample buffer solutions at pH 7.5 containing 20 mM-*N*-(2-hydroxyethyl)piperazine-*N'*2-ethanesulfonic acid (HEPES), NaCl to obtain the chosen Cl^- concentration, and either 30% (v/v) ethylene glycol or 0.4 M sucrose, as indicated; at least two resuspension and recentrifugation cycles were employed.

The procedures followed for treating PS II membrane samples with amines have been previously described [6, 12]. CH_3NH_2 was added as the free base, diluted from a concentrated stock solution (Aldrich) in the chosen sample buffer solution. The pH of the CH_3NH_2-containing treatment buffer solution was adjusted to 7.50 by addition of H_2SO_4. The protocol chosen for depletion of Cl^- from PS II membranes was adapted from that used by Ono et al. [19]; the Cl^--free buffer solution employed in the protocol consisted of 50 mM-Na_2SO_4, 20 mM-HEPES-NaOH, pH 7.50, and either 30% (v/v) ethylene glycol or 0.4 M sucrose, as indicated.

EPR spectroscopy and sample illuminations were performed with the apparatus described previously [12, 18]. Difference spectra (light–dark) were obtained through computer subtraction of the dark background spectrum from the post-illumination spectrum obtained under the same measurement conditions.

Results

Binding of halides and amines to the Type 2 binding site

In the presence of 30% (v/v) ethylene glycol, the S_2-state $g = 4.1$ EPR signal is unstable. Though it can be produced in high yield by illumination at 130 K, the $g = 4.1$ EPR signal is nearly quantitatively converted under these conditions to the S_2-state multiline EPR signal by warming the sample to 200 K in darkness [20, 21]. In a previous paper [6], however, we showed that the binding of NH_3 in the S_1 state to the Type 2 binding site stabilizes the $g = 4.1$ EPR signal produced by illumination at 210 K. In a subsequent paper [14a], we found that an increase in the Cl^- concentration decreases the intensity of the stable $g = 4.1$ EPR signal observed after illumination of NH_3-treated PS II membranes at 210 K. This result suggests that the Type 2 binding site is the site at which Cl^- binds in the O_2-evolving center. In order to test this proposal, we have examined the effect of halides on the S_2-state EPR signals. Casey and Sauer [20] observed previously that the presence of F^- suppresses the formation of the S_2-state multiline EPR signal, yielding instead the S_2-state $g = 4.1$ EPR signal. We proposed previously that the results of Casey and Sauer [20] could be explained by a binding of F^- to the Type 2 binding site in the same manner as we observed in the case of NH_3 [6].

Fig. 1*a* shows the S_2-state EPR signals that are observed in untreated PS II membranes at pH 7.50 when the Cl^- concentration is 15 mM. Only a small $g = 4.1$ EPR signal is observed under these conditions, but a large S_2-state multiline EPR signal is present, accounting for the oxidizing equivalent produced during the illumination. A $g = 1.9$ EPR signal, arising from the reduced $Fe(II)Q_A$ site on the electron acceptor side of PS II [22], is superimposed on the hyperfine line pattern of the multiline EPR signal. When the Cl^-

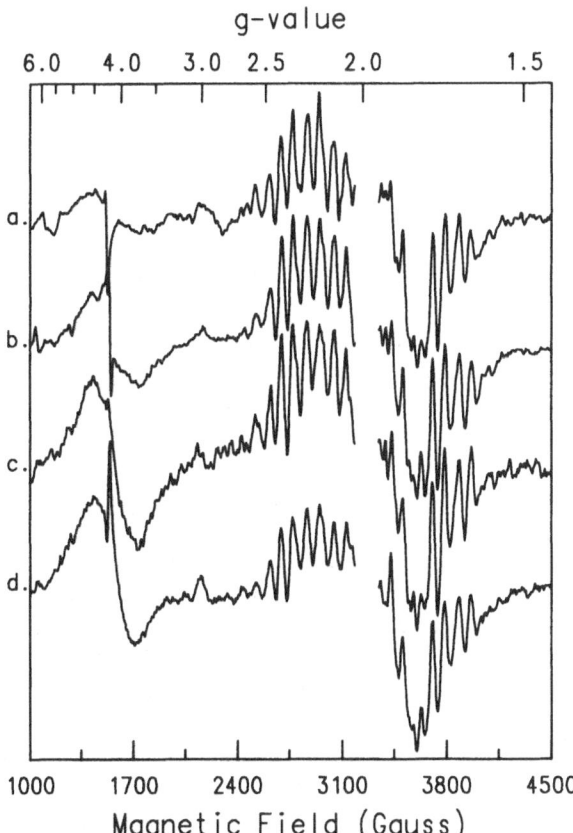

Fig. 1. S$_2$-state $g = 4.1$ and multiline EPR signals observed in spinach PS II membranes after illumination at 210 K for 120 s. The PS II membrane samples (5 mg of chlorophyll/ml) were treated with 100 μM 3-(3,4-dichlorophenyl)-1,1-dimethylurea (DCMU), and contained 30% (v/v) ethylene glycol: (a) untreated PS II membranes, total [Cl$^-$] = 15 mM; (b) untreated PS II membranes, total [Cl$^-$] = 0.2 mM; (c) 100 mM-NaF-treated PS II membranes, total [Cl$^-$] = 0.2 mM; (d) 100 mM-CH$_3$NH$_2$-treated PS II membranes, total [Cl$^-$] = 0.2 mM. EPR spectrometer conditions: microwave frequency, 9.0 GHz; microwave power, 200 μW; field modulation frequency, 100 kHz; field modulation amplitude, 20 G; sample temperature, 8.0 K. Each spectrum is the average of two, 10 min scans.

concentration is lowered to 0.2 mM, illumination at 210 K yields a slightly more intense $g = 4.1$ EPR signal (Fig. 1 b) than that observed at a Cl$^-$ concentration of 15 mM. A still larger $g = 4.1$ EPR signal is generated in a similar sample when 100 mM-F$^-$ is added (Fig. 1 c). However, 100 mM-F$^-$ has little effect on the S$_2$-state EPR signal yields when 2.5 mM-Cl$^-$ is present (data not shown). These results show that F$^-$, in fact, does bind to the Type 2 binding site in the S$_1$ state.

Similarly, when the free base form of CH$_3$NH$_2$ is added to PS II membranes at pH 7.50, at a 0.2 mM-Cl$^-$ concentration, illumination at 210 K produces a larger $g = 4.1$ EPR signal than is observed in the absence of CH$_3$NH$_2$ under similar conditions [cf Fig. 1(b and d)]. As shown previously [6], CH$_3$NH$_2$ has no effect on the yields of the S$_2$-state EPR signals when a high Cl$^-$ concentration is present. These results confirm that small primary amines, in addition to NH$_3$, can bind to the Type 2 binding site in the S$_1$ state. Even at low Cl$^-$ concentrations, however, we were previously unable to detect the binding of tris(hydroxymethyl)aminomethane or 2-amino-2-ethyl-1,3-propanediol to the Type 2 binding site [6]. Thus, the Type 2 binding site is sterically selective for relatively small Lewis bases; a similar steric selectivity at the Type 2 binding site is inferred from the kinetics of hydroxylamine reactions with the O$_2$-evolving center in the S$_1$ state [14].

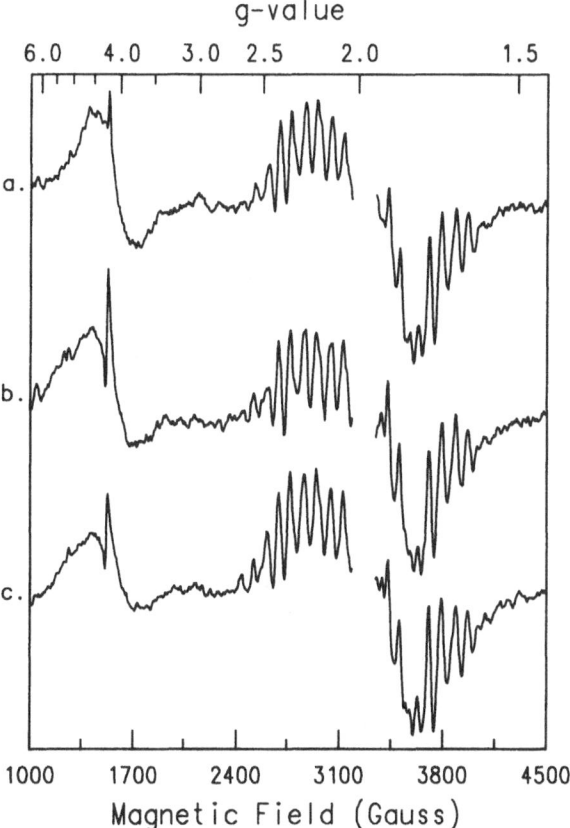

Fig. 2. S$_2$-state $g = 4.1$ and multiline EPR signals observed in 10 mM-NH$_4^+$-treated PS II membranes, after illumination at 210 K for 120 s. Sample and EPR spectrometer conditions were as described in Fig. 1: (a) 10 mM-NH$_4^+$-treated PS II membranes, total [Cl$^-$] = 0.2 mM; (b) as in (a), except with 25 mM-NaBr added; (c) as in (a), except with 25 mM-NaCl added.

When 10 mM-NH$_4^+$ is added to a PS II membrane sample containing only 0.2 mM-Cl$^-$, a substantial S$_2$-state $g = 4.1$ EPR signal is observed after illumination at 210 K (Fig. 2a). In Fig. 2(b and c), we compare the effects of added Br$^-$ and Cl$^-$ on the intensity of the stable $g = 4.1$ EPR signal induced by NH$_4^+$ in PS II membrane samples at low Cl$^-$ concentrations. Both Br$^-$ and Cl$^-$ decrease the intensity of the $g = 4.1$ EPR signal compared to that observed at low Cl$^-$ concentrations; however, Cl$^-$ causes a larger decrease in the intensity of the $g = 4.1$ EPR signal than does Br$^-$ at the same concentration. In the S$_1$ state, then, the Type 2 binding site has a higher affinity for Cl$^-$ than for Br$^-$. F$^-$ seems to bind even more weakly to the Type 2 binding site than Br$^-$ does, since the presence of 2.5 mM-Cl$^-$ completely destabilizes the $g = 4.1$ EPR signal in 100 mM-F$^-$-treated samples (data not shown).

Depletion of chloride from the Type 2 binding site

Ono *et al.* [19] found that depletion of Cl$^-$ from PS II membranes resulted in an inability to form the S$_2$-state multiline EPR signal after a single flash of light; however, the multiline EPR signal could be formed in darkness by adding Cl$^-$ following the flash of light. This result suggested that the O$_2$-evolving center is capable of advancing in the absence of Cl$^-$ to a modified form of the S$_2$ state that does not exhibit the S$_2$ state multiline EPR signal. Ono *et al.* [19] found that subsequent S-state advancement from the modified, Cl$^-$-depleted S$_2$ state was blocked. We examined the effect of depletion of Cl$^-$ from PS II membranes on the intensity of the S$_2$-state $g = 4.1$ EPR signal, since the results of Figs. 1 and 2

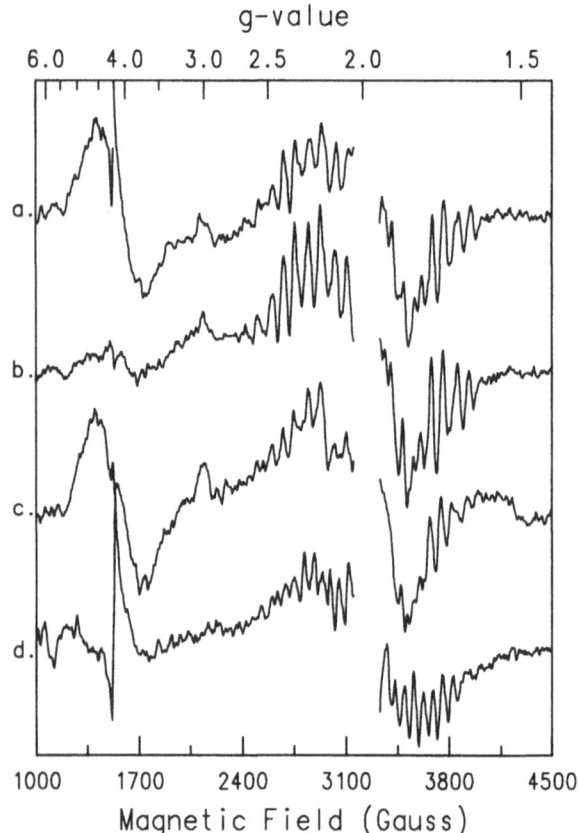

Fig. 3. S_2-state $g = 4.1$ and multiline EPR signals observed in Cl^--depleted PS II membranes after illumination at 210 K for 120 s. Sample and EPR spectrometer conditions were as described in Fig. 1: (*a*) Cl^--depleted PS II membranes, suspended in a Cl^--free buffer solution; (*b*) Cl^--depleted PS II membranes, reconstituted with 100 mM-NaCl; (*c*) Cl^--depleted PS II membranes, treated with 50 mM-$(NH_4)_2SO_4$ in a Cl^--free buffer solution; (*d*) the same sample used in (*c*), after warming in darkness to $-6\,°C$ for 1 min. In (*c*) and (*d*) the sample contained 250 μM 2,5-dichloro-*p*-benzoquinone (DCBQ). The spectrum in (*d*) is the difference between the spectrum observed after the $-6\,°C$ dark-incubation period and the spectrum obtained prior to illumination at 210 K.

suggest that displacement of Cl^- from the Type 2 binding site by an amine or F^- yields an S_2 state conformation of the Mn complex that exhibits a $g = 4.1$ EPR signal rather than a multiline EPR signal.

Figure 3*a* shows that at pH 7.50, in the presence of 30% (v/v) ethylene glycol, Cl^--depleted PS II membranes exhibit a large $g = 4.1$ EPR signal and a relatively small multiline EPR signal when the S_2 state is produced by illumination at 210 K. Addition of 100 mM-NaCl to a similar, Cl^--depleted PS II membrane sample (Fig. 3*b*) results in the production of a large S_2-state multiline EPR signal and a negligible amount of the S_2-state $g = 4.1$ EPR signal. The yield of the $g = 1.9$, $Fe(II)Q_A^-$ EPR signal in Fig. 3(*a* and *b*) is equivalent, however, showing that similar yields of the S_2 state were produced in both the Cl^--depleted and Cl^--reconstituted samples. Thus, the addition of Cl^- restored the ability to form the S_2-state multiline signal lost after Cl^- depletion, in a manner consistent with the findings of Ono *et al.* [19].

We observed similar effects (data not shown) when PS II membrane suspensions containing 0.4 M sucrose, rather than 30% (v/v) ethylene glycol, were depleted of Cl^-. However, we found that a much smaller fraction of the S_2 state exhibits the S_2-state multiline EPR signal under conditions of Cl^- depletion in the presence of 0.4 M sucrose. This result raises the possibility that PS II membranes suspended in the

presence of sucrose have a lower affinity for Cl^- than that observed when ethylene glycol is used; preliminary results based on O_2-evolution assays confirm this suggestion (Beck, W. F. and Brudvig, G. W., unpublished results). It is possible that the stable S_2-state $g = 4.1$ EPR signal observed in samples containing sucrose as a cryoprotectant [7, 19, 23] is a reflection of the lower affinity for Cl^- in these types of samples.

The results of Fig. 3(*a* and *b*) suggest strongly that the modified S_2 state present under conditions of Cl^--depletion exhibits the $g = 4.1$ EPR signal rather than the multiline EPR signal. In particular, it does not appear to be necessary to postulate an EPR-silent form of the Mn complex in the S_2 state in order to account for the results of Ono *et al.* [19]. We wondered, however, if depletion of Cl^- from PS II membranes produces a state distinguishable from that induced by the displacement of Cl^- from the Type 2 binding site by a ligand. Figure 3*c* shows that the addition of 100 mM-NH_4^+ to a Cl^--depleted PS II membrane sample does not significantly change the respective yields of the S_2-state $g = 4.1$ and multiline EPR signals after illumination at 210 K. Therefore, the condition present after depletion of Cl^- from the O_2-evolving center is not distinguishable from that resulting from binding of NH_3, for instance, to the Type 2 binding site.

Figure 3*d* shows that a brief incubation in the dark at $-6\,°C$ causes the large $g = 4.1$ EPR signal present in the NH_4^+-treated, Cl^--depleted PS II membrane sample to collapse completely, producing instead an S_2-state multiline EPR signal having a 67-G hyperfine line spacing. This altered multiline EPR signal is the same as that observed previously in NH_4Cl-treated PS II membranes [6, 12, 15] and can be attributed to the coordination of NH_3 to the Type 1 binding site on the Mn complex. The direct binding of NH_3 to the Mn complex evidently exerts a stronger influence on the conformation of the Mn complex in the S_2 state than does the nature of the ligand in the Type 2 binding site. This observation is consistent with our previous proposal that a μ-imido bridge is formed upon coordination of NH_3 to the Mn complex in the S_2 state [16].

Discussion

The results presented in this paper support the previously introduced concept of an equilibrium between the two forms of the S_2 state that exhibit either the $g = 4.1$ or the multiline EPR signals [6, 23]. In this model, a single 3 Mn(III)–Mn(IV) or 3Mn(IV)–Mn(III) tetramer complex accounts for both types of S_2-state EPR signals [5, 24]. The two conformations of the Mn complex that exhibit either the $g = 4.1$ EPR signal or the multiline EPR signal in the S_2 state are linked by an equilibrium influenced by ligand substitution at the Type 2 binding site. As shown in Scheme 1, our results indicate that the binding of a ligand L to the Type 2 binding site displaces

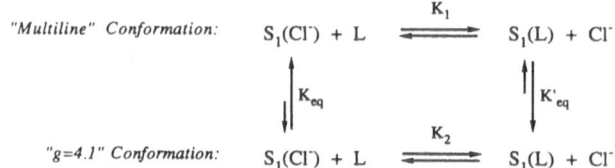

Scheme 1. Ligand-substitution reactions at the Type 2 binding site influence an equilibrium between the '$g = 4.1$' and 'multiline' conformations of the Mn complex.

Cl⁻. We propose that this ligand-substitution reaction shifts the position of the conformational equilibrium between two forms of the Mn complex, the 'multiline' and '$g = 4.1$' conformations, which exhibit the multiline and $g = 4.1$ EPR signals upon formation of the S_2 state, respectively. F⁻ and small primary amines shift this equilibrium upon binding to the Type 2 site so as to favor the '$g = 4.1$' conformation; Br⁻ and Cl⁻ favor the 'multiline' conformation upon binding to the Type 2 site. Depletion of Cl⁻ from the O₂-evolving center favors the '$g = 4.1$' conformation apparently because Cl⁻ is either lost from the Type 2 binding site or replaced by another ligand, such as OH⁻.

The size of the equilibrium constants K_{eq} and K'_{eq} in Scheme 1, is evidently also influenced by the cryoprotectant present in PS II membrane samples. In the presence of ethylene glycol, K_{eq} is apparently much larger than one since very little S_2-state $g = 4.1$ EPR signal is observed when Cl⁻ is not limiting; in contrast, the presence of sucrose causes K_{eq} to be approximately one since significant yields of both the $g = 4.1$ and multiline EPR signals are observed in the S_2 state, even when high concentrations of Cl⁻ are present. This latter phenomenon can also be explained if the presence of sucrose lowers the affinity of the O₂-evolving center for Cl⁻, effectively increasing the equilibrium constants K_1 and K_2 shown in Scheme 1.

In the light of the finding by Ono *et al.* [19] that depletion of Cl⁻ from the O₂-evolving center results in the production of a form of the Mn complex that is unable to undergo the $S_2 \to S_3$ transition, the results of this paper indicate that the '$g = 4.1$' conformation is catalytically inactive. As shown in Scheme 2, we propose that only the 'multiline' conformation of the Mn complex is able to be advanced to the S_3 state, subsequently to be stabilized by the coordination of two substrate H_2O molecules. The '$g = 4.1$' conformation, however, is not able to form the S_3 state, perhaps because the redox potential of the S_2 state has been lowered relative to that present in the 'multiline' conformation. This explanation is similar to that proposed by Ono *et al.* [19] who found that the modified form of the S_2 state observed under conditions of Cl⁻ depletion had a much longer $S_2 \to S_1$ state back-reaction time. A similar lengthening of the $S_2 \to S_1$ state back-reaction time is observed in the presence of inhibiting concentrations of NH_3 [6, 25, 26] or CH_3NH_2 [26]; our model easily accounts for these results since the '$g = 4.1$' conformation is favored in the presence of these inhibitors. Table I shows that, in general, inhibitors of photosynthetic O₂ evolution favor the '$g = 4.1$' conformation since they evidently displace Cl⁻ from the Type 2 binding site. Thus, our proposal that the '$g = 4.1$' conformation is catalytically

inactive provides a mechanism for the competitive inhibition of O₂ evolution by primary amines or F⁻, as observed by Sandusky and Yocum [13].

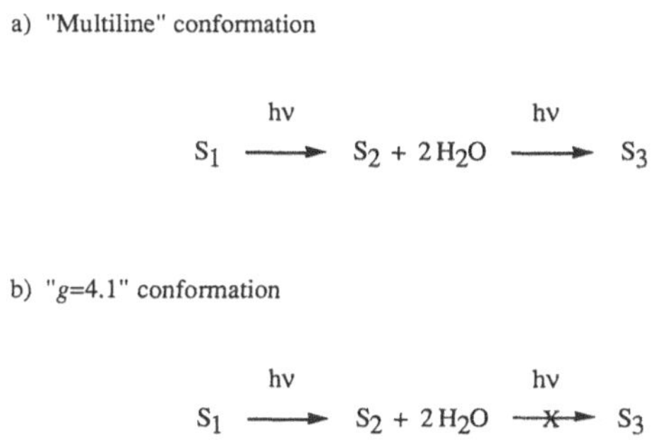

When sucrose is present as a cryoprotectant, however, Zimmermann and Rutherford [23] observed that both the S_2-state $g = 4.1$ and multiline EPR signals oscillate in yield with a period of four when produced by a series of saturating laser flashes. How can the S_2-state $g = 4.1$ EPR signal oscillate in yield when produced by a series of flashes if the '$g = 4.1$' conformation is catalytically inactive? We can account for the observations of Zimmermann and Rutherford [23] if the conformational equilibrium between the '$g = 4.1$' and 'multiline' conformations is established rapidly between flashes of light. Then, the $S_2 \to S_3$ transition would be blocked in only the fraction of the O₂-evolving centers in the S_2-state '$g = 4.1$' conformation, commensurate with the size of the equilibrium constant K_{eq} in Scheme 1. Those sites that productively form the S_3 state would go around the S-state cycle, and the S_2-state $g = 4.1$ EPR signal would be reformed in a fraction of the centers that reach the S_2 state again. This explanation accounts for the observation by Zimmermann and Rutherford [23] that the relative yields of the S_2-state $g = 4.1$ and multiline EPR signals after a given flash of light were in constant proportion.

Furthermore, the hypothesis that the '$g = 4.1$' conformation is catalytically inactive allows fitting the characteristic flash-oscillation pattern of the S_2-state EPR signals [23] without the use of an arbitrary miss coefficient; the damping of the flash-oscillation pattern with increasing flash number occurs because a fraction of the total S_2 state population present after a flash resides in the '$g = 4.1$' conformation, unable to be advanced to the S_3 state. This model affords an excellent fit to the flash oscillation data of Zimmermann and Rutherford [23] (data not shown). We note that Delrieu [26] has previously demonstrated that the presence of NH_3 increases the damping of the flash-induced O₂-yield pattern obtained in spinach thylakoid membranes; because the presence of NH_3 favors the '$g = 4.1$' conformation of the Mn complex, a larger fraction of O₂-evolving centers will not advance to the S_3 state from the S_2 state when exposed to a flash of light, increasing the effective damping of the flash-induced O₂-yield pattern. Delrieu [26] also found that an increase in the time between flashes enhanced the NH_3-induced damping of the O₂-yield oscillations. We may interpret this latter result as an indication of the kinetics involved in the establishment of the equilibrium between the '$g = 4.1$' and 'multiline' conformations in the S_2 state. Lastly, Delrieu's finding [26, 27] that better fits were obtained for the flash-induced O₂ yield pattern in spinach thylakoids when miss coefficients were postulated only for the $S_2 \to S_3$ transition is compatible with our model.

We conclude that the role of Cl⁻ in photosynthetic O₂

a) "Multiline" conformation

$$S_1 \xrightarrow{\text{hv}} S_2 + 2H_2O \xrightarrow{\text{hv}} S_3$$

b) "$g=4.1$" conformation

$$S_1 \xrightarrow{\text{hv}} S_2 + 2H_2O \xrightarrow{\text{hv}} \!\!\!\!\! \times \!\!\! S_3$$

Scheme 2. Regulation of O₂-evolution activity by ligand substitution at the Type 2 binding site occurs by inhibition of the $S_2 \to S_3$ transition.

Table I. *Effects of ligand substitution at the Type 2 binding site on the structure and function of the O₂-evolving center*

Ligand	Favored S_2 state conformation	O₂ evolution activity
Cl⁻	'Multiline'	Active
Br⁻	'Multiline'	Active
F⁻	'$g = 4.1$'	Inhibited
NH_3	'$g = 4.1$'	Inhibited
CH_3NH_2	'$g = 4.1$'	Inhibited

evolution involves regulation of the conformation of the Mn complex. It is not necessary, however, that Cl^- be directly bound to the Mn complex to accomplish a regulation of the equilibrium between the '$g = 4.1$' and 'multiline' conformations. Several lines of evidence, discussed by Yachandra *et al.* [28], argue against the presence of Cl^- in the first coordination sphere of the Mn complex. Further, the line shape, line width, and position of the $g = 4.1$ EPR signal do not depend on the ligand bound to the Type 2 binding site. If the Type 2 binding site were located on Mn, one might expect to observe a perturbation of the S_2-state $g = 4.1$ EPR signal upon substitution of ligands. The mode of action of Cl^- on the conformation of the Mn complex may be as simple as maintaining a network of hydrogen bonds between the Mn complex and the protein matrix that binds it. Such a network of hydrogen bonds has been suggested to be important in the oxido-reduction activity of Fe_4S_4 proteins such as ferredoxins [29].

References

1. For a recent review, see Babcock, G. T., in *New Comprehensive Biochemistry – Photosynthesis* (ed. J. Amesz), Elsevier, Amsterdam, chp. 6, pp. 125–158.
2. Dismukes, G. C. and Siderer, Y., *Proc. Natl. Acad. Sci. USA* **78**, 274–278 (1981).
3. Zimmermann, J.-L. and Rutherford, A. W. *Biochim. Biophys. Acta* **767**, 160–167 (1984).
4. Goodin, D. B., Yachandra, V. K., Britt, R. D., Sauer, K. and Klein, M. P., *Biochim. Biophys. Acta* **767**, 209–216 (1984).
5. de Paula, J. C., Beck, W. F. and Brudvig, G. W., *J. Am. Chem. Soc.* **108**, 4002–4009 (1986).
6. Beck, W. F. and Brudvig, G. W., *Biochemistry* **25**, 6479–6486 (1986).
7. Hansson, Ö, Aasa, R. and Vänngård, T., *Biophys. J.* **51**, 825–832 (1987).
8. Hansson, Ö, Andréasson, L.-E. and Vänngård, T., *FEBS Lett.* **195**, 151–154 (1986).
9. Brudvig, G. W. and Crabtree, R. H., *Proc. Natl. Acad. Sci. USA* **83**, 4586–4588 (1986).
10. Sheats, J. E., Czernuszewicz, R. S., Dismukes, G. C., Rheingold, A. L., Petrouleas, V., Stubbe, J., Armstrong, W. H., Beer, R. H. and Lippard, S. J., *J. Am. Chem. Soc.* **109**, 1435–1444 (1987).
11. Radmer, R. and Ollinger, O., *FEBS Lett.* **195**, 285–289 (1986).
12. Beck, W. F., de Paula, J. C. and Brudvig, G. W., *J. Am. Chem. Soc.* **108**, 4018–4022 (1986).
13. (a) Sandusky, P. O. and Yocum, C. F., *FEBS Lett.* **162**, 339–343 (1983); (b) Sandusky, P. O. and Yocum, C. F., *Biochim. Biophys. Acta* **766**, 603–611 (1984); (c) Sandusky, P. O. and Yocum, C. F., *Biochim. Biophys. Acta* **849**, 85–93 (1986).
14. (a) Beck, W. F. and Brudvig, G. W., *J. Am. Chem. Soc.* (in press); (b) Beck, W. F. and Brudvig, G. W., *Biochemistry* (in press).
15. Andréasson, L.-E. and Hansson, Ö, in *Progress in Photosynthesis Research* (ed. J. Biggins), vol. 1, pp. 503–510. Martinus Nijhoff, Dordrecht (1987).
16. Beck, W. F. and Brudvig, G. W., in *Progress in Photosynthesis Research* (ed. J. Biggins), vol. 1, pp. 499–502. Martinus Nijhoff, Dordrecht (1987).
17. Berthold, D. A., Babcock, G. T. and Yocum, C. F., *FEBS Lett.* **134**, 231–234 (1981).
18. Beck, W. F., de Paula, J. C. and Brudvig, G. W., *Biochemistry* **24**, 3035–3043 (1985).
19. Ono, T., Zimmermann, J.-L., Inoue, Y. and Rutherford, A. W., *Biochim. Biophys. Acta* **851**, 193–201 (1986).
20. Casey, J. L. and Sauer, K., *Biochim. Biophys. Acta* **767**, 21–28 (1984).
21. de Paula, J. C., Innes, J. B., and Brudvig, G. W., *Biochemistry* **24**, 8114–8120 (1985).
22. Rutherford, A. W. and Zimmermann, J.-L., *Biochim. Biophys. Acta* **767**, 168–175 (1984).
23. Zimmermann, J.-L. and Rutherford, A. W., *Biochemistry* **25**, 4609–4615 (1986).
24. de Paula, J. C., Beck, W. F., Miller, A.-F., Wilson, R. B. and Brudvig, G. W., *J. Chem. Soc. Faraday Trans.* (in press).
25. Velthuys, B. R., *Biochim. Biophys. Acta* **396**, 392–401 (1975).
26. Delrieu, M. J., *Biochim. Biophys. Acta* **440**, 176–188 (1976).
27. Delrieu, M. J., *Photochem. Photobiol.* **20**, 441–454 (1974).
28. Yachandra, V. K., Guiles, R. D., Sauer, K. and Klein, M. P., *Biochim. Biophys. Acta* **850**, 333–342 (1986).
29. Adman, E., Watenpaugh, K. D. and Jensen, L. H., *Proc. Natl. Acad. Sci. USA* **72**, 4854–4858 (1975).

Chemica Scripta 1988, **28A**, 99–104

The Spectroscopically Derived Structure of the Manganese Site for Photosynthetic Water Oxidization and a Proposal for the Protein-Binding Sites for Calcium and Manganese

G. Charles Dismukes

Princeton University, Department of Chemistry, Princeton, New Jersey 08544, USA

Paper presented at the Nobel Conference 'Biophysical Chemistry of Dioxygen Reactions in Respiration and Photosynthesis', Fiskebäckskil, Sweden, 1–4 July, 1987

Abstract

In this paper we summarize a variety of experiments from a number of laboratories which have provided evidence on how the manganese ions are organized in the catalytic site for photosynthetic water oxidation. The spectroscopic approach, in particular, has given a restricted picture of the ligand coordination environment around manganese and the distances between the four manganese ions. This has been directly aided by several new synthetic Mn complexes. A spectroscopically derived structure for the Mn cluster is proposed. The Mn core, Mn_4O_3Cl, could be comprised of a flattened right regular pyramid of four Mn ions that are capped on three faces by μ_3-oxo bridges and by a μ_3-Cl bridge opposite the apical Mn site. The second part of this paper presents a proposal for how this cluster could be coordinated to conserved carboxylate residues within the D1D2 reaction center protein, based upon the analysis of the gene sequence data from several species. A comparison of the protein sequence data for the D1 protein with the sequence data from 12 crystallographically determined calcium-binding sites from five different proteins has revealed the presence of a probable binding site for at least one Ca^{2+} in D1. This site is very close to the proposed manganese-binding site.

The proteins comprising the water-oxidizing complex of Photosystem II

Four manganese ions, all required for water oxidation, are located on one or more integral membrane proteins of the Photosystem II (PS II) complex and are exposed to the inner aqueous space. In addition, 1–3 calcium ions per PS II and an undetermined number of chloride ions are physiological co-factors. The smallest protein complex which binds manganese in a physiologically functional form is an O_2-evolving reaction center particle that contains 7 polypeptides and about 50 chlorophyll molecules, in the case of spinach (reviewed in ref. [1]). The functions of these subunits is a matter of current research. The emerging data support a picture in which two of the subunits, designated D1 and D2, form the reaction center core where the primary photochemistry and charge separation occurs [2]. These subunits exhibit about 25–30% homology, both between themselves and with the L and M subunits of the analogous bacterial reaction centers from several organisms [3]. This homology extends to the secondary structure level where both D1 and D2 have been proposed to contain five membrane-spanning

α-helical segments based upon hydropathy plots of their primary sequences [3,4]. These proposed helices coincide with the positions of the five membrane-spanning helices observed in the crystallographically determined structure of the bacterial reaction center proteins from *Rb. sphaeroides* and *Rb. virides* [5–7]. Trebst's model for the folding of the five helical domains in D1 and D2 from the higher plant sequence data is given in Fig. 1. This uses the numbering scheme given in refs [4] for the 353 amino acids found in both polypeptides. D1 is given on the left side and D2 on the right side of this figure. This model is based upon the DNA sequence data for six different species, including spinach [8,9], pea [10] and *Chlamydomonas reinhardtii* [11,12]. The general folding pattern through the membrane is supported by studies of the binding of antibodies raised against synthetic polypeptides corresponding to regions of D1 and D2 which were anticipated to be exposed on one or the other side of the membrane [13]. By removing three peripheral polypeptides of the water-oxidizing complex, which are bound to the reaction center proteins on the lumenal surface of the membrane, there is an enhancement by 2- to 3-fold in the binding of antibodies to the exposed segments of the D1 polypeptide which loop between helices A and B, C and D and between E and the carboxy terminus, but not between D and E, which is thus proposed to be exposed only to the matrix side.

Shown in Fig. 1 are three pairs of histidine residues from conserved, symmetrically related, positions on the D and E pairs of helices, which were proposed to bind the two magnesium atoms of the chlorophyll special pair and the non-heme ferrous ion on the basis of their conserved positions in various bacterial reaction center proteins [3,4]. The two accessory bacteriochlorophylls found in bacterial reaction centers are ligated to histidines L153 and M180 between the C and D helices in *Rb. virides*, while no corresponding pair was found in D1 and D2 [3]. The model given in Fig. 1 reveals that histidine residues 190D1 and 190D2 could be involved in the ligation of the two accessory chlorophylls, should these exist, since the histidines are symmetrically positioned between the C and D helices, if the folding model of Trebst is accepted [4]. In addition to conservation of their location, the downstream flanking residue in both cases is conserved. Alternatively, His118D1 and His118D2, which are located in

---

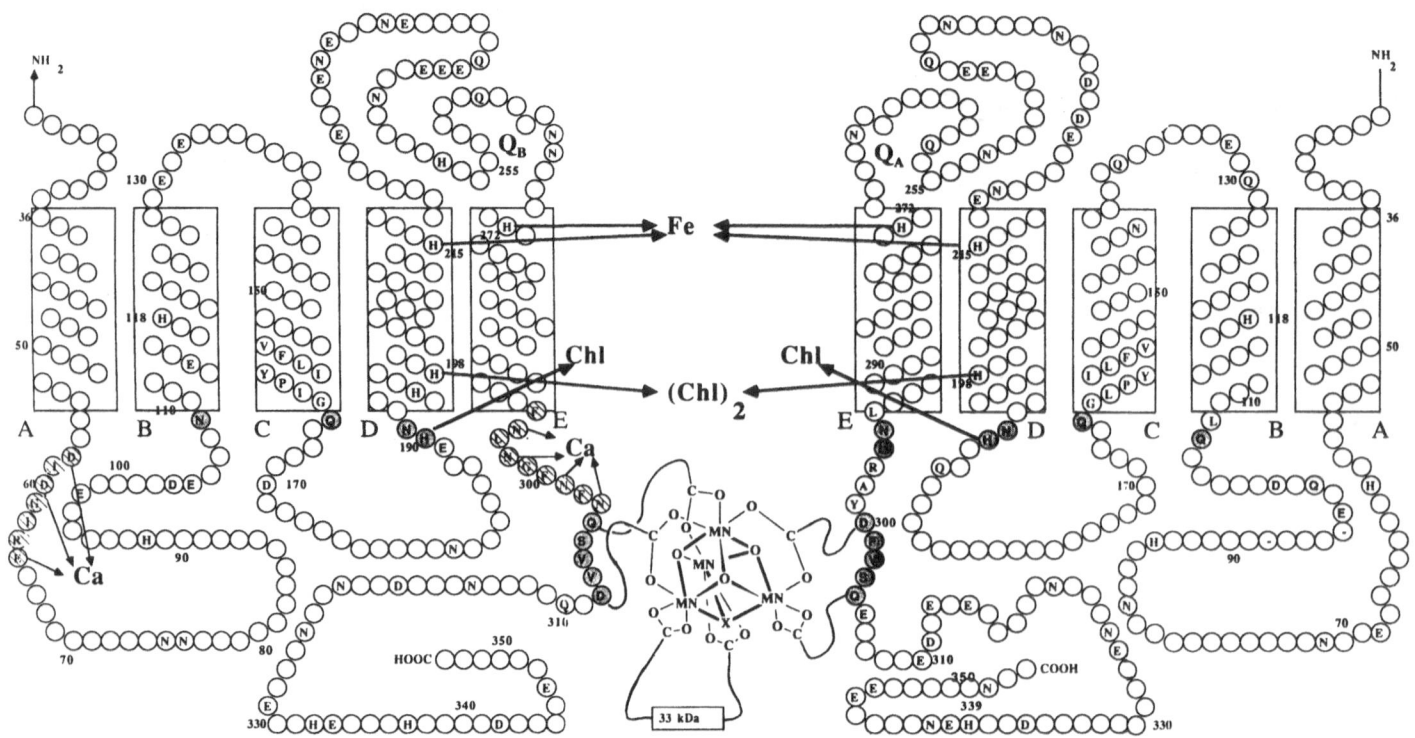

Fig. 1. The proposed structure of the Mn_4O_3X cluster comprising the catalytic site for water oxidation and its ligation to the D1D2 reaction center and the 33 kDa protein is given. $X = Cl$ for the native site and F for F-inhibited samples. The sequences for two proposed calcium sites in the D1 polypeptide are given in hatching. The folding pattern in the membrane of the D1D2 protein is taken from ref [4]. Sequences of conserved residues exposed to the lumen that are candidates for metal binding and which occur as symmetrical pairs between D1 and D2 are shown in gray. The locations of all of the histidine (H), aspartate (D), glutamate (E), asparagine (N) and glutamine (Q) residues are shown.

conserved positions [12] in the middle of the B helices, may be able to function as ligands to the Mg atoms of the accessory chlorophylls. These would appear to be too distant from the presumed location of the special pair chlorophylls between the D helices. However, the packing of helices is unknown and there is sufficient flexibility in the loops between them to allow permutation of the order of the helices, and thus the B and D helices could be adjacent.

The D1D2 core, although apparently forming the reaction center complex, is unable to oxidize water without other protein components. The other protein components found in O_2 evolving core particles include (reviewed in ref. [1]: two chlorophyll binding proteins of 47 and 43 kDa, believed to function as antenna complexes specific for PS II; a 20 kDa polypeptide of unknown function that appears to interact with the acceptor side, and so may be the counterpart of the H subunit seen in the bacterial reaction center; a pair of small subunits of 9 and 4.5 kDa that comprise cytochrome b_{559}, a heme protein of unknown function; and last, a water-soluble peripheral membrane protein of 33 kDa, first characterized by Kuwabara and Murata [14], that plays an essential role in stabilizing the functional binding of two of the four Mn ions to the core polypeptides [15–18].

Manganese and calcium coordination in Photosystem II

Only circumstantial evidence so far has pointed to the possible binding sites for manganese and calcium on the lumenal surface of the D1D2 reaction center, so the metal binding model given in Fig. 1 should be considered a proposal that can be tested and not a statement of observed

facts. It is not yet known, for example, if the 47 and 43 kDa chlorophyll binding polypeptides of the O_2-evolving reaction center complex are also needed for metal binding.

The rebinding of four Mn(II) ions to Mn-depleted (Tris/MgCl$_2$ washed) PS II membranes has been shown to occur in pairs [19]. These Mn(II) ions are coupled by an electron spin-exchange interaction, indicating the presence of bridging ligand atoms between the Mn atoms. This indicates that the Mn(II) ions are arranged either as two uncoupled pairs or as a single tetranuclear site. Mn reconstitution under similar conditions has recently been shown to produce recovery of O_2 evolution provided calcium and light are present [20, 21], suggesting a correlation with the functional Mn site for water oxidation. Photoactivation of O_2 evolution in leaves and isolated PS II membranes has been previously observed to require both Mn and Ca [22, 23]. This pairing of Mn ions in a common site can also be observed during the release of Mn upon increasing the pH during the washing of PS II membranes in 1 M-NaCl [24]. Alkaline pH releases all four Mn ions, with the first two or three being released cooperatively as a unit. These PS II membranes, so-called BBY type, are more complex than the core preparations described above, differing primarily by the presence of additional antenna subunits and contain about 200–250 chlorophyll molecules.

A high-affinity binding site has been found for Mn(II) on Tris-washed PS II membranes [25, 26]. This competes with artificial electron donors for light driven electron transport through PS II [25]. The affinity of this site for Mn(II) was shown to be enhanced by Cl$^-$, Br$^-$ and NO$_3^-$, cofactors which support O_2 evolution, relative to F$^-$ and CH$_3$COO$^-$, which inhibit O_2 evolution in the native membranes [25]. This was taken as evidence that Cl$^-$, the physiological anion required for water oxidation, is required for stabilizing the binding of Mn and that this is the native site for endogenous Mn functioning. Manganese EXAFS studies on PS II membranes have not shown significant changes upon halide exchange, which has been taken as evidence against chloride binding to Mn [27]. This view is not widely shared [28–30], based on a variety of studies summarized elsewhere [1, 30]. Especially compelling is the observation that the multiline EPR signal for the S_2 state is not formed upon addition of F^- [32] – though oxidation of Mn still occurs to form the S_2 state [33] – nor upon Cl^- extraction [34], but in its place an EPR signal at $g = 4.1$ appears. This signal can be interpreted in terms of the $S = \frac{3}{2}$ ground state produced by a modified spin coupling pattern imposed by F^- coordination to Mn within either a trinuclear or tetranuclear cluster [35, 36]. An alternative view is that this could be due to a monomeric Mn(IV) [37]. Possible functions for Cl^- have included charge neutralization of the S_2 and S_3 states [38], and, as a ligand to Mn, the co-factor which controls electron transfer [29] and both the Mn oxidation potential and the binding of substrate water to Mn [36].

Both the identity and the location of the protein amino acids responsible for ligation of the manganese and calcium ions are unknown. EPR linewidth studies on PS II membranes have suggested the possibility that ^{17}O–water exchanges with some labile oxygen ligands which could be μ-oxo ligands [39]. Electron spin-echo spectroscopy has shown that protons are coordinated to the Mn ligands, but no other magnetic nuclei are seen [40].

Manganese EXAFS data has given compelling evidence for three different Mn–X bond lengths, one at 1.8 Å attributed to μ_2-O, another at 2.0 Å attributed to O or N ligands, and a longer distance of 2.7–2.8 Å that coincides with a single transition metal atom, such as a second Mn atom [41]. This latter distance $(2.7 \pm 0.1$ Å$)$ is observed in the synthetic dimanganese complexes having either di-μ_2-oxo bridges [42] or di-μ_2-oxo-μ_2-carboxylato bridges [43], but not in the μ_2-oxo-di-μ_2-carboxylato bridges, which have considerably longer Mn–Mn separations of 3.1 Å [44, 45]. The model complexes thus indicate a di-μ_2-oxo or a di-μ_2-oxo-μ_2-carboxylato linkage between the closest Mn ions in the S_1 and S_2 states. A third shell of scattering atoms around the manganese ions at 3.3 Å has also been suggested on the basis of Mn EXAFS data [41]. While the data are noisy at this separation, it has been proposed that this could represent a longer Mn–Mn separation arising from a single Mn atom within an asymmetric cluster of more than two Mn atoms. Recently these results have been extended with data from oriented thylakoid membranes at 10- to 20-fold improvement in signal/noise [46]. The shell of strongly scattering atoms that appears at 3.3 Å is indicative of the presence of the mono-μ_2-oxo bridge between Mn neighbors. This distance is only slightly longer than the 3.1–3.2 Å separation observed in the synthetic μ_2-oxo-di-μ_2-carboxylato bridged dimanganese(III,III) and (III,IV) complexes [44, 45]. It is very close to the 3.3–3.4 Å separation found between Mn(III) ions in the μ_3-oxo-centered planar triangles, $[Mn_3O(O_2Cr)_6]^+$ [47], and in the analogous asymmetric Mn isosceles triangles within the tetrameric μ_3-oxo-bridged cation $[Mn_4O_2(O_2CR)_7]^+$ [48]. The former cation has in addition two μ_2-carboxylato bridges between all Mn pairs, while the latter cation has mono-μ_2-carboxylatos between the longest Mn pairs (3.4 Å)

and di-μ_2-carboxylatos between the closest Mn pairs (3.3 Å). This μ_3-oxo-centered M_3 structure is the fundamental building unit of the so-called basic carboxylates of the trivalent metals Cr, Mn and Fe, and represent the stable form of Mn(III) in mild acidic solutions of carboxylic acids [49]. Similar conditions prevail in the Mn-binding environment of PS II. Thus, there is both structural evidence and chemical precedent for the presence of either the mono-μ_2-oxo or mono-μ_3-oxo linkage between 2 or 3 Mn pairs in the water-oxidizing complex.

The first and still compelling evidence for the organization of the manganese into a cluster of 2–4 Mn ions is the multiline EPR signal [31, 50, 51]. Two spectral forms of this signal have been observed [36]. In samples that are partly denatured by freezing and thawing a '16 line' form is observed, while more lines (17–19) are found in native samples that exhibit the '19 line' form. The '16 line' form has spectral and thermal characteristics typical of a variety of dimanganese(III, IV) complexes, while the more complex '19 line' form has not yet been satisfactorily modelled chemically. The latter form has characteristics that have been predicted for trinuclear and tetranuclear manganese complexes [36]. The temperature dependence of this form has been found to be non-Curie in some cases and not in others, a feature which remains unexplained. Attempts to explain the temperature dependence have considered models with both ferromagnetic and antiferromagnetic coupling between the six possible pairs within a Mn tetramer [35].

In dimanganese(III, IV) complexes with di-μ_2-oxo bridges (2.7 Å), with or without a μ_2-carboxylato bridge, strong antiferromagnetic coupling occurs between the Mn ions [43, 45, 58]. This contrasts with dimanganese(III, III) complexes with mono-μ_2-oxo-di-μ_2-carboxylato bridges (3.3 Å) for which the coupling is weak or essentially zero [44, 45]. Combining three pairs of each type to form a Mn tetramer could produce a complex having magnetic properties similar to that observed for the photosynthetic Mn site.

In summary, the core of the tetrameric Mn site is most probably organized in one of two limiting structures – either as a pair of spin coupled dimers having di-μ_2-oxo bridges within each pair (2.7 Å) and single-μ_2-oxo ligands between pairs (3.3 Å); or as a trigonal pyramidal cluster in which the unique apical Mn is linked to the basal set of three Mn by di-μ_2-oxo bridges (2.7 Å), and within the basal set they are linked by either mono-μ_2-oxo or mono-μ_3-oxo bridges (3.3 Å). Coordination to protein carboxylates in each of these cases could be envisioned, since the structural constraints imposed by the Mn EXAFS allows the bridging of mono-μ_2-carboxylatos between all di-μ_2-oxo bridged Mn pairs (short) and either mono- or di-μ_2-carboxylatos between all mono-μ_2-oxo bridged pairs (long). Coordination sites for chloride binding to Mn are also believed to exist, but these must be few given the lack of a dominant feature in the Mn EXAFS [27]. A proposal for the structure of the Mn core consistent with the latter of these two limiting cases is given in Fig. 1. In its most symmetric form, the core Mn_4O_3Cl, could be comprised of a compressed right regular pyramid of 4 Mn atoms facially capped by three μ_3-oxo and a μ_3-chloro ligand opposite the apical Mn. This would be the presumed site for fluoride inhibition of S_2 [30]. It contains three μ_2-carboxylates, one each between the three apical-basal Mn pairs, as well as three terminally coordinated bidentate carboxylates to basal Mn sites. In the next section we present evidence for the existence of a small number of conserved proteins residues that are candidates for the carboxylate ligands which could bind this Mn core to the PS II complex.

Proposed manganese-binding sites on the D1D2 reaction center

Combining this information on the organization of manganese with the sequence data on the D1 and D2 polypeptides, we can suggest the location of some possible metal ligands from D1 and D2. Sulfur donor ligands to Mn, such as cystine and methionine, have been previously excluded [31]. Deprotonated alkoxyl type ligands, which model serine, threonine and tyrosine residues, coordinate to Mn(II), Mn(III) and Mn(IV) producing stable complexes. However, these are stable only under alkaline conditions, [52, 53] in contrast to the acidic conditions at the Mn site, and so these may be excluded. We are left with nitrogen donor ligands such as histidine and lysine, carboxylate donors such as aspartate and glutamate and primary amides such as asparagine and glutamine. The locations of these in the D1 and D2 sequences are given in Fig. 1. There is only a single lysine residue on the lumenal surface of the D2 polypeptide, hence we exclude it from consideration. In choosing among these we restrict consideration to those residues in D1 and D2 which are located in regions exposed to the lumenal side and which lie within conserved sequences. In the case of the D1 polypeptide, the primary sequence data is derived from the gene and so includes an additional 1.5 kDa fragment at the C terminus which is removed in the mature functional protein [54]. The precise location of the site of detachment is uncertain, but is thought to occur between 12 and 16 residues from the C terminus. This would eliminate residues above 338 in D1. For D2 the gene sequence homology between spinach and pea is 97%, with only 9 out of 353 residues different. The homology with the *C. reinhardtii* is 83% and this increases to 96% upon exclusion of 30 N-terminal and 28 C-terminal residues [8]. The last 28 C-terminal residues beginning at Glu315D2 are thus not conserved and so we exclude these from consideration as metal-binding sites.

Within the conserved regions we apply the initial restriction that amino acids in D1 which bind metals involved in water oxidation ought to be exposed to the lumen and have identical residues symmetrically located in D2. In the evolution of oxygenic organisms this criterion would enable the assembly of a tetranuclear Mn cluster from a simple genetic blueprint in which protein ligands for only two Mn ions needed to be encoded in the sequence. Gene duplication and subsequent divergence would then produce the evolutionary precursors of the D1 and D2 polypeptides. This view extends the 2-fold symmetry model proposed for the binding of chlorophyll and the non-heme iron to include the manganese ions. There are four sequence regions within each polypeptide, containing amino acids suitable for manganese binding, which strictly fulfill this symmetry criterion. These are shaded in gray in Fig. 1. His190 is part of a conserved pair, His190-Asn191, at the base of the D helices which was discussed above as a candidate for both chlorophyll and manganese ligation. The analogy to the bacterial reaction center suggests this pair is more likely involved in chlorophyll binding. The second region occurs as a pair, Asn297-Leu298,

at the base of the E helices. This could provide a pair of carboxylate ligands derived from the asparagine side-chain amides following hydrolysis to remove NH_3. In the case of D1, this pair is part of a larger sequence which we propose may be a calcium-binding site (next section), and so we suggest it is not directly involved in Mn binding. As a potentially significant general feature, four of the five helices, B to E in both D1 and D2, terminate with either Asn or Gln at the juncture with the lumen. This places Gln165 at the base of the C helices as a conserved pair of residues that could provide two carboxylate ligands to Mn.

It should be realized that all primary amides including asparagine and glutamine could be hydrolyzed to the carboxylate if coordinated to a strong Lewis acid metal, especially in the acidic environment such as the lumenal space of thylakoid membranes [55]. Carboxylates form complexes with Mn that are stable in high oxidation states in aqueous solutions, while amides do not, presumably due to hydrolysis. With this in mind, all Asn and Gln residues deduced from the gene sequence which are involved in metal binding on the lumenal surface may actually occur as the free carboxylate side chain in the functional protein. Such a mechanism could actually be essential for the insertion of a sequence that is abundant in charged residues through the thylakoid membrane – by initially disguising them as neutral residues. Subsequent hydrolysis of the amides could then activate these for metal binding, for signalling the completion of insertion of the newly synthesized protein through the membrane and in subsequent protein folding.

The fourth region of D1D2 symmetry occurs as an apparent inverted sequence of five residues D1: Asp308-Val-Val-Ser-Gln304 and D2: Gln304-Ser-Val-Phe-Asp300. The change of Val for Phe adjacent to the Asp site is a conservative one, representing a single base change in the codon for this position. This conserved sequence could provide two pairs of carboxylate ligands from Asp and Glu (hydrolyzed Gln) as suggested in Fig. 1. As shown, three of these are proposed to bind as μ_2-carboxylato bridges between the apical Mn and (three) basal Mn atoms, with the fourth acting as a terminal bidentate ligand. This region appears to be the strongest candidate for the site of binding of the putative Mn_4O_3Cl core to the D1D2 reaction center.

A fifth region which may qualify for metal binding, yet does not fully qualify as a symmetric D1D2 sequence, is D1: Glu98-Ala-Ala-Ser-Val-Asp103 and D2: Glu97-Ala-Gln-Gly-Asp101, both located at the base of the B helices. Only the pair of carboxylate donors are preserved in this sequence and could provide another pair of ligands for the manganese site, if the location of the B helices is close enough to the other ligands from the E helices. We do not favor this interpretation.

This leaves two of the four Mn ions coordinatively unsaturated. We propose that two terminal carboxylate ligands to Mn are provided by the extrinsic 33 kDa protein. This is then capable of completing the normal six-coordination environment for Mn, while additionally explaining the observation that 1–2 Mn atoms bind to the 33 kDa protein when extracted in the presence of chemical oxidants to suppress Mn reduction [17, 18, 26, 56]. In this model the 33 kDa protein provides only terminal ligands and thus could be extracted from the D1D2 core without extraction of Mn, as has been found using concentrated $CaCl_2$ as a selective denaturant [15]. Facile chemical exchange of oxo and

Table I. *Sequence alignment of Ca²⁺-binding sites in GBP and EF-hand loops compared to a proposed site in PS II D1*

The sequence of the proposed calcium-binding site in the D1 protein of PS II is compared to the crystallographically determined calcium sites in GBP (galactose-binding protein), the EF loops of (parvalbumin), troponin C, ICaBp (vitamin-D-dependent intestinal calcium-binding protein) and calmodulin. Data taken from ref. [61]. Calcium ligands, determined crystallographically or predicted on the basis of sequence homology, are marked by asterisks. The ligand at position 7 is provided by the peptide carbonyl oxygen while the others are from side-chain oxygens.

		1 *	2	3 *	4	—	5 *	6	7 *	
PS II D1	296	Asn	Leu	Asn	Gly	Phe	Asn	Phe	Asn	303
GBP										
Site	134	Asp	Leu	Asn	Lys		Asp	Gly	Gln	140
Parvalbumin										
Loop CD	51	Asp	Gln	Asp	Lys		Ser	Gly	Phe	57
Loop EF	90	Asp	Ser	Asp	Gly		Asp	Gly	Lys	96
Troponin C										
Loop I	30	Asp	Ala	Asp	Gly		Gly	Gly	Asp	36
Loop II	66	Asp	Glu	Asp	Gly		Ser	Gly	Thr	72
Loop III	106	Asp	Lys	Asn	Ala		Asp	Gly	Phe	112
Loop IV	142	Asp	Lys	Asn	Asn		Asp	Gly	Arg	148
ICaBP										
Loops III–IV	54	Asp	Lys	Asn	Gly		Asp	Gly	Glu	60
Calmodulin										
Loop I	20	Asp	Lys	Asp	Gly		Asn	Gly	Thr	26
Loop II	56	Asp	Ala	Asp	Gly		Asn	Gly	Thr	62
Loop III	93	Asp	Lys	Asp	Gly		Asn	Gly	Tyr	99
Loop IV	129	Asn	Ile	Asp	Gly		Asp	Gly	Glu	135

carboxylato ligands to Mn with free ligands in solution has been demonstrated with dimanganese (III, III) and (III, IV) complex [45, 57].

Proposed calcium-binding sites in Photosystem II

There appears to be one high-affinity and 1–2 lower-affinity specific Ca sites per PS II in wheat [58], spinach [59] and the cyanobacterium *Synechococcus* [60]. Although the precise location of these sites is not known, they reside on one or more of the six integral polypeptides of the O_2-evolving core preparations, and are exposed on the lumenal side upon removal of the peripheral 23 kDa protein of the water-oxidizing complex. There now exists several crystallographically determined sequences for calcium-binding sites involved in signal proteins and contractile proteins which enable predictions to be made for the calcium-binding sites in PS II [61]. In table I the sequence data for five calcium-binding proteins having a total of 12 calcium sites are compared to a proposed calcium site in the PS II D1 protein. An unusual sequence between Asn296 and Asn303 is comprised of eight residues which bear a striking similarity to part of the calcium-binding domains in all of these proteins. Four asparagines (aspartates if hydrolyzed) are located in three conserved positions (1, 3, 5) found to coordinate via the side-chain carboxylates. This leaves two coordination sites undetermined, since six ligands are invariably found. Two other oxygen-donor protein ligands make up the complete coordination environment for these calcium sites, with a Glu comprising one of these in all 12 sites contiguous with the first 7 residues in Table I. Because of this the location of these two remaining residues in D1 is not clear. Also, the possibility that water or chloride serve as ligands to calcium can not be excluded for the remaining ligands. The close proximity of the proposed calcium site at the base of the E helix to the proposed manganese-binding domain suggests a possible direct role for calcium in structural regulation of the manganese site. A second region within D1 which exhibits somewhat less similarity to known calcium sites is located between Asp58 and Glu64 at the base of the A helix. This contains three carboxylate side-chains in the proper locations for Ca binding, and so may qualify as a lower-affinity site. The proposed sequences of the calcium sites in D1 are identified in Fig. 1 with hatching of the residues.

Acknowledgements

Our work was supported by grants from the US Department of Energy Soleras program DE-FG02-84CH10199, the New Jersey Commission on Science and Technology and the Exxon Educational Foundation.

References

1. Babcock, G. T., in *New Comprehensive Biochemistry Photosynthesis* (ed. J. Amesz,). Elsevier, Amsterdam, pp. 125–158 (1987).
2. Satoh, K., *FEBS Lett* **204**, 357 (1986).
3. Michel, H. and Deisenhofer, J., *Progress in Photosynthesis Research* (ed. J. Biggins), vol. 1, p. 353. Martinus Nijhoff, Amsterdam (1986).
4. Trebst, A., *Naturforsch.* **41**c, 240 (1986).
5. Allen, J. P., Feher, G., Yeates, T. O. and Rees, D. C., *Progress in Photosynthesis Research.* vol. 1, p. 375, Martinus Nijhoff, Amsterdam (1986).
6. Chang, C.-H., Tiede, D., Tang, J., Norris, J. R. and Schiffer, J. *Progress in Photosynthesis Research*, vol. 1, p. 371, Martinus Nijhoff, Amsterdam (1986).
7. Deisenhofer, J., Epp, O., Miki, K., Huber, R. J. and Michel, H., *J. Mol. Biol.* **180**, 385 (1984).
8. Zurawski, G., Bohnert, J. J., Whitfeld, P. R. and Bottomley, W., *Proc. Nat. Acad. Sci. USA* **79**, 76 (1982).
9. Holschuh, K., Bottomley, W. and Whitfeld, P. R, *Nucleic Acid Res.* **12**, 8819 (1984).
10. Alt, J., Morris, J., Westhoff, P. and Herrmann, R. G., *Curr. Genet.* **8**, 597 (1984).
11. Rasmussen, O. F., Bookjans, G. Stumann, B. M. and Henningsen, K. W., *Plant Mol. Biol.* **3**, 371 (1984).
12. Rochaix, J.-D., Dron, M., Rahire, M. and Malnoe, P., *Plant Mol. Biol.* **3**, 363 (1984).
13. Sayre, R. T., Andersson, B. and Bogorad, L., *Cell* **47**, 601 (1986).
14. Kuwabara, T. and Murata, N., *Biochim. Biophys. Acta* **680**, 210 (1982).

15. Ono, T and Inoue, Y., *FEBS Lett.* **170**, 281 (1984).
16. Miyao, M. and Murata, N., *FEBS Lett* **170**, 350 (1984).
17. Abramowicz D. A. and Dismukes, G. C., *Biochim. Biophys. Acta* **765**, 318 (1984).
18. Yamamoto, Y., Shinkai, H., Isogai, K., Matsurra, K. and Nishimura, M., *FEBS Lett* **175**, 429 (1984).
19. Kulikov, A. C., Bogatyrenko, V. R., Likhtenstein, G. I., Allakhverdiev, S. I., Klimov, V. V., Shuvalov, V. A. and Krasnovsky, A. A., *Biofizica* **28**, 357 (1983).
20. Klimov, V. V., Ganago, I. B., Allakhverdiev, S. I., Shafiev, M. A. and Ananyev, G. M., *Progress in Photosynthesis Research*, vol. 1, p. 581, Martinus Nijhoff, Amsterdam (1986).
21. Tamura, N. and Cheniae, G., *Biochim. et Biophys. Acta* **890**, 179 (1987).
22. Tamura, N. and Cheniae, G., *Progress in Photosynthesis Research.*, Vol. 1, p. 621. Martinus Nijhoff, Amsterdam (1986).
23. Ono, T. and Inoue, Y., *Biochim. Biophys. Acta* **723**, 191 (1983).
24. Hunziker, D., Abramowicz, D., Damoder, R. and Dismukes, G. C., *Biochim. Biophys. Acta* **890**, 6 (1987).
25. Hsu, B.-D., Lee, J.-Y. and Pan, R.-L., *Biochim. Biophys. Acta* **890**, 89 (1987).
26. Hunziker, D., Abramowicz, D., Damoder, R. and Dismukes, G. C., *Progress in Photosynthesis Research*, vol. 1, p. 597, Martinus Nijhoff, Amsterdam (1986).
27. Yacchandra, V. K., Guiles, R. D., McDermott, A., Cole, J., Britt, R. D., Dexheimer, S. L., Sauer, K. and Klein, M. P., *Progress in Photosynthesis Research*, vol. 1, p. 557, Martinus Nijhoff, Amsterdam (1986).
28. Critchley, C., *Biochim. Biophys. Acta* **724**, 1, (1983).
29. Sandusky, P. O. and Yocum, C. F., *Biochim. Biophys. Acta* **766**, 603 (1984).
30. Damoder, R., Klimov, V. V. and Dismukes, G. C., *Biochim. Biophys. Acta* **848**, 378 (1986).
31. Dismukes, G. C., Ferris, K. and Watnick, P., *Photobiochem. Photobiophys* **3**, 243 (1982).
32. Casey, J. L. and Suer, K., *Biochim. Biophys. Acta* **767**, 21 (1984).
33. Ono, T., Zimmerman, J.-L., Inoue, Y. and Rutherford, A. W., *Biochim. Biophys. Acta* **851**, 193 (1986).
34. Dismukes, G. C. and Rutherford, A. W. (unpublished).
35. de Paula, J. C., Beck, W. F. and Brudvig, G. W., *J. Am. Chem. Soc.* **108**, 4002 (1986).
36. Dismukes, G. C., *Photochem. and Photobiol.* **43**, 99 (1986).
37. Hansson, O., Aasa, R. and Vanngard, T., *Biophysical J.* **51**, 825 (1987).
38. Preston, C. and Pace, R. J., *Biochim. Biophys. Acta* **810**, 388 (1985).
39. Hansson, O., Andreasson, L.-E., Vanngard, T., *FEBS Lett.* **195**, 151 (1986).
40. Britt, R. D., Sauer, K. and Klein, M. P., *Progress in Photosynthesis Research*, vol. 1, p. 573. Martinus Nijhoff, Amsterdam (1986).
41. Guiles, R. D., Yachandra, V. K., McDermott, A., Britt, R. D., Dexheimer, S. L., Sauer, K., and Klein, M. P., *Progress in Photosynthesis Research*, vol. 1, p. 561, Martinus Nijhoff, Amsterdam (1986).
42. Stebler, M., Ludi, A. and Burgi, H.-B., *Inorg. Chem.* **25**, 4743 (1986).
43. Wieghardt, K., Bossek, U., Zsolnai, L., Huttner, G., Blondin, G., Girerd, J.-J. and Babonneau, F., *J. Chem. Soc. Chem. Commun.* 651 (1987).
44. Wieghardt, K., Bossek, U. Bonvoisin, J., Beauvillain, P., Girerd, J.-J., Nuber, B., Weiss, J. and Heinze, J., *Angew Chem.* **98**, 1026 (1986).
45. Sheats, J. E., Czernuszewicz, R. S., Dismukes, G. C., Rheingold, A. L., Petrouleas, V., Stubbe, J. A., Armstrong, W. H. Beer, R. H. and Lippard, S. J., *J. Am. Chem. Soc.* **109**, 1435 (1987).
46. Prince, R. C., George, G. and Cramer, S. P., (unpublished).
47. Blake, A. B. and Fraser, L. R., *Dalton Trans.* 193 (1977).
48. Vincent, K. B., Christmas, C., Huffman, J. C., Christou, G., Chang, H.-R. and Hendrickson, D. N., *Chem. Commun.* **236** (1987).
49. Cotton, F. A. and Wilkinson, G., *Advanced Inorganic Chemistry*, p. 154, J. Wiley, New York (1980).
50. Dismukes, G. C. and Siderer, Y., *Proc. Natl Acad. Sci. USA* **78**, 274 (1981).
51. Hansson, O. and Andreasson, L.-E., *Biochim. Biophys. Acta* **679**, 261 (1982).
52. Pecoraro, V. L., Kessissoglou, D. P, Li, X. and Butler, W. M., *Progress in Photosynthesis Research*, vol. 1, p. 725, Martinus Nijhoff, Amsterdam (1986).
53. Mathur, P., Crowder, M. and Dismukes, G. C., *J. Am. Chem. Soc.* **109**, 5227 (1987).
54. Marder, J. B., Goloubinoff, P. and Edelman, M., *J. Biol. Chem.* **259**, 3900 (1984).
55. Noller, C. R., *The Chemistry of Organic Compounds*, p. 269. W. R. Saunders (1965).
56. Yamamoto, Y., *Progress in Photosynthesis Research*, vol. 1, p. 593, Martinus Nijhoff, Amsterdam (1986).
57. Sheats, J. E., Unni Nair, B. C., Petrouleas, V., Artandi, S., Czernuszewicz, R. S., and Dismukes, G. C., *Progress in Photosynthesis Research*, vol. 1, p. 721, Martinus Nijhoff, Amsterdam (1986).
58. Cammarata K. and Cheniae, G., *Progress in Photosynthesis Research*, vol. 1, p. 617, Martinus Nijhoff, Amsterdam (1986).
59. Ghanotakis, D. F., Babcock, G. T. and Yocum, C. F., *FEBS Lett* **167**, 127 (1984).
60. Katoh, S., Satoh, K., Ohno, T., Chen, J.-R., and Kasino. Y., *Progress in Photosynthesis Research*, vol. 1, p. 625, Martinus Nijhoff, Amsterdam (1986).
61. Vyas, N. K., Vyas, M. N. and Quiocho, F. A., *Nature* **327**, 635 (1987).

Chemica Scripta 1988, **28A**, 105–109

On the Mechanism of Photosynthetic Water Oxidation to Dioxygen

Gernot Renger

Max Volmer Institut für Biophysikalische und Physikalische Chemie der Technischen Universität, Straße des 17 Juni 135, D-1000 Berlin 12, Germany

Paper presented at the Nobel Conference 'Biophysical Chemistry of Dioxygen Reactions in Respiration and Photosynthesis', Fiskebäckskil, Sweden, 1–4 July, 1987

Abstract

Flash-induced absorption changes and the oxygen yield pattern were measured in dark-adapted samples. Information was obtained about (*a*) the activation parameters of the univalent oxidation of the catalytic site; (*b*) the proton release coupled with these reactions in $5.5 < \text{pH} < 8.9$; (*c*) the resting state of the catalytic site at alkaline pH and (*d*) the action of hydroxylamine. Based on these findings and data gathered from the literature the reaction coordinate and the molecular mechanism of photosynthetic water oxidation are analyzed. The structure of the catalytic site is discussed.

Introduction

Photosynthetic water cleavage into dioxygen and metabolically bound hydrogen is the essential step in order to permit a sufficiently efficient exploitation of solar radiation as the free energy source for the development and sustenance of all higher forms of life. The crucial reaction of the overall process is the oxidation of water to dioxygen. The 'invention' of a device to perform this reaction made the redox couple H_2O/O_2 'manageable' for biological organisms not only in the oxidative pathway of photosynthesis but with dioxygen as its 'waste' product it also opened the road to use the system in the oppositive direction of respiration [1]. A tremendous bioenergetic advantage was achieved with this 'discovery'. However, it had to be paid for by the 'toxicity' of dioxygen and its precursor states. Therefore proper defense mechanisms are required in order to survive under aerobic conditions. In respect to the variety of oxygen enzymology [2] a very interesting evolutionary aspect arises if one takes into account the widely accepted idea that the availability of dioxygen at the earth surface is a consequence of photosynthetic water oxidation [3]. The water-oxidizing enzyme system developed 2 to 3 billion years ago at the level of cyanobacteria could be considered as the phylogenetic ancestor of all biological systems dealing with the redox couple H_2O/O_2 and its intermediates. The implications of this possibility have to be analyzed in future work. In this brief communication the mechanistic aspects of photosynthetic water oxidation will be considered. Generally, two prerequisites have to be satisfied in order to permit the performance of this reaction: (*a*) the light-induced generation of redox equivalents of sufficient oxidizing power (holes); (*b*) the co-operative reaction of four holes with two water molecules leading to dioxygen and four protons. The former goal is achieved within the reaction center complex of system II by electron ejection from the excited singlet state of a special chlorophyll a molecule referred to as P680. The latter process takes place via a four-step univalent redox reaction

sequence at a manganese-containing catalytic site (for recent review see ref. [4]).

Each reaction center complex is functionally connected with only one catalytic site for water oxidation and the stepwise electron abstraction by $P680^+$ occurs via a further redox component Z. Despite of its characterization by the difference spectrum Z^{ox}/Z in the UV and of Z^{ox} by EPR the chemical identity of Z still remains to be clarified. It becomes oxidized by $P680^+$ in the ns time domain [5–7]. The reduction kinetics of Z^{ox} coincide with those of the oxidation of the catalytic site of water oxidation depending on its redox states S_i [8–10]. The half-life times of the univalent electron-transfer reactions $Z^{ox}S_i \rightarrow ZS_{i+1}$ were determined by measurements of time-resolved UV-absorption changes [9, 10]:

$$t_{\frac{1}{2}}(S_0 \rightarrow S_1) = 30\text{–}50 \ \mu s, \ t_{\frac{1}{2}}(S_1 \rightarrow S_2) = 100\text{–}150 \ \mu s$$
$$t_{\frac{1}{2}}(S_2 \rightarrow S_3) = 200\text{–}300 \ \mu s \text{ and } t_{\frac{1}{2}}(S_3 \rightarrow (S_4) \rightarrow S_0) = 1\text{–}1.2 \text{ ms}$$

The states S_i formally represent the redox states of photosynthetic water oxidation where S_0 corresponds with the redox level of water and S_4 with that of dioxygen. Accordingly, the central problem for understanding the mechanism of this process is the unravelling of the electronic structure and the nuclear geometry of the catalytic site in the redox states S_i. It is now well established that the catalytic site contains at least two manganeses that undergo valence changes during the redox cycle leading from water to dioxygen (for review see ref. [4]). Water as substrate is bound to the first coordination sphere [11]. The protein matrix also contributes to the ligation of manganese, a possible participation of further components remains to be clarified.

The existence of protonizable groups in the surrounding of the catalytic manganese raises the question of their involvement in proton release during each redox transition $S_i \rightarrow S_{i+1}$. In this respect it is mechanistically very important to distinguish the stoichiometry of deprotonation at the catalytic site itself, referred to as the *intrinsic* proton release pattern, from that of the whole enzyme complex which could be affected by Bohr type effects and is therefore designated as the extrinsic pattern [12].

Different spectroscopic methods (NMR, EPR, IR, UV–VIS, XAES, EXAFS) have been used in order to analyze the properties of the redox states S_i (for review see ref. [4]). Despite of many interesting results obtained a number of essential problems still remain to be clarified without ambiguity:

(1) What is the number of manganeses establishing the catalytic site?

(2) What is the valence of the manganeses in the different S_i-states?

(3) At which state S_i and at which redox level of the substrate is the O–O bond formed?

(4) Why is S_1 the resting state of the catalytic site in dark-adapted samples?

(5) Do there exist at the catalytic site other redox active components beyond manganese and the water substrate?

(6) What is the intrinsic proton release pattern?

(7) How does the protein matrix regulate the reaction coordinate of each individual redox step $S_i \rightarrow S_{i+1}$?

In this study new experimental data and theoretical considerations are presented that address the above mentioned questions.

Materials and Methods

The preparations of samples and the equipments for measuring flash-induced absorption changes and oxygen evolution have been described in previous reports [13–17].

Results

The general features of the reaction coordinates of the univalent redox transitions at the catalytic site of water oxidation were analyzed by measurements of the electron transport rates as a function of temperature for the reactions $P680^+Z \rightarrow P680\ Z^{ox}$ and $Z^{ox}S_i \rightarrow ZS_{i+1}$. In order to cope with a larger temperature range the experiments were performed with PS II fractions from the thermophilic cyanobacterium *Synechococcus vulcanus* Copeland. The experimental data obtained are depicted in Fig. 1. At the left side, flash-induced absorption changes obtained under repetitive excitation are presented. The signal measured a 33 °C exhibits almost the same relaxation kinetics as that observed at 0 °C (represented by the dashed curve). In the presence of 3 mM-NH_2OH the ns kinetics are completely transformed into μs kinetics (dotted curve). The relaxation of the 830 nm absorption changes reflect the $P680^+$ reduction by Z. Accordingly, the close similarity of the decay kinetics at 0 and 33 °C indicates an

almost temperature-independent electron transport from Z to $P680^+$.

A more complex pattern arises for the electron abstraction from the catalytic site of water oxidation by Z^{ox}. The half-life time of this reaction was determined by measurements of flash-induced absorption changes at 350 nm as described recently [10, 14]. In Fig. 1, right side, the reciprocal half times are presented as a function of the reciprocal temperature for the reactions $Z^{ox}S_i \rightarrow ZS_{i+1}$ with $i = 1, 2$ and 3 (the problems arising for the $S_0 \rightarrow S_1$-transition prevented unambiguous conclusions about the T-dependence of this reaction). The data reveal two interesting features. (*a*) The activation energy for the electron abstraction from S_2 (26.8 kJ/mol) derived from the Arrhenius plot is about three times larger than that for S_1 oxidation (9.6 kJ/mol), (*b*) The activation energy of S_3 oxidation exhibits a characteristic break point at about 16 °C, it is markedly smaller (15.5 kJ/mol) than for S_2 oxidation above this temperature but significantly larger (59.4 kJ/mol) below that point. As the oxidation of Z by $P680^+$ is only slightly dependent on temperature the above-mentioned data are indicative for structural changes at the catalytic site of water oxidation due to electron abstraction (see Discussion). This finding raised questions about a possible effect of pH upon the oxidation kinetics of the different S_i-states because regulatory subunits of the water-oxidizing enzyme system were inferred to undergo structural changes between pH = 6.0 and 7.5 [17, 18]. Experiments performed in the range of $5.5 < \text{pH} < 8.0$ did not reveal drastic effects on the oxidation kinetics. A suspension of the sample in D_2O caused detectable changes in the kinetics (data not shown) as will be outlined elsewhere.

In respect to the protein matrix another mechanistically important aspect arises for the interpretation of the extrinsic proton release pattern due to the possibility of pK-shifts of protonizable groups in the polypeptide(s) surrounding the catalytic site caused by the redox transitions $S_i \rightarrow S_{i+1}$. In order to analyze this point the protolytic reactions were measured in the range of $5.5 < \text{pH} < 8.9$ in dark adapted PS II membrane fragments excited with a train of saturating 10 μs flashes. In the range of $6.2 < \text{pH} < 7.7$ the data obtained (see Fig. 2) can be interpreted by a 1, 0, 1, 2 stoichiometry (the H^+-release of the 1st flash has been omitted because of its unknown origin, see ref. [15]). This result corresponds with previous findings in isolated thylakoids [19] and inside-out vesicles [20]. However, it has to be emphasized that considering enhanced double-hit probability in the 1st flash due to Q-400 oxidation (for details see ref. [15]) sometimes a better fit could be achieved within the framework of Kok's model [21] with stoichiometries of 0, 0, 1, 3 and/or 0, 0, 2, 2. This phenomenon suggests some variability of the stoichiometry that might be caused by pK-shifts in the protein matrix.

Significant changes of the oscillation patterns of proton release are observed at pH = 5.5 and at pH > 8.5. In the acidic range the oscillation tends to level off. A similar phenomenon arises in the alkaline region, except for the release due to the 2nd flash that exhibits about the double extent compared with the other flashes. The interpretation of the data obtained in the alkaline region is complicated because of the sharp decline of the oxygen-evolving capacity between pH = 8.0 and 9.0 [22, 23]. On the other hand at pH = 5.5 the oxygen-evolving capacity is hardly affected,

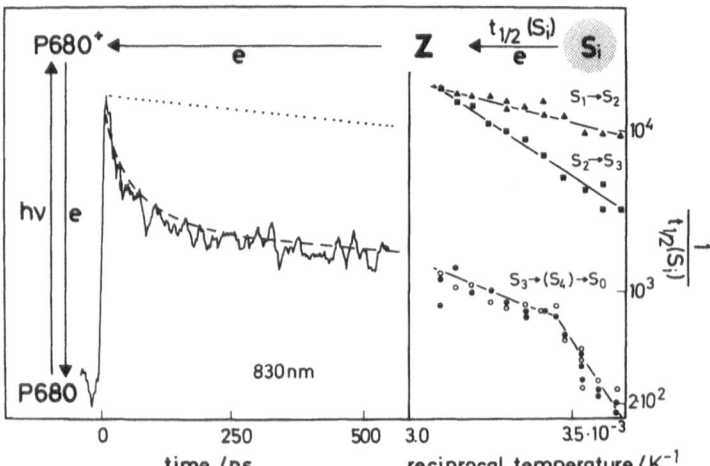

Fig. 1. Absorption changes at 830 nm (left side) as a function of time and reciprocal half-life times as a function of reciprocal temperature (right side) in PS II fragments from *Synechococcus vulcanus* Copeland. The signal on the left side represents a trace monitored at 33 °C, the dashed curve symbolizes the data at 0 °C. The dotted curve was obtained in the presence of 3 mM-NH_2OH. Repetitive excitation (1 Hz) with laser flashes. For further details see refs. [14] and [17].

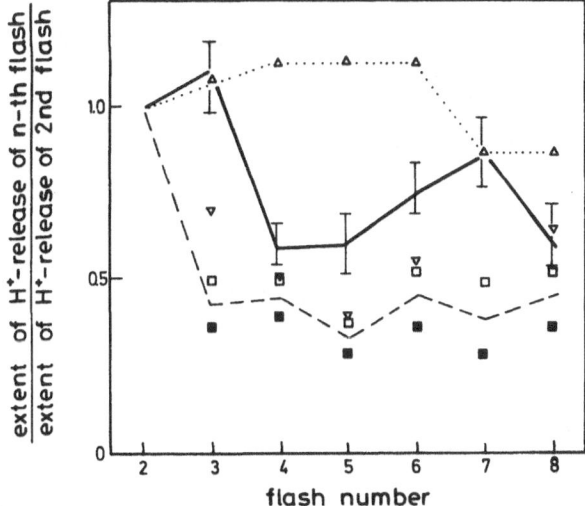

Fig. 2. Extent of H^+ release due to the nth flash normalized to the extent due to the 2nd flash as a function of flash no. (n) in dark adapted PS II membrane fragments from spinach at different pH. The data at pH < 7.0 were obtained with bromocresol purple, at pH > 7.0 with phenol red as indicator; signals obtained in the presence of buffer were subtracted. For further details see ref. [15]. The following symbols are used (pH): △, 5.5; ▽, 8.1; □, 8.5 and ■, 8.9, the full curve represents data obtained in the range of 6.0 < pH < 7.7. The bars do not represent error bars but reflect the variation of the data in the range 6.2 < pH < 7.7.

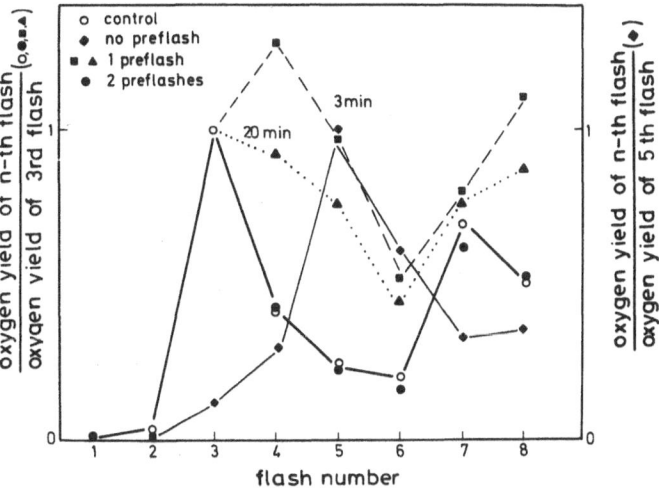

Fig. 3. Oxygen yield of nth flash normalized to that of the 3rd flash as a function of flash no. (n) in dark-adapted spinach thylakoids in the absence (○) and presence (■, ▲, ●, ◆) of 50 μM-NH_2OH. The measurements were started after 3 or 20 min dark incubation, preflashes as indicated in the figure. For further details see ref. [32].

whereas the stoichiometry of proton release markedly differs from that in the range 6.2 < pH < 7.7. These findings support the idea that protonizable groups in the protein matrix probably contribute to the extrinsic proton release pattern. In this respect it has to be emphasized that so far no method is available to detect the intrinsic proton release pattern (for discussion see ref. [24]). The protein matrix cannot only affect the protolytic reactions but it also determines the redox properties of the functional manganeses in the catalytic site. In this respect it appears worthwhile to analyze possible structural effects of polypeptide(s) on the resting state of the water-oxidizing enzyme system. It was recently shown that the Triton X-100 solubilization step used in the isolation procedure of PS II membrane fragments modified the properties of the PS II acceptor side [25]. Therefore slight structural changes of functional relevance could also arise at the water oxidizing enzyme system. This possibility was analyzed by determining the oxygen yield oscillation pattern in dark-adapted thylakoids and PS II membrane fragments at alkaline pH. A marked change (especially of the oxygen yield ratio of the 3rd and 4th flash) was observed in PS II membrane fragments within the range of 7.6 < pK < 8.6, whereas the pattern in thylakoids remained almost invariant up to pH = 9.0 (data not shown, see ref. [26]). These results can be consistently explained by an increased probability of misses rather than by a thermodynamic stabilization of S_0 at higher pH in PS II-membrane fragments. The assignment of the elevated misses to donor and/or acceptor side components remains to be clarified (for details see ref. [26]). A suitable approach for analyzing the properties of the S_i-states is the use of specifically interacting substances. The most powerful species are certain ADRY-agents [27] which destabilize at substoichiometric concentrations the redox states S_2 and S_3 but do not affect S_0 and S_1 [28]. Another modification of the reaction pattern of the water oxidizing enzyme system can be achieved by low concentrations of NH_2OH [29]. In this case formally a redox state 'S_{-1}' is formed. It was assumed that

NH_2OH directly interacts as a substrate analogue with the manganese at the catalytic site but latest data question this idea [30].

In order to analyze this problem experiments were performed in dark-adapted thylakoids incubated with 50 μM-NH_2OH corresponding to a stoichiometry of > 10 NH_2OH molecules per PS II reaction center. The oscillation pattern of the oxygen yield is shifted by two flashes and remains the same after 3 or 20 min dark incubation. If however, the samples are illuminated by two flashes after 3 min dark incubation the oxygen yield pattern measured 17 min after the preflashes exhibit almost the same feature as the control without NH_2OH (see Fig. 3). After one preflash a pattern for S_0 accumulation is observed that slowly shifts towards the control due to S_1 formation in the dark.

Furthermore, excess amounts of NH_2OH were found to become degraded during dark incubation of thylakoids (Hanssum and Renger, unpublished results). Therefore, a stoichiometric and exclusive reaction with the manganese of the catalytic site can be excluded. However, it is clear that the overall reaction supplies two reducing redox equivalents at the catalytic site. Taking into account previous findings about the co-operativity [31, 32] of NH_2OH and NH_2NH_2 at least two molecules are assumed to bind in the neighborhood of the catalytic site. This state formally represents 'S_{-1}'. The electronic structure of 'S_{-1}' remains to be clarified.

Discussion

The present study provided further pieces to the huge puzzle which resolution will eventually lead to a detailed picture of the mechanism of photosynthesic water oxidation. Despite lack of essential information (e.g. about the ligands of the 1st coordination sphere, the existence and possible role of hydrogen bridges, the protein dynamics of the apoenzyme and its functional role) an attempt will be made to extend our previous model [12, 33] on the basis of recent data from the literature and results presented here. Two essential assumptions were taken from the original model:

(a) The O–O bond formation occurs at a binuclear manganese center at the redox level of a peroxide.

(b) There exists another redox center M that participates in the overall reaction at the level of $S_0 \to S_1$ and $S_3 \to (S_4) \to S_0$.

Component M is still not yet substantiated (for detailed discussion see ref. [34]). Latest data favour the idea that M is a special mononuclear manganese center [35] which undergoes a $Mn(II) \to Mn(III)$ oxidation during the $S_0 \to S_1$ transition. Based on EPR measurements, component M is assumed to act as a two-electron redox species becoming oxidized up to $Mn(IV)$ [35, 36]. Beyond these redox groups of the catalytic site there exists another component D that interacts with the S_i-states. The reduced form, D, causes the rapid decay of S_2 and S_3 in a fraction of PS II whereas D^{ox} (EPR-signal II slow) slowly oxidizes S_0 to S_1 [37, 38].

Taking into account the existence of μ-oxo-bridge(s) between the two manganeses of the catalytic binuclear cluster [39] a preliminary cartoon can be drawn about the structure of the catalytic site that is presented in the lower insert of Fig. 4. The other two manganeses located close to the catalytic site are not explicitly shown because the nature of the component M which could be a manganese [35] is not yet unequivocally clarified.

For the discussion of the reaction mechanism two redox equilibria depicted in Fig. 4 (top, left) have to be taken into consideration. It is assumed that S_3 satisfies the structural and energetic requirements for the formation of the crucial O–O bond. Whether this condition is already achieved at S_2 remains to be clarified. Based on the results of this study (Fig. 1) a reaction coordinate is presented for photosynthetic water oxidation. A very interesting aspect is the finding that above a critical temperature, T_c, the activation energy for the rate-limiting step of the complex reaction $S_3 \to (S_4) \to S_0$ is comparatively small whereas below T_c this quantity is rather large. This indicates gross structural changes of state S_3 which might be indicative for a peroxide-like configuration

above T_c. Taking into account the properties of thermophilic cyanobacteria, T_c could be lower in nonthermophilic organisms [14]. If one assumes that the transmission coefficients of the redox transitions $S_i \to S_{i+1}$ are close to one the following activation entropies can be determined: $\Delta S^*(S_1 \to S_2) = -140$ J/mol K; $\Delta S^*(S_2 \to S_3) = -85$ J/mol K and $\Delta S^*(S_3 \to (S_4) \to S_0) = -140$ J/mol K $(T > T_c)$ and -10 J/mol K $(T < T_c)$.

The implications of these considerations will be outlined elsewhere. A molecular mechanism of photosynthetic water oxidation was presented recently [4, 12].

Acknowledgements

The author would like to thank H. J. Eckert, B. Hanssum, Y. Inoue, H. Koike and U. Wacker for their contributions and A. Bowe-Gräber for drawing the figures. The financial support by Deutsche Forschungsgemeinschaft (Re 354/8-1 and Sfb 312) is gratefully acknowledged.

References

1. Renger, G., in *Biophysics* (ed. W. Hoppe, H. Lohmann, H. Markl and H. Ziegler). Springer, Berlin (1983).
2. Malmström, B. G., *Annu. Rev. Biochem.* **51**, 21 (1982).
3. Berghorn, E. A., *Sci. Am.* **224**, 30 (1971).
4. Renger, G., *Angew. Chem., Int. Ed.* **26**, 643 (1987).
5. Renger, G., Eckert, H. J. and Weiss, W., in *The Oxygen Evolving System in Photosynthesis* (ed. Y. Inoue, A. R. Crofts, Murata N. Govindjee, G. Renger and K. Satoh). p. 73, Academic Press, Japan (1983).
6. Brettel, K. and Witt, H. T., *Photobiochem. Photobiophys.* **6**, 253 (1983).
7. Eckert, H. J., Renger, G. and Witt, H. T., *FEBS Lett.* **167**, 316 (1984).
8. Babcock, G. T., Blankenship, R. E. and Sauer, K., *FEBS Lett.* **61**, 286 (1976).

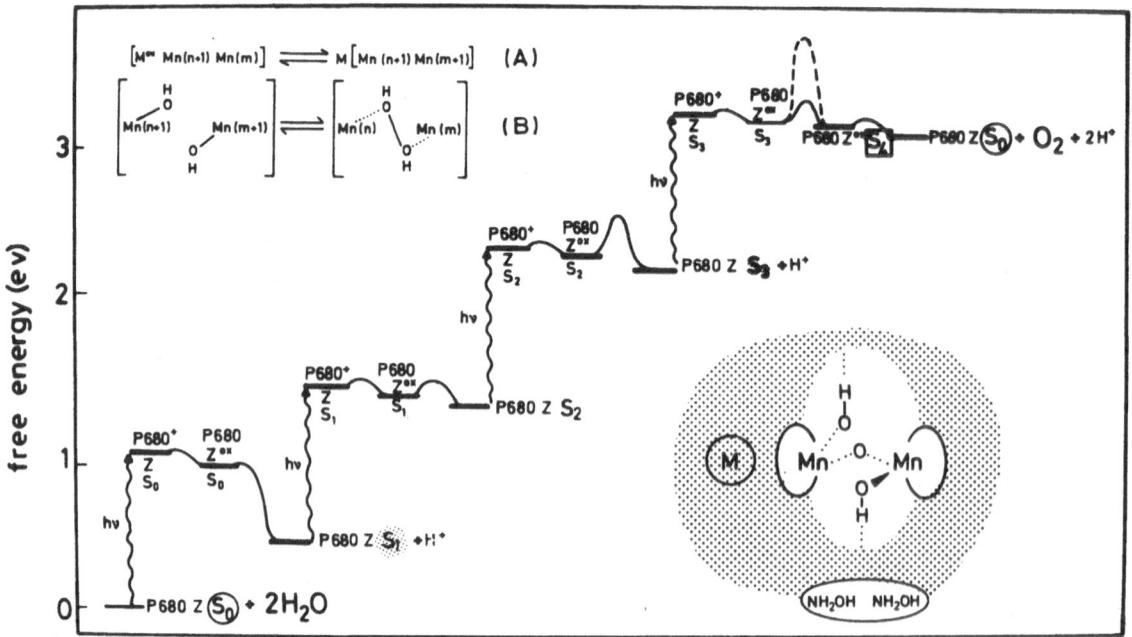

Fig. 4. Overall reaction coordinate of photosynthetic water oxidation and structural scheme of the catalytic site (lower insert, right side). The reaction coordinate is based on the extrinsic proton release pattern, the activation energies determined in this study and data taken from the literature (for details see refs. [4] and [12]). In the structural scheme the unknown coordination spheres of the binuclear manganese cluster and of component M are symbolized by half or full circles, respectively. Component D is not shown. The binding site of NH$_2$OH close to the catalytic site is represented by an oval in the protein matrix. Further details are outlined in the text. In the upper left, equations for two essential redox equilibria are depicted: (A) describes the possibility of electron transfer between M and the binuclear manganese; (B) describes the redox equilibrium between the ligated substrate and the binuclear manganese center giving rise to peroxide-like configuration.

9. Dekker, J. P., van Gorkom, H. J., Wessink, J. and Ouwehand, L., *Biochim. Biophys. Acta* **77**, 176 (1984).
10. Renger, G. and Weiss, W., *Biochem. Soc. Trans.* **14**, 17 (1986).
11. Hansson, Ö., Andreasson, L. E. and Vänngard, T., *FEBS Lett.* **195**, 151 (1986).
12. Renger, G., in *Photosynthetic Oxygen Evolution* (ed. H. Metzner), p. 229, Academic Press, London (1978).
13. Völker, M., Ono, T., Inoue, Y. and Renger, G., *Biochim. Biophys. Acta* **806**, 25 (1985).
14. Koike, H., Hanssum, B., Inoue, Y. and Renger, G., *Biochim. Biophys. Acta* **893**, 524 (1987).
15. Renger, G., Wacker, U., Völker, M., *Photosynth. Res.* **13**, 167 (1987).
16. Weiss, W. and Renger, G., *Biochim. Biophys. Acta* **850**, 173 (1986).
17. Völker, M., Eckert, H. J. and Renger, G., *Biochim. Biophys. Acta* **890**, 66 (1987).
18. Völker, M. and Renger, G., in *Advances in Photosynthesis Research* (ed. C. Sybesma), vol. 1, p. 605, Martinus Nijhoff/Dr W. Junk, The Hague (1984).
19. Förster, V. and Junge, W., *Photochem. Photobiol.* **41**, 183 (1984).
20. Glaubitz-Dietrich, R., Völker, M., Renger, G. and Gräber, P., in *Progress in Photosynthesis Research* (ed. J. Biggins), vol. 1, p. 519, Martinus Nijhoff, Dordrecht (1987).
21. Kok, B., Forbush, B. and McGloin, M. P., *Photochem. Photobiol.* **11**, 457, (1970).
22. Renger, G., Thesis, Technical University, Berlin (1969).
23. Renger, G., Gläser, M. and Buchwald, H. E., *Biochim. Biophys. Acta* **461**, 392 (1977).
24. Renger, G., *Photosynthetica* **21**, 203 (1987).
25. Renger, G., Hagemann, R. and Fromme, R., *FEBS Lett.* **203**, 210 (1986).
26. Renger, G., Hanssum, B. *Photosynth Res.* (in press).
27. Renger, G., *Biochim. Biophys. Acta* **256**, 428 (1972).
28. Hanssum, B., Dohnt, G. and Renger, G., *Biochim. Biophys. Acta* **806**, 210 (1985).
29. Bouges, B., *Biochim. Biophys. Acta* **234**, 103 (1971).
31. Brudvig, G. W., Beck, W. F., de Paula, J. C. and Zewert, T., This volume.
31. Förster, V. and Junge, W., *Photosynth, Res.* **9**, 197 (1986).
32. Hanssum, B. and Renger, G., *Biochim. Biophys. Acta* **810**, 225 (1985).
33. Renger, G., *FEBS Lett.* **81**, 223 (1977).
34. Renger, G. and Govindjee, *Photosynth Res.* **6**, 33 (1985).
35. Hansson, Ö., Thesis, Chalmers University Göteborg, (1986).
36. Hansson, Ö., Aasa, R. and Vänngård, T., *Biophys. J.* **51**, 825 (1987).
37. Vermaas, W. J. F., Renger, G. and Dohnt, G., *Biochim. Biophys. Acta* **767**, 194 (1984).
38. Styring, S. and Rutherford, A. W., *Biochemistry* **26**, 2401 (1987).
39. Sauer, K., Guiles, R. D., McDermott, A. E., Cole, Y. L., Yachancha, V. K., Zimmermann, J.-L., Klein, M. P., Docheimer, S. L. and Britt, R. D., *Chemica Scripta* **28**A, 87 (1988).

Chemica Scripta 1988, 28A, 111–116

Protolytic Reactions of the Photosynthetic Water Oxidase in the Absence and in the Presence of Added Ligands

Verena Förster and Wolfgang Junge

Biophysik, Universität Osnabrück, PF 4469, D-4500 Osnabrück, Germany (FRG)

Paper presented by Wolfgang Junge at the Nobel Conference 'Biophysical Chemistry of Dioxygen Reactions in Respiration and Photosynthesis', Fiskebäckskil, Sweden, 1–4 July, 1987

Abstract

Photosynthetic water oxidation proceeds by a sequence of four one-electron abstractions by photosystem II (PS II) from the catalytic manganese centre, which accumulates four oxidizing equivalents. This is formally described by transitions $S_0 \rightarrow S_1$, $S_1 \rightarrow S_2$, $S_2 \rightarrow S_3$, and $S_3 \rightarrow S_4 \rightarrow S_0$, with state S_1 most stable in the dark. Dioxygen is produced and protons are liberated into the thylakoid lumen. Flash-spectrophotometric techniques have previously revealed the stoichiometry and kinetics of proton release per univalent transition.

Via the pattern of proton release as function of flash number we have studied the reactions of NH_2OH and of NH_2NH_2 as putative ligands and electron donors to the Mn centre. Our results suggest reversible binding of these agents to the centre in state S_1, completed after 5 min incubation. After excitation of PS II by a first flash of light a two-step reaction is initiated. The first step is apparent via the release of two protons into the lumen with approximately 3 ms half-rise-time. The second, with 30 ms, is apparent by relaxation of the centre into state S_0. The effect of NH_2OH is not influenced by chloride in the concentration range between 100 μM and 30 mM. This excludes competitive binding of hydroxylamine to chloride-binding sites. The concentration dependence is very steep with Hill coefficients 2.4 (NH_2OH) and 1.5 (NH_2NH_2). This can be understood in terms of co-operative binding to at least 3, if not 4, binding sites for hydroxylamine. If a binuclear manganese centre accommodated four molecules of hydroxylamine (bidentate complexes) or two molecules of hydrazine as bridging ligands, relatively large Mn-Mn distances were expected (up to 5.4 Å). These have not been observed in EXAFS. We show that the compared with optical studies, 100-fold more concentrated samples used in EXAFS are heterogeneous with regard to the occupation of centres with 0, 1, 2, 3, 4 molecules of hydroxylamine. They have therefore to be reinterpreted.

Introduction

The photosynthetic water oxidase is a complex enzyme which integrates a photochemical reaction centre, photosystem II (PS II), and a manganese cluster as catalytic centre (reviewed in refs. [1, 2]). One quantum of light drives a very rapid charge separation, half rise time about 400 ps, between a donor chlorophyll a, P680, and a bound plastoquinone, Q_a. This reaction is directed from the lumenal side of the thylakoid membrane to the stroma side. It fully spans the membrane dielectric [3]. The manganese cluster acts as an accumulator of oxidizing equivalents which are produced one by one in PS II. Once four oxidizing equivalents are accumulated two molecules of water are oxidized to one molecule of dioxygen [4, 5]. The step-by-step accumulation of oxidizing equivalents is formally described by transitions between five states, S_0 to S_4, where S_1 is most stable in the dark, and S_4 spontaneously decays to S_0 under liberation of dioxygen [6].

Because the catalytic manganese centre and the bound quinones are placed at different sides of the membrane the water oxidase acts as an electrogenic proton pump. The photochemical electron transfer charges the lumenal side positively. Water oxidation releases protons into the lumen, and quinone reduction is followed by proton uptake from the stroma. This generates a pH-difference [3]. It is unique to photosynthetic membranes that the elementary electrochemical reactions are measurable by flash spectrophotometric techniques at high time resolution [3]. Proton release into the thylakoid lumen, in particular, is a sensitive indicator of the sequential reactions of the water oxidase.

Water is substrate and solvent of the water oxidase. That its concentration cannot be varied hampers studies on its interaction with the catalytic site. Hydroxylamine and hydrazine interact with the water oxidase. They are of great interest under the aspect that they may bind to and react with the catalytic manganese centre proper. We investigated their highly co-operative binding to the oxidase.

Experimental

Thylakoids were prepared from pea seedlings as described in [14]. Concentrated stock suspensions were stored on ice until dilution of aliquots in a medium containing 25 mM-KCl, 3 mM-$MgCl_2$, 2.6 g/l bovine serum albumin, 4 μM-DNP-INT, 2 mM potassium hexacyanoferrate(III). Spectrophotometric experiments were carried out with the dilute suspension at 20 μM chlorophyll in a cuvette of 1 cm optical pathlength and at room temperature. PS II was excited with single turnover flashes of red light (Xenon, 1 μs or Ruby laser, 40 ns FWHM). pH transients in the lumen were detected photometrically by way of transient absorption changes of neutral red (13 μM), bovine serum albumin acted as selective buffer for the suspending medium [8]. The response of neutral red to pH transients resulted from a subtraction of a transient absorption change at a wavelength of 548 nm obtained in the absence of neutral red from another one obtained in its presence [8].

The submicroscopic dimensions of the lumen, radius about 300 nm and width about 5 nm, imply that there is only 0.1 free proton around at pH 7. Nevertheless is the pH in each thylakoid well defined as a time average with a standard deviation depending on the number of buffering groups inside this volume, their protonation/deprotonation time and the time interval under inspection. To give an example: at pH 7 there are about 2×10^4 effective buffering groups in a thylakoid of the above dimensions (see ref. [8]) which according to Gutman's work on the kinetics of acid/base reactions at protein surfaces [9] undergo diffusion-controlled protonation. This implies a protonic on/off relaxation time in the order of $200 \, \mu s$ for each of these groups. Taken together they provide a protonic relaxation time of some 10 ns in the lumen. In other words the concentration of free protons when averaged over a time interval of $100 \, \mu s$ is precise within less than 3% error and the pH is defined within less than 0.02 units around pH 7.

Neutral red is sensitive for pH transients in the narrow lumen phase because of its membrane solubility. The free base has a distribution coefficient between membrane and aqueous bulk phase of 880 in favour of the membrane (ref. [10] in correction to ref. [11]). Neutral red, adsorbed to the membrane surface, is a true [8] indicator of the surface pH [11].

In response to the flash-induced pH difference, protonation of membrane-bound neutral red is very fast, for example $100 \, \mu s$ as found for proton release in Tris-treated thylakoids [7]. Redistribution of neutral red across the membrane occurs in the second time range and thus it only interferes with slower pH transients [10].

Results and discussion

Stoichiometry and kinetics of proton release during the sequential reactions of the water oxidase

The kinetic parameters of electron transport from the site of catalytic water decomposition via the intermediate carrier, Z, into P680 and the kinetics of concomitant proton release are summarized in Fig. 1. In dark-adapted pea thylakoids, the catalytic centre is predominantly in state S_1. Under excitation with a series of flashes the yield of rapid (< 3 ms) proton release proceeds as $0:1:2:1$ per flash [12–14]. (The extent of rapid proton release after damping out of the period-of-four oscillations after many flashes is taken as standard, which, according to several lines of evidence, represents one proton per flash and PS II (α-centre) (reviewed in ref. [3]).) The lack of proton extrusion during transition $S_1 \rightarrow S_2$ leaves a positive net charge in the catalytic site which is not compensated until two protons are liberated during the transition $S_3 \rightarrow S_4$. The

transients of the net charge, detected by Saygin and Witt [15] via electrochromic absorption changes, are consistent with this notion. The rate of electron transfer between Z and $P680^+$ is slowed down under conditions where the centre carries a positive net charge. This is attributed to coulombic attraction of the electron by the net charge in the centre [16].

In brief, the catalytic centre of the water oxidase acts as an accumulator of four oxidizing equivalents. Its net charge is controlled by the release of protons into the thylakoid lumen. During the four transitions $S_0 \rightarrow S_1$ to $S_3 \rightarrow S_4 \rightarrow S_0$ the extent of proton release oscillates like $1:0:1:2$ protons per flash. This leaves a net charge of $0(S_0):0(S_1):+1(S_2):+1(S_3)$ in the catalytic centre.

The ultimate source of liberated protons is the oxidation of water. Transient proton release after a distinct flash, however, does not necessarily reflect the oxidation of 'bound water'. This becomes obvious by comparison of the half-rise time of electron transfer between the catalytic site and Z [17, 18] and the half-rise time of proton liberation [14]. Proton release follows in time electron abstraction from the catalytic site with one exception (see Fig. 1); in $S_2 \rightarrow S_3$ the half-rise time of proton release ($200 \, \mu s$ [14]) is shorter than the one of electron abstraction from the catalytic site ($350 \, \mu s$ [18]). In consequence, when the catalytic centre is in state S_2 proton release at $200 \, \mu s$ has to be attributed to a transient deprotonation/protonation of the intermediate electron carrier Z or of an acidic group in its vicinity [14]. We consider it as likely that protons are transiently liberated from amino-acid side groups which are only reprotonated during the reaction of the catalytic centre with water.

How to get hold of the so far ill-defined groups involved in the storage and channelling of protons? A possible clue might be expected from work by Peter Jahns and Andrea Polle in our laboratory. They found the proton-pumping activity of PS II short-circuited after covalent modification by N,N'-dicyclohexylcarbodiimide of two particular polypeptides in the range of 20–24 kD molecular mass. Treatment with this notorious blocker of proton channels indeed blocked proton release and also proton uptake but it also opened a channel which conducted protons, produced during water oxidation, across the membrane to the freshly reduced quinone. It will be interesting to identify the respective polypeptides among the about 15 ones [19] which are contained in the photosynthetic water oxidase.

The reaction of hydroxylamine and of hydrazine with the water oxidase: kinetics, binding parameters, structure and products

Bouges [20] was the first to report that hydroxylamine shifts the pattern of oxygen release as function of flash number by two digits to higher flash numbers. This 'delay' of the water oxidase is also seen in the pattern of proton liberation [23]. For hydroxylamine it is documented in Fig. 2. Upon flash no. 1 two protons per PS II are released with a half-rise time of 3.1 ms. The following flashes produce the usual pattern oscillating with period of four but as if starting from state S_0 upon flash no. 2 [23]. Thus the maximum of oxygen evolution and of proton release is shifted from flash no. 3 in the absence to flash no. 5 in the presence of hydroxylamine

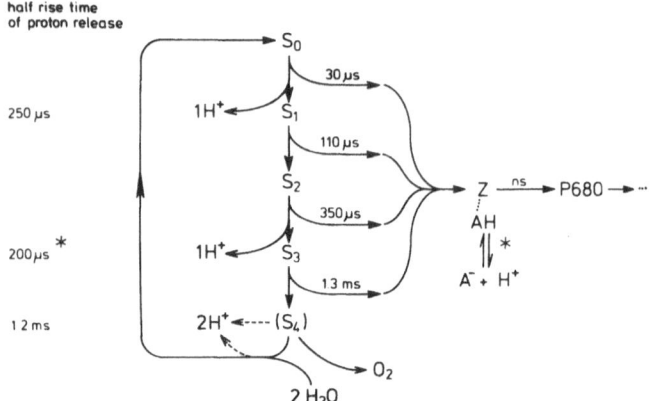

Fig. 1. Electron transfer and protolytic reactions at the donor side of PS II (for refs. see text).

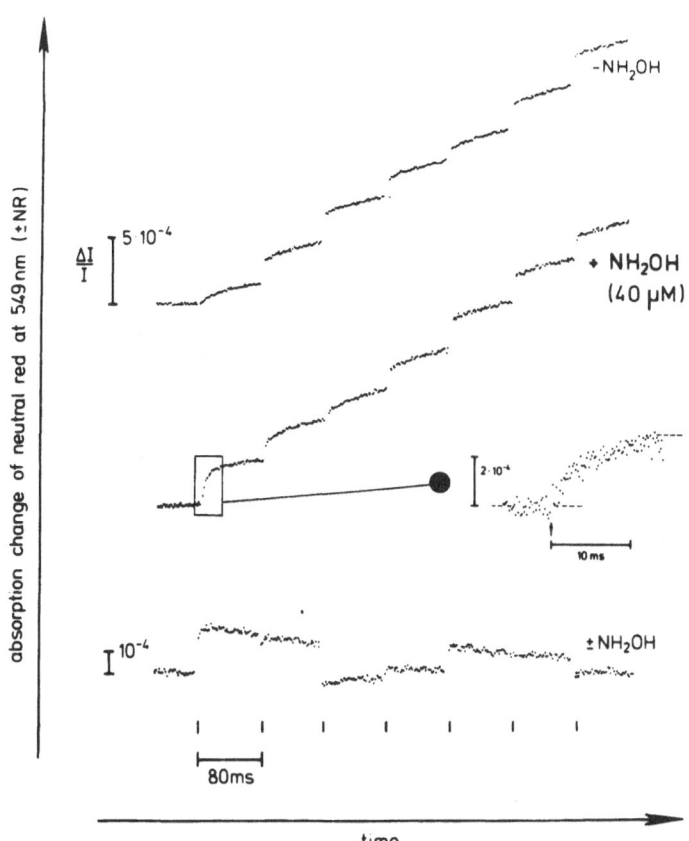

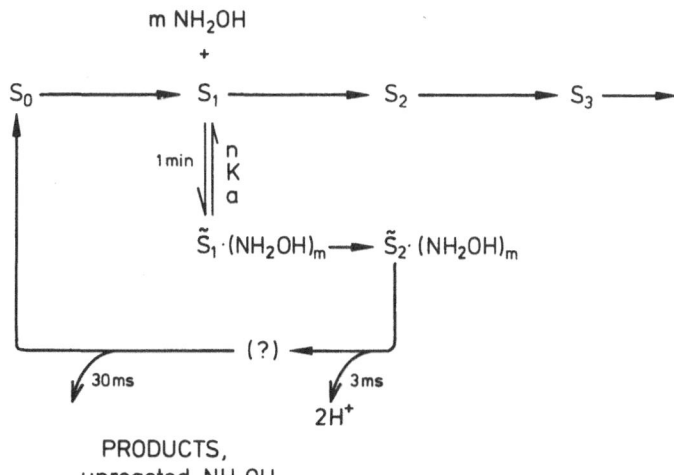

Fig. 3. Reactions of the water oxidase with hydroxylamine and their approximate half-rise times.

Fig. 2. Pattern of proton release into the thylakoid lumen observed upon a series of flashes in dark-adapted thylakoids, monitored by neutral red, without and with hydroxylamine (upper and middle trace). The bottom trace represents the difference between the middle and the upper trace. Time resolution of the pattern was 2 ms/point, and in the expanded region around the first flash of the middle trace 100 μs/point.

[20–24]. The same behaviour is observed in the presence of hydrazine and of *O*-sulphonyl-hydroxylamine [26].

It is under debate whether hydroxylamine reacts with the water oxidase already in the dark or whether it binds in the dark to react only after the centre is oxidized by input of one quantum of light. Judging from the pattern of proton release, we found that three washes of thylakoids in hydroxylamine-free medium abolished the effect of a previous incubation with hydroxylamine widely [23]. (Incubation plus washes were carried out all in total darkness!) This result was at variance from a report by Bouges who did not observe a reversal [20]. But it was in line with a report from Hanssum & Renger [24] who observed reversibility of the action for hydrazine. Thus we tend to assume that these agents bind reversibly to the centre in the dark (see also below and Fig. 3).

The kinetic parameters of hydroxylamine binding and its flash-induced reactions are given in Fig. 3. When dark-adapted thylakoids are incubated with 30 μM hydroxylamine the two-digit shift in the proton release pattern was fully expressed after 5 min, incubation for further 20 min did not enhance the effect. After firing of a single laser flash the release of two protons was half-completed in about 3 ms. If the second flash was fired 80 ms after the first one and a series of flashes at 80 ms intervals thereafter the pattern of proton release showed the period-of-four oscillation as documented in Fig. 2. The pattern was out of order, however, if the spacing between the first and the second flash was shorter. The time delay required for half-relaxation of the centre, supposedly to yield state S_0, was approximately 30 ms. Thus,

we postulate two reaction steps following the first flash, one releasing protons and an about tenfold slower step completing the relaxation to S_0 (see Fig. 3). It is noteworthy that we found no protolytic reaction concomitant with or subsequent to the second reaction. This is documented in Fig. 2. The bottom trace is the difference between the proton release pattern in the presence and that in the absence of hydroxylamine. After flash no. 1 there is no extra proton produced or consumed except from the two which appear with 3 ms half-rise time (time resolution of the patterns was 2 ms/point).

In summary, the reaction of the water oxidase with hydroxylamine requires 5 min of incubation of dark-adapted thylakoids (at room temperature). One flash causes the release of two protons into the lumen at 3 ms half-rise time. The unperturbed state S_0 is reached with a half-rise time of approximately 30 ms. It is conceivable that this time interval is required for the release of products and of unreacted ligand molecules from the centre (see Fig. 3). Together with Friedhelm Lendzian in Klaus Möbius' laboratory we looked for an EPR-detectable intermediate. Unfortunately, no such intermediate could be detected in the time interval between 500 μs and some milliseconds after the flash. Either it was not produced, or it was too shortlived to be detected.

Which are the binding parameters of hydroxylamine and hydrazine to the catalytic centre? Figure 4*a* shows the extent of rapid proton release upon flash no. 1 in dark-adapted thylakoids as function of the hydroxylamine concentration (data points taken from ref. [25]). The effect *vs.* concentration behaviour is sigmoidal. A Hill plot reveals a Hill coefficient at half maximum of 2.43 for hydroxylamine (see Fig. 5 in ref. [5]) and of 1.48 for hydrazine [26]. For hydrazine the same degree of cooperativity was found by Hanssum and Renger [24]. At first sight this implies the existence of at least three binding sites for hydroxylamine and at least two for hydrazine, which are populated in a cooperative manner. A prerequisite for this interpretation is the reversible binding of these agents to the water oxidase. Before discussing the implications of the concentration dependence for the interaction of the agents with the catalytic centre in more quantitative terms we scrutinize alternative interpretations, in particular that of Radmer and Ollinger [21, 22] and of Beck and Brudvig (see this volume) who favour the idea that hydroxylamine irreversibly reacts with the water oxidase in state S_1 in the dark.

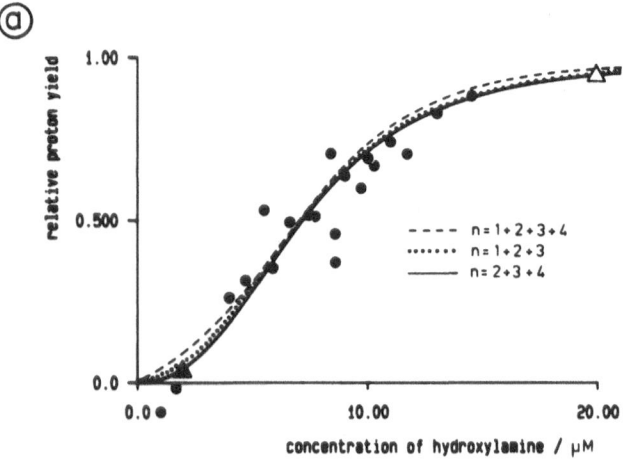

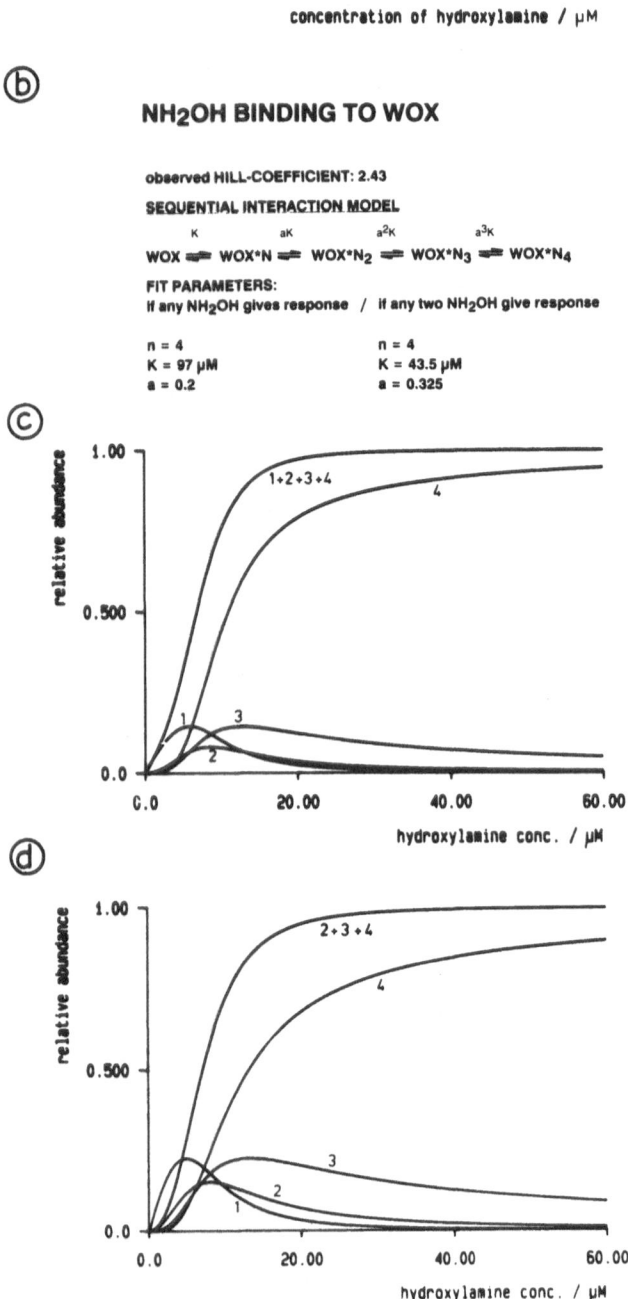

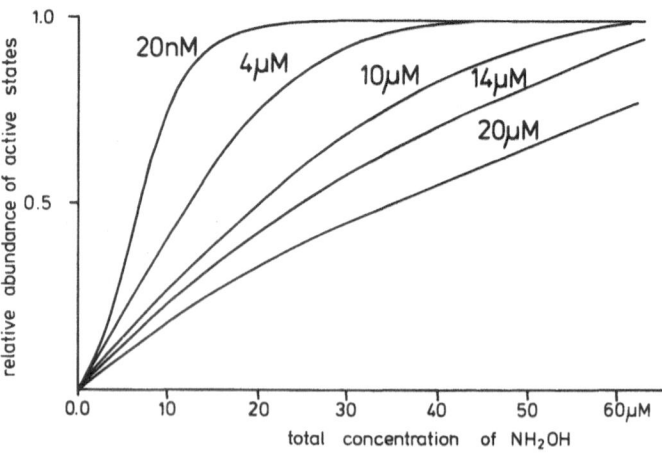

Fig. 5. Concentration dependence of the two-digit shift by hydroxylamine, as expected for different concentrations of the water oxidase. Note that the sigmoidal behaviour is converted into a Langmuir binding isotherm at high concentrations of the water oxidase (see text).

Fig. 4. Concentration dependence of the two-digit shift in the proton release pattern by hydroxylamine. (*a*) Hydroxylamine-mediated proton yield upon the first flash as a function of the concentration of hydroxylamine. Full circles: in the presence of 30 mM chloride. Open triangles: in the presence of 100 μM chloride. Lines: fit curves by model for sequential co-operative interaction (see Fig. 4*b* and text). (*b*) Reaction scheme for the co-operative reaction of the water oxidase with hydroxylamine and two parameter sets to fit the data in Fig. 4*a*. (*c*) and (*d*) Occupancy of states as function of hydroxylamine concentration for the two parameter sets (*n, K, a*) given in (*b*).

Can the observed concentration dependence be explained under the assumption of an irreversible reaction of hydroxylamine in the dark? The induction up to about 3 μM can. If there were hydroxylamine-trapping groups with a capacity of say 3 μM and with a very high affinity, there was indeed no effect expected before saturation of these groups. Beyond this, however, an irreversible reaction would proceed unless all water oxidase centres were saturated. At low concentrations this may go on slowly but it proceeds to completion. Since this consumed only some multiple of 40 nM, the concentration of the water oxidase in our samples, we expected a concentration dependence after the induction which was steeper than the observed one (see Fig. 4*a*). Note that the data shown in Fig. 4*a* were obtained after the effect was stationary. This and the above described washing experiment showed that the reaction was reversible.

What if the reversibility was indirect? Imagine an irreversible reaction of the centre with hydroxylamine, e.g. a two-electron reduction from state S_1 to S_{-1}, followed by a slower reoxidation of S_{-1} back to S_1 caused by some other agent, for example dioxygen. Then the effect *vs.* concentration data would not reflect a binding isotherm but rather the stationary level of the intermediate, S_{-1}, of a consecutive reaction. No such oxidation reaction is known so far, the only known spontaneous oxidation of the water oxidase is from state S_0 to S_1 caused by the intrinsic electron acceptor D^+ and proceeding in 20 min [27], which is too slow to be relevant in our context (see Fig. 3).

Based on the foregoing considerations we fitted the effect *vs.* concentration data for hydroxylamine by an isotherm for an enzyme with several co-operative binding sites. For the sake of simplicity we chose the Pauling–Adair model of sequential interaction with only three fit parameters, namely *n*, the number of binding sites, *K*, the dissociation constant of the first ligand, and *a*, the interaction factor, the same factor valid for all occupation states. This is illustrated in Fig. 4*b*. As common, these three parameters are sufficient to provide considerable ambiguity. Three fit curves are plotted in Fig. 4*a*. For two of these curves it is assumed that the effect is the same independent of whether 1, 2, 3, or more sites are occupied. (Fig. 4*a*, dashed and dotted lines.) They resulted in the following parameter sets (*n, K, a*): 4, 97 μM, 0.2; and 3, 152 μM, 0.05 [25]. For the third fit we assume that the minimum requirement is an occupation by 2 molecules of hydroxylamine, but three and four giving the same effect as two (Fig. 4*a*, solid line). The fit parameters are 4, 44 μM, and 0.325. We do not wish to overstress the better quality of the last fit. The common features of all fits are as follows:

The water oxidase in its oxidation state S_1 contains 3–4 sites for the reversible binding of hydroxylamine. These are cooperatively populated (Hill coefficient 2.4). The effect of bound hydroxylamine which is observed after excitation of PS II with one flash of light is the same once the occupation number exceeds a minimum value (1 or 2 mol).

Which are the sites of hydroxylamine binding? According to Sandusky and Yocum [28] and Beck and Brudvig [29] there are at least two classes of binding sites for primary and secondary amines. The first class is possibly associated with the manganese centre proper and the second one with the sites of chloride binding in the vicinity of the centre. We examined the possibility that hydroxylamine is bound to the latter class. The effect *vs.* concentration behaviour was redetermined under variation of the chloride concentration. As documented by the open triangles in Fig. 4*a* there was no influence of chloride on the binding curve in the concentration range between 100 μM and 30 mM. If hydroxylamine was replacing chloride at its binding site an effect of chloride was to be expected.

For the binding of hydroxylamine to the manganese centre we have previously proposed a model where two intrinsic bridges, e.g. μ-oxo bridges between a binuclear manganese centre, are replaced either by coordination of 4 molecules of hydroxylamine or, alternatively, by 2 molecules of hydrazine as bridging ligand [26]. Only amino groups interact with manganese. This model most naturally incorporates the different Hill coefficients, 2.43 indicative of three if not four sites for hydroxylamine, and 1.48 indicative of two (or more) sites for hydrazine. It makes falsifiable predictions on Mn–Mn distances, 4.2 Å with hydrazine instead of 2.7 Å in controls and about 5.7 Å with hydroxylamine [26]. The enlarged distances are based on the known crystal structure of manganese–hydroxylamine complexes [30].

The larger Mn–Mn distances have not been detected in EXAFS studies with hydroxylamine treated samples ([31,32] and Sauer *et al.*, this issue). This may indicate that there is indeed no enlargement of distances. However, a word of caution may be worthwhile. Very different sample concentrations are used in optical spectroscopy (our approach) and in EPR or in EXAFS. We work with thylakoid suspensions containing only 20 μM chlorophyll, the concentration of water oxidases is about 40 nM. Out of the 10 μM of hydroxylamine required for the expression of the effects only a very small portion will be bound. So the total concentration of hydroxylamine is very close to the concentration of free hydroxylamine. This is totally different in concentrated samples of approximately 10 mM-Chl and 4 μM water oxidases as used in EPR and EXAFS studies. Using the fit parameters based on our experiments with diluted samples (see Fig. 4) we calculated the concentration dependence of the effect of hydroxylamine in samples with micromolar concentrations of the water oxidase. In that calculation we assumed that there were no binding sites for hydroxylamine other than the cooperative ones (Fig. 4*b*). It is intuitively apparent that for a concentrated sample the concentration of free hydroxylamine over the total concentration will be

flattened the most in the range where the cooperative binding curve is steepest. This 'buffering' also flattens the effect *vs.* concentration curves as documented by the simulated curves in Fig. 5. This particular simulation was carried out for the parameter set {(n; K; a) = (4; 97 μM; 0.2)} and under consideration of the relation between free and total hydroxylamine. It is important to note that the effect *vs.* concentration dependence approaches the one of a simple first-order binding isotherm in concentrated samples, although the effect *vs.* free concentration behaviour, is highly cooperative and sigmoidal. The published EXAFS work was carried out with samples containing about 14 μM of water oxidases. The authors observed an influence of 40 μM hydroxylamine on about 65% of the centres as measured by the disappearance of the multiline EPR signal [32]. This agrees with our prediction (see Fig. 5, " 14 μM "). Fig. 4(*c* and *d*) shows the frequency of occupation numbers, 0, 1, 2, 3, 4. With the percentage of modified centres being 65% (i.e. $1+2+3+4$ or $2+3+4 = 65\%$ depending on the respective parameter set) we found all occupation states, broadly speaking, about equally probable. This implied that the EXAFS work might have been carried out with an ensemble of water oxidases which was strongly heterogeneous with respect to the ligand state of the water oxidase.

Acknowledgements

We wish to thank Peter Jahns and Andrea Polle for communicating their unpublished work and Friedhelm Lendzian for sharing the up to now fruitless adventure to chase transient intermediates of hydroxylamine oxidation. We are grateful to Hella Kenneweg for the graphs. This work was supported by a grant (TP A2) from the DFG (Sonderforschungsbereich 171).

References

1. Babcock, G. T. *New Compreh. Biochem.* **15**, 125–158 (1987).
2. Dismukes, C. G. (1986) in: *Manganese Metab. Enzyme Funct.* (ed. V. Schramm and F. C. Wedler), pp. 275–309, Academic, New York.
3. Junge, W., *Curr. Top. Membr. Transp.* **16**, 431–465 (1982).
4. Radmer, R. and Ollinger, O., *FEBS Lett.* **195**, 285–289 (1986).
5. Hansson, O., Andréasson, L.-E. and Vänngård, T., *FEBS Lett.* **195**, 151–154 (1986).
6. Kok, B., Forbush, B. and McGloin, M., *Photochem. Photobiol.* **11**, 457–475 (1970).
7. Förster, V. and Junge, W., in *Advances in Photosynthesis Research*, vol. 2 (ed. Ch. Sybesma), pp. 305–308. Martinus Nijhoff/Dr W. Junk, The Hague (1984).
8. Junge, W., Ausländer, W., McGeer, A. and Runge, T., *Biochim. Biophys. Acta* **646**, 121–141 (1979).
9. Gutman, M. and Nachliel, E., *Biochemistry* **24**, 2941–2946 (1985).
10. Junge, W., Schönknecht, G. and Förster, V., *Biochim. Biophys. Acta* **852**, 93–99 (1986).
11. Hong, Y.-Q. and Junge, W., *Biochim. Biophys. Acta* **722**, 197–208 (1983).
12. Saphon, S. and Crofts, A. R., *Z. Naturforsch.* **32 C**, 617–626 (1977).
13. Wille, B. and Lavergne, J., *Photobiochem. Photobiophys.* **4**, 131–144 (1982).
14. Förster, V. and Junge, W., *Photochem. Photobiol.* **41**, 183–190 (1985).
15. Saygin, Ö. and Witt, H. T., *FEBS Lett.* **176**, 83–87 (1984).
16. Brettel, K., Schlodder, E. and Witt, H. T., *Biochim. Biophys. Acta* **766**, 403–415 (1984).
17. Babcock, G. T., Blankenship, R. E. and Sauer, K., *FEBS Lett.* **61**, 286–289 (1976).
18. Dekker, J. P., Plitter, J. J., Ouwekand, L. and van Gorkom, H. J., *Biochim. Biophys. Acta* **767**, 1–9 (1984).
19. Murata, N. and Miyao, M., in *Progress in Photosynthesis Research*, vol. 1. (ed. J. Biggins), pp. 453–462, Martinus Nijhoff, Dordrecht (1987).

20. Bouges, B., *Biochim. Biophys. Acta* **234**, 103–112 (1971).
21. Radmer, R. and Ollinger, O., *FEBS Lett.* **144**, 162–166 (1982).
22. Radmer, R. and Ollinger, O., *FEBS Lett.* **152**, 39–43 (1983).
23. Förster, V. and Junge, W., *Photochem. Photobiol.* **41**, 191–194 (1985).
24. Hanssum, B. and Renger, G., *Biochim. Biophys. Acta* **810**, 225–234 (1985).
25. Förster, V. and Junge, W., *FEBS Lett.* **186**, 153–157 (1985).
26. Förster, V. and Junge, W., *Photosynthesis Research* **9**, 197–210 (1986).
27. Styring, S. and Rutherford, W. A., *Biochemistry* **26**, 2401–2405 (1987).
28. Sandusky, P. and Yocum, C. F., *Biochim. Biophys. Acta* **766**, 603–611 (1984).
29. Beck, W. F. and Brudvig, G. W. *Chemica Scripta* 28A, 93 (1988).
30. Ferrari, A., Braibanti, A., Bigliardi, A. and Dallavalle, F., *Z. Krist.* **119**, 284 (1963).
31. Yachandra, V. K., Guiles, R. D., McDermott, A., Britt, R. D., Dexheimer, S. L., Sauer, K. and Klein, M. P., *Biochim. Biophys. Acta* **850**, 324–332 (1986).
32. Guiles, R. D., Yachandra, V. K., McDermott, A. E., Britt, R. D., Dexheimer, S. L., Sauer, K. and Klein, M. P., in *Progress in Photosynthetic Research*, vol. 1 (ed. J. Biggins), pp. 561–564. Martinus Nijhoff, Dordrecht (1987).

Chemica Scripta 1988, **28A**, 117–122

Characterization of the Tyrosine Radical Involved in Photosynthetic Oxygen Evolution

Bridgette A. Barry and Gerald T. Babcock

Department of Chemistry, Michigan State University, East Lansing, MI 48824-1322, USA

Paper presented by Gerald T. Babcock at the Nobel Conference 'Biophysical Chemistry of Dioxygen Reactions in Respiration and Photosynthesis', Fiskebäckskil, Sweden, 1–4 July, 1987

Abstract

The process of photosynthetic oxygen evolution involves two organic co-factors other than the reaction center chlorophyll. One of these co-factors, called Z, is involved in electron transport from the site of water oxidation to the reaction center of Photosystem II. The other species, D, gives rise to the stable EPR signal known as Signal II. Z^+ and D^+ have identical EPR spectra and are generally assumed to arise from species with the same chemical structure. Here, we present the evidence that has led us to conclude that D^+, and probably Z^+, are tyrosine radicals and not radicals of plastoquinone, as had previously been suggested. We also propose a model for the geometry of the methylene protons in the D^+ tyrosine radical which can explain the unique EPR properties of D^+. Finally, we show that we can roughly simulate the D^+ spectrum through the use of this model and the use of previous EPR studies of tyrosine radical model compounds.

Photosynthetic oxygen evolution in plants and algae involves the transfer of electrons from water to plastoquinone. This process takes place in the protein complex known as Photosystem II (PS II); the driving force for the reaction arises from the photochemical charge separation that takes place in the chlorophyll reaction center upon absorption of light (for review, see refs. [1,2]). One intermediate in the electron-transport chain between water and the PS II reaction center is an organic co-factor, called Z [3]. The EPR spectrum and kinetic behavior of Z^+ are well characterized [1,4–6]. In addition to the Z^+ EPR signal, PS II preparations also give rise to a stable EPR spectrum with the same linewidth, partially resolved hyperfine couplings, and g-value as the Z^+ spectrum [7]. The radical that is detected by the stable signal is called D^+. Recent data suggest that this species may be responsible for maintaining the integrity of a manganese complex, which is also involved in the catalysis of oxygen evolution [2,8]. Because of the similarity of their EPR spectra, D^+ and Z^+ are usually considered to arise from the same chemical species. The hyperfine couplings in the D^+/Z^+ spectra are known to arise from protons, since the D^+ spectrum narrows in D_2O [9,10].

We have recently shown that D^+, and probably Z^+, are tyrosine radicals [11], not radicals of plastoquinone, as had previously been suggested [10,12–17]. However, the lineshape of the D^+ signal does not resemble that of the enzyme, ribonucleotide reductase (RDPR), which also contains a tyrosine radical [18,19]. Here, we review the results that have led us to the conclusion that D is a tyrosine moiety, and we discuss the physical mechanisms that can alter the lineshape of these radicals.

Materials and methods

The methionine mutant (Met-27) of *Anabaena variabilis* [20] and wild-type *Synechochystis* 6803 were grown under sterile conditions on BG-11 media [21]; all amino acids were added by sterile filtration. The BG-11 media for the methionine mutant was supplemented with 90 μM methionine. Methionine was obtained from Sigma; deuterated methionine, $D_3CSCH_2CH_2CH(NH_2)COOH$, was from MSD Isotopes (98% d_3). *Synechocystis* was grown in the presence of 0.5 mM phenylalanine, 0.25 mM tryptophan, and 0.25 mM tyrosine. All the aromatic amino acids were from Sigma; deuterated tyrosine, $HOC_6D_4CO_2CD(NH_2)COOH$, was from MSD isotopes (98% d_7).

Cells were pelleted for the EPR studies when the cultures had reached either $OD_{720} = 0.22$ (*Anabaena*) or $OD_{720} = 0.44$ (*Synechocystis*). The pellets were resuspended in buffer containing 7.5% polyethylene glycol-3400 (Aldrich), 20 mM Hepes, pH 7.5, 10 mM $CaCl_2$, and 1 mM $MgCl_2$, and the cells were pelleted again. The final chlorophyll concentration was between 1.0 and 1.5 mg/ml for *Anabaena* and between 0.3 and 0.8 mg/ml for *Synechocystis*. EPR spectra were recorded at room temperature on a Bruker ER200D spectrometer operating at X-band and using a Varian TM cavity. The power was 10 mW, the modulation amplitude was 5 G, the scan time was 200 s, and the time constant was 500 ms. A Hewlett-Packard 5245L electronic counter with a 5255A frequency converter plug-in and a Bruker ERO35M gauss-meter were used to measure microwave frequency and magnetic field, respectively, in order to determine g-values. Simulation of the powder EPR spectrum was performed by using the program described in [22].

After the EPR studies, plastoquinone-9 was isolated from cell cultures by the method of Schoeder and Lockau [23]. Cells were extracted with chloroform and methanol, the solvent phases were separated, and the chloroform phase was dried and concentrated. The chloroform extract was chromatographed on thin-layer plates (EM Reagents, pre-coated PLC plates, silica gel 60 F254), and the quinone-containing band was eluted in ethanol and rechromato-graphed. Purification was obtained with a Waters HPLC and a C18 Bondapak column. The absorbance of the mobile phase was monitored at 254 nm; the retention time for plastoquinone-9 was 20 min (100% methanol). The absorption spectrum of the purified plastoquinone-9 was recorded and found to have a maximum at 255 nm [24] and to be identical to that of a plastoquinone standard. The plasto-quinone used for the standard was a gift from Hoffman–LaRoche.

Mass spectra were recorded on a Jeol JMS HX 110 mass spectrometer. The sample was ionized by fast atom bombard-ment (FAB) by using a matrix of triethanolamine and a 6 keV Xenon beam. The accelerating voltage was 8 keV, and negative ions were detected. Electron impact mass spectra were obtained by using a Finnigan 4000 E1/CI mass spectrometer. The electron voltage was 70 eV. The molecular weight of plastiquinone is 748 [25]. In FAB, the PQ-9 standard gives a strong molecular ion at $(M+H)^-$ or 749 and weaker peaks at $(M)^-$ and $(M-H)^-$. Reduction of the positive and negative ions of quinones in FAB mass spectrometry has been described previously [26].

Results

A model for the D^+ EPR spectrum had been proposed in which the major hyperfine couplings arise from the protons of one of the methyl groups of a plastoquinone cation radical [15, 16]. This model predicts that deuteration of these methyl groups should narrow the D^+ spectrum.

Figure 1 shows that the methyl groups of plastoquinone can be deuterated by feeding a methionine auxotroph of the cyanobacteria, *Anabaena variabilis*, deuterated methionine. In the mass spectrum of plastoquinone from cultures fed protonated methionine, the most intense molecular ion appears at $M/Z = 749$ (Fig. 1 A). By contrast, in cultures fed deuterated methionine (Fig. 1 B), approximately 48 % of the plastiquinone is labeled at both methyl groups, which produces a molecular weight of 755. An additional 40 % of the plastiquinone is labeled at one methyl groups (MW = 752); only 12 % is unlabeled. In a separate labeling experiment, a similar pattern of deuterium incorporation was observed with electron impact mass spectrometry (Table I).

Table I. *Electron impact mass spectral data on plastoquinone-9 isolated from* Anabaena *cultures that were fed either protonated or deuterated methionine.*

The 189 base peak of the PQ mass spectrum is attributed to formation of the pyrylium ion by cleavage of the bond δ to the ring and cyclization [25].

Relative Abundance		
M/Z	Protonated methionine	Deuterated methinone
189	100	23
192	—	21
195	—	56

The intensities of the $M/Z = 189$ base peak of plastoquinone were analyzed in this experiment [25], since the large amount of fragmentation caused by our system made the molecular ion unobservable.

With this amount of specific deuterium incorporation, the D^+ EPR signal should narrow if the model described in refs [15, 16] is correct. In Fig. 2, we present EPR spectra of D^+ in cultures fed protonated methionine (Fig. 2A) and deuterated methionine (Fig. 2B). As expected, in the control (Fig. 2A), the D^+ spectrum is approximately 20G wide and has a g-value of 2.0043. The D^+ spectrum of algae grown on deuterated methionine (Fig. 2B) is identical in linewidth and g-value to the control, in spite of the large amount of deuterum incorporation into plastoquinone in these cultures. We conclude that the methyl groups of plastoquinone are not

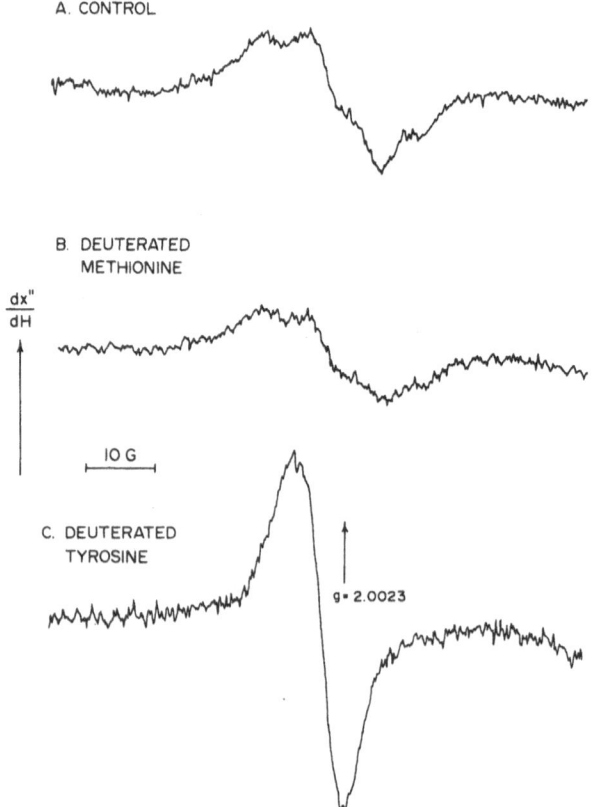

Fig. 2. EPR spectra of D^+ in cyanobacteria grown in the presence of (A) 90 μM protonated methionine, (B) 90 μM deuterated methionine, and (C) 0.5 mM phenylalanine, 0.25 mM tryptophan, and 0.25 mM per-deuterated tyrosine. The spectra were obtained 2 min after illumination. The gain in (A) is 3.2×10^6, in (B) 3.2×10^6, and in (C) 5×10^6. Other instrument settings and experimental conditions are given in the Methods section.

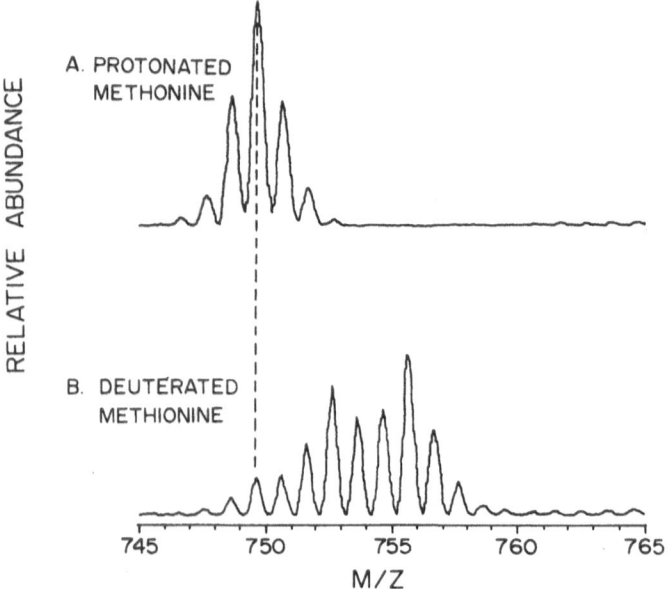

Fig. 1. FAB mass spectra of plastoquinone isolated from *Anabaena* cultures grown on BG-11 supplemented with (A) 90 μM protonated methionine and (B) 90 μM deuterated methionine. In the text, the molecular weights are rounded to the next lower integer (i.e. 749.6 to 749). The observed ion distribution from 749 to 753 was reproduced by a computer program, supplied by the manufacturer of the mass spectrometer, that uses the natural isotope abundance and the molecular formula as inputs.

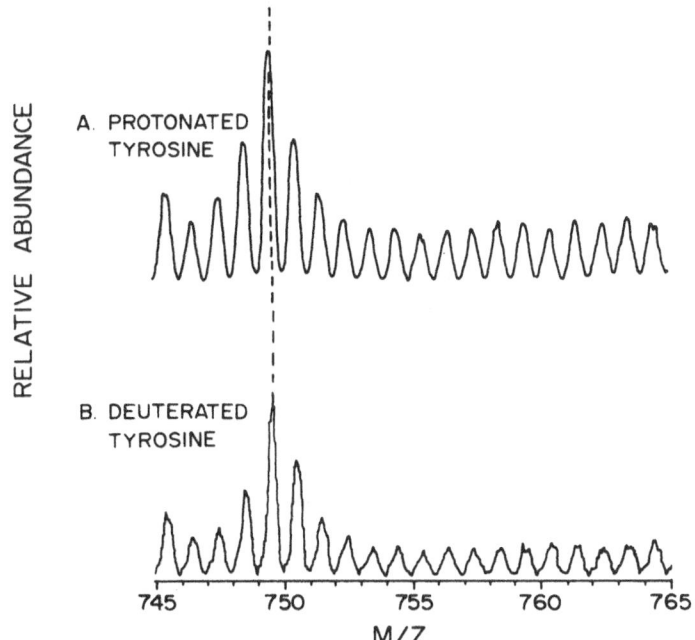

Fig. 3. FAB mass spectra of plastoquinone isolated form *Synechocystis* cultures grown on BG-11 supplemented with 0.5 mM phenylalanine, 0.25 mM tryptophan and (A) 0.25 mM protonated tyrosine or (B) 0.25 mM per-deuterated tyrosine. Smaller sample size makes background peaks evident in these spectra. For example, M/Z = 745 is a peak from the triethanolamine matrix.

responsible for the hyperfine couplings observed in the D^+ spectrum.

To test the possibility that D^+ may be a tyrosine radical, we used feedback inhibition of the tyrosine biosynthetic pathway to label tyrosine completely in the cyanobacterium, *Synechocystis* 6803 [11]. When *Synechocystis* 6803 is grown in the presence of 0.5 mM phenylalanine, the algae is dependent on the import of tryptophan and tyrosine for growth.

In Fig. 3, we demonstrate that deuterated tyrosine does not label plastoquinone in these algae. Fig. 3A shows the mass spectrum of plastoquinone from cultures fed protonated tyrosine. In Fig. 3B, we show the mass spectrum of plastoquinone from algae grown on deuterated tyrosine. No significant amount of deuterium incorporation is observed, as the two spectra are essentially identical.

However, Fig. 2C shows that the D^+ EPR signal narrows to 7 G in cultures fed deuterated tyrosine. The control D^+ spectrum, taken of cells grown under identical conditions, except in the presence of protonated tyrosine, is identical to Fig. 2A (not shown). Although the D^+ spectrum in the deuterated cultures has narrowed, the g value of the spectrum is still 2.0043. Moreover, 7 G is approximately the D^+ linewidth in algae grown on D_2O [9, 10]. So, in these cultures that were fed deuterated tyrosine, all protons with significant hyperfine couplings have been replaced with deuterium. We conclude, therefore, that D^+ is a tyrosine radical.

In Fig. 4 we show a simulation (Fig. 4b) of an unoriented D^+ spectrum (Fig. 4a), where the parameters for the simulation are based on our conclusion that D is a tyrosine. The program used to produce the powder EPR spectrum [22] takes as parameters the anisotropic components of the D^+ g tensor, which have been measured by Brok and co-workers [10]. Also, the values of the anisotropic hyperfine tensors are supplied to the program, along with the orientations of these tensor components with respect to the molecular axes. The

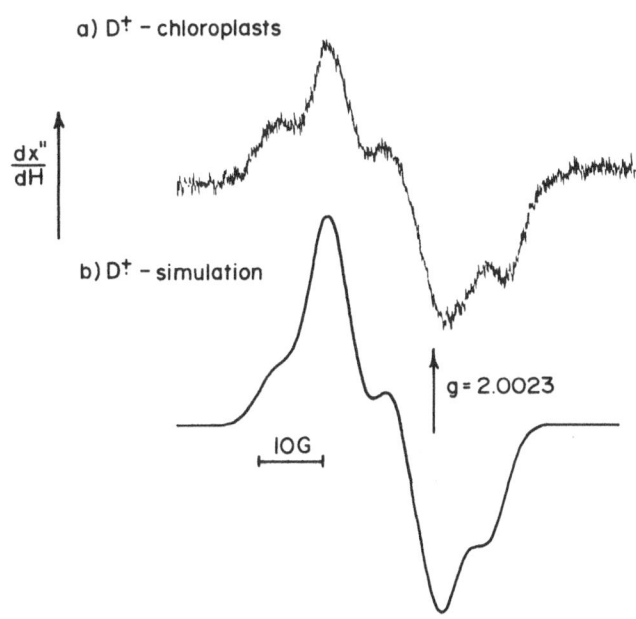

Fig. 4. Computer simulation (b) of an unoriented D^+ spectrum (a) taken from Babcock and Sauer [37]. The parameters used in the simulation are given in Table II.

Table II. *Components of the proton hyperfine tensors and their orientations*

These values were used in the computer simulation of an unoriented D^+ spectrum (Fig. 4). N is the number of protons in the group. θ is the angle between the Z-axis of the hyperfine tensor and the molecular Z-axis; ϕ is the angle between the Y-axes.

	X (G)	Y (G)	Z (G)	N	θ	ϕ
$A(CH_2)$	11.2	9.7	9.7	1	39°	50°
$A(H)$	9.0	3.1	6.5	2	0°	30°

parameters used to generate the simulation (Fig. 4b) are listed in Table II; the method by which the hyperfine values and angle were chosen is described in the discussion.

Discussion

The assignment of Z^+ and D^+ to tyrosine radicals with the unpaired electron delocalized over the phenol substituent is consistent with the majority of the physical properties of these cofactors. The absorption spectrum of Z^+ bears a marked resemblance to that of a phenoxy radical [17, 27]. The redox potential of a tyrosine radical is greater than 0.8 V at pH 7 [28], in agreement with the redox potentials of Z^+ and D^+ [3, 29]. The isotropic g-values of the D^+ and Z^+ EPR spectra are essentially the same as the g-value of a neutral phenoxy radical [30]. Moreover, the principal components of the D^+ g-tensor, measured as 2.0074, 2.0044 and 2.0023 [10], agree well with the g-tensor components, 2.0067, 2.0045 and 2.0023, measured for the tyrosine radical by Fasanella and Gordy ([31], see Table III). These tensor components are also similar to those that have been estimated by simulation for the tyrosine radical in RDPR (Table III), although for this species these values have not been determined with high accuracy.

The EPR spectrum of D^+, however, has a different linewidth and lineshape than that of the tyrosine radical in RDPR [18, 19]. For example, the RDPR signal is more than 40 G wide, while the D^+ spectrum has a linewidth of less than

Table III. *Tyrosyl radical g-values*

radical species	g_{iso}	g_x	g_y	g_x	Ref.
1. Solution tyrosyl radical (neutral)	2.0044	—	—	—	[30]
2. Solution tyrosyl radical (cation)	2.0033	—	—	—	[30]
3. Single crystal tyrosyl radical[a]	2.0045	2.0067	2.0045	2.0023	[31]
4. Tyrosyl radical in RDPR	2.0049	2.008	2.003	2.003	[19]
5. D⁺ tyrosyl radical in PS II	2.0047	2.0074	2.0044	2.0023	[10]

[a]g_x lies along the C_1–C_4 axis, g_y is in the phenol plane and perpendicular to g_x, g_z is the out-of-plane component.

20 G. In order to understand these observations, previous work done on tyrosine radicals must be reviewed.

The EPR spectra of tyrosine radicals in single crystals and in solution have been studied in detail [31–33]. In these compounds, there is significant spin density at the 1, 3 and 5 carbons (see Fig. 5 for the tyrosine numbering scheme used here). This spin density distribution produces an anistropic coupling to the equivalent α-protons at the 3,5-ring positions and a more isotropic coupling to one or both of the β-methylene protons. The hyperfine coupling constants for the 3,5-protons are relatively insensitive ($\sim$6–7 G) to the mode of preparation and physical state of the tyrosine radical. For the β-methylene protons, however, there is considerable variation in the hyperfine parameters. In the crystalline state, one of the two protons is strongly coupled to the unpaired electron spin whereas the second has only negligible coupling [31,32]. For tyrosine radicals in solution the methylene proton splittings observed in the EPR spectrum are temperature dependent [33]. Restricted rotation about the single bond between C_1 and C_β (see Fig. 5) is regarded as the cause of this temperature dependence [33]. At room temperature, the hyperfine coupling to each of the β-methylene protons is estimated to be 7.5 G. At 60 °C, a good simulation of the spectrum is obtained with hyperfine splittings to the two inequivalent β-methylene protons of 7.1 and 8.5 G ([33], see Table IV).

These variations in the hyperfine couplings can be used to determine the geometry of the C_β methylene protons. For β-protons such as these, the isotropic hyperfine coupling is determined as $a_{iso} = B\rho \cos \theta^2$, where θ is the angle between the H–C_β bond and the C_1 p_z orbital when projected in a plane perpendicular to the C_1–C_β bond direction. B is a proportionality constant, usually taken as approximately 50 G, and ρ is the unpaired electron spin density at the C_1 position. The above equation indicates, then, that with a knowledge of ρ,

the orientation of the methylene protons with respect to the ring can be deduced. For the tyrosyl radical, Sealy and co-workers [33] took the $B\rho$ product as 25 G, which implies an unpaired electron-spin density of 0.5 at C_1. However, lower values for this spin density have been reported. For example, Fasanella and Gordy [31] give an experimental value of 0.32 and an SCF calculated value of 0.31 for the spin density at C_1; Box *et al.* [32] found a spin density of only 0.185 in calculations in which simple Huckel theory was used. In the absence of quantitative agreement about the C_1 spin density in tyrosyl radicals, the Sealy *et al* values of $\rho = 0.5$ and $B = 50$ G seem appropriate. These values were estimated from the splittings observed for the 1-methyl phenoxy radical, where $a_{iso} = 12.5$ G. Thus, by using $B\rho = 25$ G, the conformation of each of the methylene protons in the tyrosyl radical in solution was deduced to be approximately 60° [33]. Sealy and co-workers argued that this orientation represents a low energy conformation.

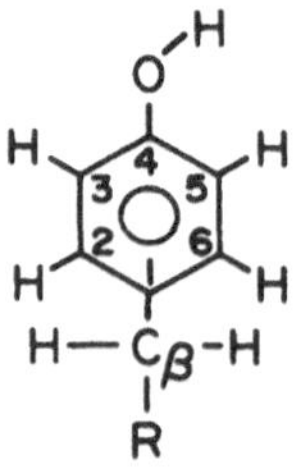

The geometry of the methylene protons in crystalline tyrosine radical is shifted markedly from this solution conformation. From the crystal structure and EPR spectrum of tyrosine, Fasanella and Gordy found that one proton is oriented at 30° with respect to the ring and is strongly coupled to the C_1 unpaired electron-spin density. The second β-methylene proton had an 80° orientation and interacts only weakly. Table IV summarizes the hyperfine coupling values determined for the β-methylene protons in these studies as well as for the tyrosyl radicals in RDPR and in PS II (see below); Fig. 6 presents a diagrammatic representation of the geometries of the methylene protons in the various species.

For RDPR, experiments with specifically deuterated tyrosines have indicated that the spectrum is dominated by a large, 19 G coupling to one methylene proton. The second methylene proton contributes negligible hyperfine coupling to the spectrum (Table IV). Also, these experiments have shown that the splittings due to the 3,5-protons are similar to those observed for the tyrosine radical, $\sim$7 G [18]. In order

Table IV. *Tyrosine radical methylene proton hyperfine couplings and dihedral angles*

Radical species	A_{iso}	A_x	A_y	A_z	θ	Ref.
1. Solution tyrosyl radical room temperature						
β-CH₁	7.5	—	—	—	$\sim$60°	[33]
β-CH₂	7.5	—	—	—	$\sim$60°	
2. Single crystal tyrosyl radical						
β-CH₁	14	14	14	14	$\sim$30°	[31]
β-CH₂	0.7	—	—	—	$\sim$80°	
3. Single crystal tyrosyl cation radical						
β-CH₁	18.8	20.3	18.3	17.8	—	[32]
β-CH₂	—	—	—	—	—	
4. Tyrosyl radical in RDPR						
β-CH₁	18.2	19.5	17.5	17.5	$\sim$28°	[18,33]
β-CH₂	—	—	—	—	$\sim$90°	
5. D⁺ tyrosyl radical in PS II						
β-CH₁	10.2	11.2	9.7	9.7	$\sim$47°	(this work)
β-CH₂	1.9	2.1	1.85	1.75	$\sim$73°	

Fig. 5. The phenol moiety of tyrosine showing the carbon numbering system used in the text.

β - Methylene Geometries

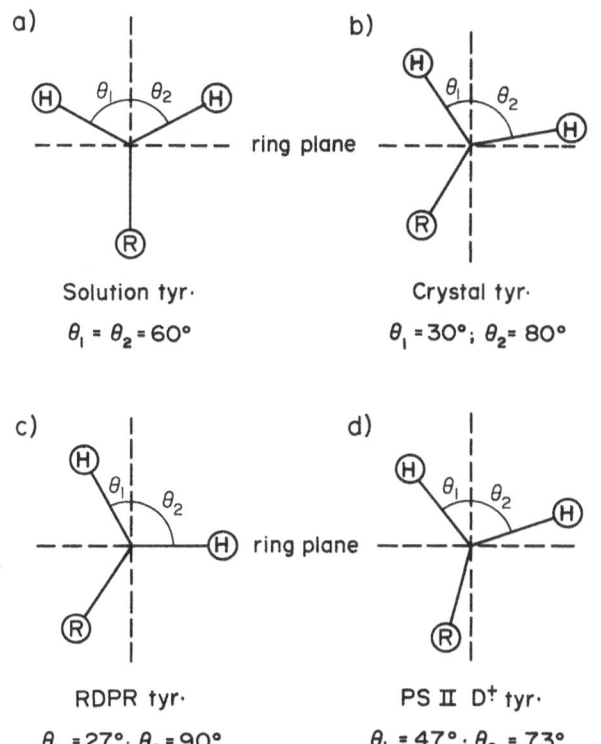

Fig. 6. The conformation of the β-methylene group with respect to the phenol ring plane in the indicated tyrosyl radical species. Both (*a*) and (*c*) are derived from data in Sealy *et al.* [33] and (*b*) is from data in Fasanella and Gordy [31]. The conformation shown in (*d*) is discussed in the text.

to explain such a large methylene proton splitting, it has been proposed that the protein constrains the ring to a high energy conformation [18, 19, 33] such that $\theta_1 \neq \theta_2 \neq 60°$. Sealy *et al.* have estimated that the dihedral angle for the strongly coupled proton is $\sim 27°$ in RDPR isolated from *E. coli* and that the weakly coupled proton lies at $\theta \cong 90°$ with respect to the $C_1 p_z$ orbital. Figure 6 shows that the postulated conformation of the β-methylene protons in the RDPR tyrosyl radical is very similar to that found for the radical in single crystals.

In carrying out the preliminary simulations of the D^+/Z^+ EPR spectrum in Fig. 4, we have relied on these earlier studies of the tyrosyl radical. The *g*-tensor for the D^+ species has been determined by Brok *et al.* [10], and we used the single crystal data of Fasanella and Gordy to orient the tensor components with respect to the tyrosyl radical molecular axes. Thus, g_x is taken along the C_1–C_4 direction, g_y is in the plane of the ring and perpendicular to g_x, and g_z is out-of-plane. Couplings to the methylene and to the 3,5-ring protons are assumed to dominate the spectrum. We expect that the methylene-proton splitting will be large and nearly isotropic. We assume, therefore, that the largest coupling observed in the D^+ ENDOR spectrum can be assigned to a methylene proton. These couplings are 11.2, 9.7 and 9.7 G [16], giving an isotropic coupling of 10.2 G. For this β-proton, the direction from C_1 to the methylene proton is the A_x direction, which should be the largest tensor component. Therefore, we assign the 11.2 G value to the *x*-direction in the methylene proton hyperfine tensor (Tables II and IV).

This value of a_{iso} for one of the methylene protons allows us to estimate the geometry of the β-methylene group. If we use the Sealy *et al.* value of $\rho = 0.5$, we find that the dihedral angle for the strongly coupled proton, θ_1, is 51°. This geometry predicts that the second methylene proton will have an isotropic coupling of ~ 3 G. In the ENDOR spectrum of D^+, however, we do not observe a β-proton with this coupling. The ENDOR spectrum does contain a set of lines with $a_{iso} = 1.9$ G, which have previously been attributed to β-protons ([34], see Table IV). Assuming that this resonance does arise from the second, more weakly coupled β-methylene proton, we find that we are able to reproduce the observed couplings with $\theta_1 = 47.1°$ and $\theta_2 = 72.9°$ and $\rho = 0.44$.

The couplings expected for the 3,5-ring protons have not yet been clearly observed in the ENDOR spectrum. Therefore, in the simulation, we have used the 3,5-ring proton hyperfine tensor components and orientations that have been measured on tyrosine radicals in single crystals [31].

In spite of the preliminary nature of our assumptions, the computer simulation (Figure 4B) that was performed with the parameters in Table II is a relatively good fit to the D^+ spectrum. We are now engaged in refining the preliminary model we have developed by carrying out experiments where cyanobacteria are labelled with specifically deuterated tyrosine. EPR, ENDOR, and spectral simulation studies of these samples should define the contributions of different protons to the D^+ EPR spectrum more precisely.

One final question to be addressed here is the extent of orientation of the cyanobacteria during the EPR measurements. The shape of the D^+ spectrum depends on the orientation of chloroplast membranes with respect to the applied field [16, 35]. For example, the D^+ spectrum of flowing and static chloroplast samples, which are believed to be oriented and unoriented, respectively, are quite different [36]. Comparison of our algae D^+ spectra with our chloroplast D^+ spectra implies that the algae may be partially oriented in these room-temperature experiments. Therefore, in order to simulate these spectra successfully, we will need to take the extent of sample orientation into account.

Acknowledgements

We thank Professor C. F. Yocum, Mr Ivan Rodriguez and Mr Mohammed El-Deeb for useful discussions. We are also grateful to P. Wolk for the gift of the methionine mutant and to L. McIntosh, S. Ferguson-Miller and D. McConnel for the use of their laboratory facilities. Mass spectral data were obtained from the Michigan State University Mass Spectrometry Facility supported by a grant (RR-00480) from the Biotechnology Resources Branch, Division of Research Resources, NIH. Bridgette Barry is a NIH postdoctoral fellow; this work was supported by the National Institute of Health (EM 37300) and the Photosynthesis Program of Competitive Research Grants Office of the U.S. Department of Agriculture.

References

1. Babcock, G. T., *New Comprehensive Biochemistry: Photosynthesis* (ed. J. Amesz). Elsevier, Amsterdam 125 (1987).
2. Dismukes, G. C., *Photochem. Photobiol.* **43**, 99 (1986).
3. Bouges-Bocquet, B., *Biochim. Biophys. Acta* **594**, 85 (1980).
4. Babcock, G. T. and Sauer, K., *Biochim. Biophys. Acta* **376**, 315 (1975).

5. Blankenship, R. E., Babcock, G. T., Warden, J. T. and Sauer, K., *FEBS Lett.* **51**, 287 (1975).

6. Hoganson, C. W., Demetriou, Y. and Babcock, G. T., in *Progress in Photosynthesis Research* (ed. J. Biggins), vol. 1, p. 479. Martinus Nijhoff, Dordrecht (1987).

7. Commoner, B., Heise, J. J. and Townsend, J., *Proc. Natl. Acad. Sci. USA* **42**, 710 (1956).

8. Styring, S. and Rutherford, A. W., *Biochemistry*, **26**, 2401 (1987).

9. Kohl, D. H., Townsend, J., Commoner, B., Crespi, H. L., Dougherty, R. C. and Katz, J. L., *Nature* **206**, 1105 (1965).

10. Brok, M., Ebskamp, F. C. R. and Hoff, A. J., *Biochim. Biophys. Acta.* **809**, 421 (1985).

11. Barry, B. A. and Babcock, G. T., *Proc. Natl. Acad. Sci. USA* **84**, 7099 (1987).

12. Weaver, E. C., *Arch. Biochem. Biophys.* **99**, 193 (1962).

13. Kohl, D. H. and Wood, P. M., *Plant Physiol.* **44**, 1439 (1969).

14. Hales, B. J. and das Gupta, A., *Biochim. Biophys. Acta* **637**, 303 (1981).

15. O'Malley, P. J. and Babcock, G. T., *Biochim. Biophys. Acta* **765**, 370 (1984).

16. O'Malley, P. J., Babcock, G. T. and Prince, R. C., *Biochim. Biophys. Acta* **766**, 283 (1984).

17. Dekker, J. P., van Gorkom, H. J., Brok, M. and Ouwehand, L., *Biochim. Biophys. Acta* **764**, 301 (1984).

18. Sjoberg, B.-M., Reichard, P., Graslund, A. and Ehrenberg, A., *J. Biol. Chem.* **253**, 6863 (1978).

19. Sahlin, H., Graslund, A., Ehrenberg, A. and Sjoberg, B.-M., *J. Biol. Chem.* **257**, 366 (1982).

20. Currier, T. C., Haury, J. F. and Wolk, C. P., *J. Bacteriol.* **129**, 1556 (1977).

21. Rippka, R., Deruelles, J., Waterbury, J. B., Herdman, M. and Stanier, R. Y., *J. Gen. Microbiol.* **111**, 1 (1979).

22. Brok, M., Babcock, G. T., deGroot, A. and Hoff, A. J., *J. Mag. Res.* **70**, 368 (1986).

23. Schoeder, H.-U. and Lockau, W., *FEBS Lett.* **199**, 23 (1986).

24. Barr, R. and Crane, F. L., *Meth. in Enzymology* **23**, 372 (1971).

25. Das, B. C., Lounasmaa, M., Tendelle, C. and Lederer, E., *Biochem. Biophys. Res. Communication* **21**, 318 (1965).

26. Cooper, R. and Unger, S., *J. Antibiotics* **38**, 24 (1985).

27. Land, E. J., Porter, G. and Strachan, E., *Trans. Faraday Soc.* **57**, 1885 (1961).

28. Jovanovic, S. V., Harriman, A. and Simic, M. G., *J. Phys. Chem.* **90**, 1935 (1986).

29. Boussac, A. and Etienne, A. L., *Biochim. Biophys. Acta* **766**, 576 (1984).

30. Dixon, W. T. and Murphy, D., *J. Chem. Soc., Faraday Transactions II*, **72**, 1221 (1976).

31. Fasanella, E. I. and Gordy, W., *Proc. Natl. Acad. Sci. USA* **62**, 299 (1969).

32. Box, H. C., Budzinski, E. E. and Freund, H. G., *J. Chem. Phys.* **61**, 2222 (1974).

33. Sealy, R. C., Harman, L., West, P. R. and Mason, R. P., *J. Am. Chem. Soc.* **107**, 3401 (1985).

34. Chandrasekar, T. K., O'Malley, P. J., Rodriguez, I. and Babcock, G. T., in *Progress in Photosynthesis Research* (ed. J. Biggins), vol. 1, p. 475. Martinus Nijhoff, Dordrecht (1987).

35. Rutherford, A. W., *Biochim. Biophys. Acta* **807**, 189 (1985).

36. Berthold, D. A., Babcock, G. T. and Yocum, C. F., *FEBS Lett.* **134**, 231 (1981).

37. Babcock, G. T. and Sauer, K., *Biochim. Biophys. Acta* **325**, 509 (1973).

Chemica Scripta 1988, **28A**, 123–126

S-state Formation after Ca²⁺ depletion in the Photosystem II Oxygen-evolving Complex

A. Boussac and A. William Rutherford

Service de Biophysique, Departement de Biologie, CEN de Saclay 91191 Gif-sur-Yvette, France

Paper presented by A. William Rutherford at the Nobel Conference 'Biophysical Chemistry of Dioxygen Reactions in Respiration and Photosynthesis', Fiskebäckskil, Sweden, 1–4 July, 1987

Abstract

The site of inhibition in oxygen evolution induced by the release of Ca²⁺ has been investigated by studying the advancement of the S-states using luminescence and electron paramagnetic resonance. We conclude that the three charges stored in the absence of Ca²⁺ can be used in oxygen evolution after its readdition and that Ca²⁺ is not essential for the formation of the multiline signal characteristic of the S_2-state.

Introduction

Oxygen evolution by plants requires the storage of four positive charges on a cluster of four Mn atoms located on the donor side of Photosystem II (PS II) before a pair of water molecules is oxidized. The five redox states are denoted S_n, n varying from 0 to 4 according to the model of Kok *et al.* [1]. The molecular mechanism of this reaction is not understood, nevertheless we know that it requires a minimum structure which includes three extrinsic polypeptides (33, 24 and 18 kDa) in physiological conditions (see ref. [2] for a review). Although these polypeptides are not essential in non-physiological conditions, some important functions can be attributed to them. The 33 kDa protein stabilizes the Mn cluster in its site [2] and seems to increase the efficiency of oxygen evolution [3]. The 24 and 18 kDa proteins act as concentrators of the essential co-factors, Ca²⁺ and Cl⁻ [2]. Although the role of Cl⁻ is not yet understood it has been shown that in its absence the formation of the S_2-state is abnormal, showing no characteristic EPR multiline signal [4] and that an inhibition of the $S_2 \rightarrow S_3$ transition occurs [5]. The level at which Ca²⁺ acts is more ambiguous. Addition of Ca²⁺ restores oxygen evolution after the release of the 24 and 18 kDa proteins by NaCl-washing of PS II particles [6]. Several studies have been performed to determine the step inhibited in the absence of Ca²⁺. In repetitive flash experiments it has been shown that the fast reduction of Z^+ (the electron donor to the oxidized primary donor of PS II), is reversibly inhibited in the absence of Ca²⁺ [7]. However experiments with a flash sequence using dark adapted samples have allowed the determination of the S-state transition which is inhibited. In an earlier paper it was shown by luminescence that Ca²⁺ was essential for the $S_3Z^+ \rightarrow S_0$ transition [8]. This conclusion was also supported by Ono and Inoue [9]. However several authors have concluded that the $S_1 \rightarrow S_2$ transition was inhibited by Ca²⁺ depletion in NaCl-washed PS II particles. For example, Blough and Sauer [10] and more recently de Paula *et al.* [11] did not observe the characteristic EPR multiline signal at low temperature

which arises from the S_2-state [12]. By monitoring the kinetics of Z^+ reduction by EPR, Cole and Sauer also concluded that electron transfer from S_1 to Z^+ was inhibited [13]. Franzen *et al.* [14] under comparable conditions still showed rapid rereduction of Z^+ in the ms range. In this paper we have investigated the role of Ca²⁺ in oxygen evolution using EPR at low temperature to detect the signal from the S_2 state and luminescence at room temperature to monitor the characteristic period-four oscillation of charge storage on the donor side of PS II.

Materials and methods

PS II particles from spinach chloroplasts were prepared as in ref. [15] and were stored at -80 °C in 0.3 M sucrose, 30 mM-NaCl, 25 mM-MES pH 6.5, 30 % (v/v) ethylene glycol. NaCl-washing of PS II particles was done in room light at 4 °C at 0.5 mg chlorophyll (Chl)/ml in 0.3 M sucrose, 1.2 M-NaCl, 25 mM-MES pH 6.5. After 30 min incubation 50 μM ethylene glycol bis (β-aminoethyl ether)-N-N'-tetraacetic acid (EGTA) was added and the NaCl-washed particles were pelleted by 15 min centriguation at 40 000 g, washed once in 30 mM-Nacl, 25 mM-MES pH 6.5, 50 μM-EGTA, pelleted again and resuspended at 5–10 mg Chl/ml in the same medium in some cases with 10 mM-CaCl₂ or SrCl₂. After 10 min of dark adaptation the samples were put in calibrated quartz tubes, frozen and stored at 77 K. For NaCl-washing in the dark, the same protocol was used except that particles were incubated in 0.3 M sucrose, 1.2 M-NaCl, 25 mM-MES pH 6.5, 5 mM-EGTA. SrCl₂ was obtained from Merck and the CaCl₂ contamination did not exceed 0.0005 %. EPR measurements were done as in ref [15]. The samples were illuminated at 200 K in a solid CO₂-ethanol bath in a non-silvered dewar as in ref. [15] just before the recording of the spectra. It was observed that in NaCl-washed particles the multiline signal was not fully stable at 77 K. The initial rate of oxygen evolution was measured with a Clark-type electrode. Phenyl-p-benzoquinone (PPBQ) at 0.45 mM was used as an electron acceptor. The luminescence experiments were already described in ref. [8]. The particles were dark adapted in 30 mM-NaCl, 25 mM-MES pH 6.5, 50 μM-EGTA at 10 μg Chl/ml with 10 mM-CaCl₂ or SrCl₂ when added. 50 μM 2,5-dichloro-p-benzoquinone (DCBQ) was added after dark adaptation as the electron acceptor. Mixing with an equal volume of the same medium with 20 mM-CaCl₂ after the preflashes was done as in ref. [16]. The time between the

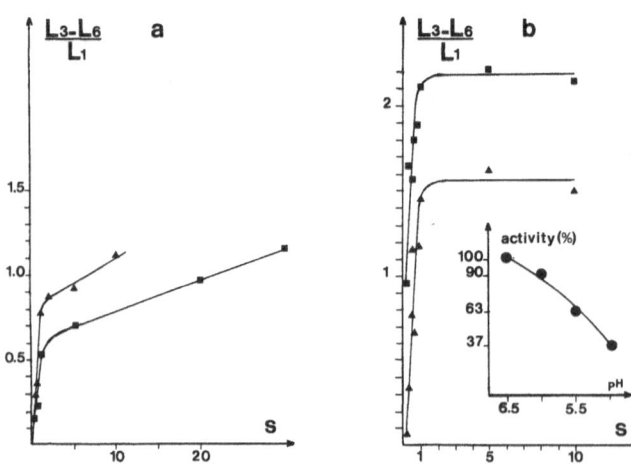

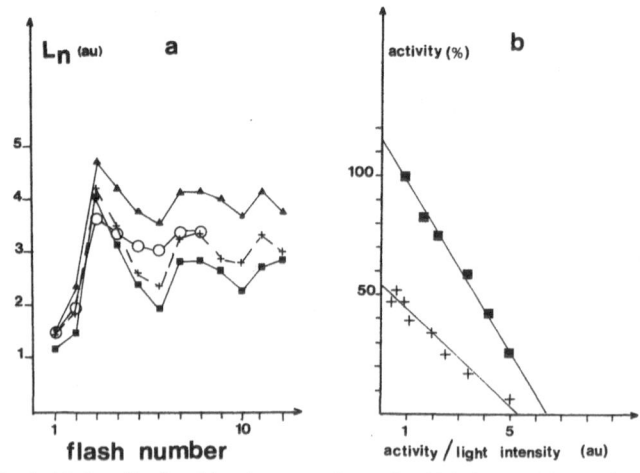

Fig. 1. Reactivation by Ca^{2+} of the oscillations detected in luminescence, measured arbitrarily by the ratio $(L3-L6)/L1$, *vs.* the time between the mixing with $CaCl_2$ (10 mM final) and the first flash of the sequence. The light NaCl-washed particles were incubated in a medium containing 30 mM-NaCl, 25 mM-MES pH 6.5 (■) or 5.5 (▲), 50 μM-EGTA, 50 μM-DCBQ with (*a*) or without (*b*) 0.3 M sucrose. The inset in (*b*) shows the amplitude of oxygen evolution *vs.* the pH of the resuspending medium with 10 mM-$CaCl_2$ and 0.45 mM-PPBQ.

Fig. 2. (*a*) Amplitude of luminescence intensity (L_n) detected 2 ms after each flash of a sequence (0.5 s spaced) in 30 mM-NaCl, 25 mM-MES pH 6.5, 50 μM-EGTA, 50 μM-DCBQ. ▲—▲, no addition ■—■, 10 mM-$CaCl_2$; +---+, 10 mM-$SrCl_2$. The trace (○—○) corresponds to PS II which do not evolve oxygen (see ref. [8]). (*b*) Effect of 10 mM-$CaCl_2$ (■), or $SrCl_2$ (+) on light-intensity dependence of oxygen evolution activity under continuous illumination. Same medium as in (*a*) with 0.45 mM-PPBQ. Particles NaCl-washed in the light. 100% activity corresponds to 450 μM-O_2/mg Chl x h.

last preflash and the first detecting flash was 1 s, the time between the start of the mixing and the detecting flash was 0.9 s.

Results

To test if the S-states formed in the absence of Ca^{2+} are competent in further charge accumulation, experiments were performed in which Ca^{2+} was added rapidly after preflash illumination. Firstly, however it was necessary to have conditions in which Ca^{2+} reconstitution would occur rapidly enough so that the S-states, formed by the preflashes, did not significantly decay during the reconstitution time. It was found that a damping of oscillations occurred if the time between the preflash and the measuring flash was greater than 1.5 s (not shown). Figure 1 shows that in buffer containing sucrose the reactivation of the oscillations was biphasic with a significant slow phase with a half time greater than 15 s. In buffer with no sucrose reconstitution occured within 1 s. The damping was more pronounced at pH 5.5 than at pH 6.5, this correlates well with the measurement of the oxygen evolution *vs.* the pH (inset of Fig. 1*b*).

Figure 2*a* shows the luminescence intensity recorded 2 ms after each flash of a sequence in NaCl-washed PS II particles adapted in the dark with (■) or without $CaCl_2$ (▲). The trace corresponding to PS II centres unable to go beyond the S_3Z^+ state is also plotted (○) see (ref. [8]). Figure 3 shows the luminescence intensity recorded 2 ms after each flash of a sequence in NaCl-washed particles adapted in darkness with (Fig. 3*a*) or without (Fig. 3*b*) $CaCl_2$. In both cases the sample was mixed with an equal volume of the same medium containing 20 mM-$CaCl_2$ after 0, 1 or 2 preflashes. The luminescence was recorded after each flash of the second series fired after the mixing. In the control experiment (Fig. 3*a*) the same oscillatory pattern was observed with 0, 1 or 2 preflashes. In Fig. 3*b* we can see that when Ca^{2+} was added after 0, 1 or 2 preflashes the luminescence intensity oscillated with the same characteristics as in Fig. 3*a*. This shows that the charges stored in absence of Ca^{2+} can be used for a period

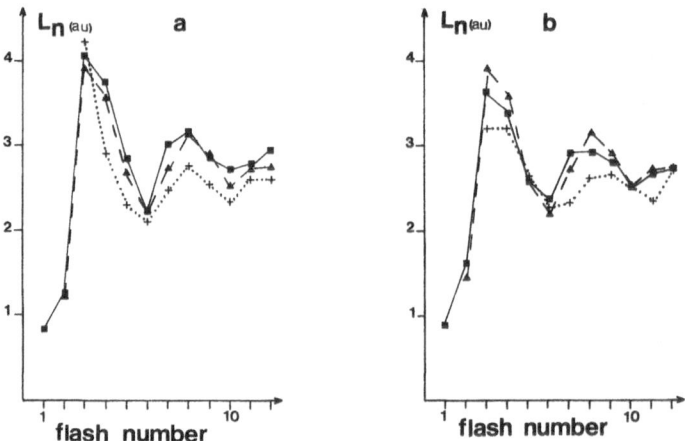

Fig. 3. Amplitude of the luminescence intensity (L_n) detected 2 ms after each flash of a sequence in dark-adapted PS II particles NaCl-washed in light; ■—■, no preflash, ▲—▲, one preflash, +···+, two preflashes. The particles were suspended either in (*b*) 30 mM-NaCl, 25 mM-MES pH 6.5, 50 μM-EGTA, 50 μM-DCBQ or (*a*) the same medium with 10 mM $CaCl_2$. In (*a*) and (*b*) the sample was rapidly mixed 0.1 s after the last preflash with an equal volume of the same medium containing 20 mM-$CaCl_2$. Flash frequency was 1 Hz.

four charge accumulation mechanism. The smaller intensity observed after the flash following two preflashes (Fig. 3*b*) probably reflects a partial deactivation during the mixing of the S_3-state since it has a shorter life-time (4 s) in the absence of Ca^{2+} [8].

The discrepency between the luminescence data indicating charge accumulation in the absence of Ca^{2+} and the EPR data showing a decrease in the yield of the multiline signal has been investigated. Figure 4 and Table I show that the multiline signal is formed to a comparable extent in Ca^{2+}-depleted and Ca^{2+}-reconstituted PS II membranes. Ca^{2+} depletion in the experiment of Fig. 4 was done by salt-washing in the light. In the previous EPR work salt-washing was done in the dark in the presence of 5 mM-EGTA [11]. We have washed PS II particles in the dark with 5 mM-EGTA. This sample was rewashed and resuspended in media containing 50 μM or 5 mM-EGTA. In both cases the mutiline

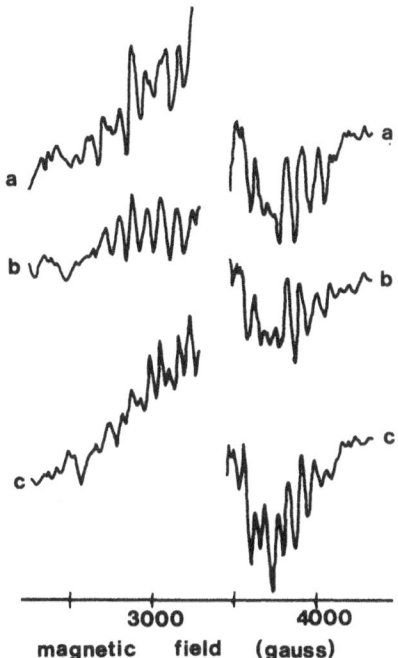

Fig. 4. Light minus dark spectra induced by an illumination at 200 K of NaCl-washed PS II particles in the light. (*a*) No addition; (*b*) with 10 mM-CaCl₂, (*c*) with 10 mM-SrCl₂. Other conditions were described in materials and methods. Modulation amplitude, 22 G; temperature, 10 K; microwave power, 20 mW. The residual activity in oxygen evolution after NaCl-washing was 20% of the activity of the Ca²⁺-reconstituted sample.

Table I. *Formation of multiline spectrum*

Three types of particles are resuspended in media as indicated in materials and methods. ML is the total amplitude of the three lines of the right part of the multiline spectrum recorded as in Fig. 4, normalized to the amplitude of Signal II slow. Signal II was recorded at 15 K with a modulation amplitude of 2.8 G and a microwave power of 2.10^{-4} mW. Each values is the average of 3–5 experiments.

Type of particles	Addition	ML/Signal II
Untreated	50 μM-EGTA	1.33
Untreated	5 mM-EGTA	1.13
NaCl-washed in light	50 μM-EGTA	0.95
NaCl-washed in light	10 mM-Ca²⁺	1.21
NaCl-washed in dark	50 μM-EGTA	1.02
NaCl-washed in dark	5 mM-EGTA	0.70
NaCl-washed in dark	10 mM-Ca²⁺	1.18

signal could be induced by an illumination at 200 K (Table I). The spectra (not shown) have the same characteristics as in Ca²⁺-reactivated samples. However with 5 mM-EGTA, the multiline signal was smaller than with 50 μM-EGTA. This quenching by EGTA was more pronounced at low power since the saturation characteristics of the signal were slightly different (not shown). The extent of this effect varies greatly from preparation to preparation and can be also observed in untreated-particles (Table I).

In Fig. 4 some slight differences in the multiline signal are apparent when the signal in the Ca²⁺-reconstituted sample is compared to that from Ca²⁺-depleted sample. It was of interest, then, to test whether replacement of Ca²⁺ by another metal would lead to a further modification of the EPR spectrum. It has already been shown that Sr²⁺ can partially reactivate oxygen evolution [17]. However Fig. 1a shows that the luminescence oscillations are reactivated to the same extent with Sr²⁺ as with Ca²⁺. This discrepancy is explained

by the experiment in Fig. 1b which shows that, when the light intensity is limiting, Sr²⁺ reactivates oxygen evolution as efficiently as Ca²⁺. Under saturating light the reactivation with Sr²⁺ is only partial, as already observed [17]. This indicates that Sr²⁺ binding reconstitutes oxygen evolution but that the limiting step of the reaction is slower. We have performed EPR measurements on NaCl-washed particles reconstituted with Sr²⁺. The preliminary results shown in Fig. 4 indicate that the multiline signal is modified when Ca²⁺ is replaced by Sr²⁺.

Discussion

Figure 3 shows that the charges stored in absence of Ca²⁺ can be used in a four step charge storage mechanism when Ca²⁺ is rapidly added after the last preflash. This suggests that the states formed in absence of Ca²⁺ are normal or very quickly transformable into normal S-states upon Ca²⁺ readdition. This is in contradiction with other reports [10, 11, 13, 17, 18]. Firstly Cole and Sauer [13] concluded that Ca²⁺-depletion resulted in an inhibition of the reduction of Z⁺ by the S-states. However this conclusion was based on the absence of a period four oscillation in the amplitudes of Signal II very fast in NaCl-washed PS II particles. In untreated particles, such oscillations are due to the reduction of Z⁺ by S₀, S₁ and S₂ being faster than the time resolution of the apparatus while donation from S₃ is in the ms range and Z⁺ is thus detectable. After NaCl-washing reduction of Z⁺ by the S₃-state becomes slower [7]. Therefore an oscillation in the amplitude of Signal II very fast is expected to be replaced by a partial oscillation in its life-time. Indeed in ref. [13] the life-times of Z⁺ oscillates with a period four after NaCl-washing and therefore the results seem not to be in contradiction with an inhibition between S₃Z⁺ and S₀. Secondly, Blough and Sauer [10] and, more recently, de Paula *et al.* [11] observed a decrease in the multiline signal after NaCl-washing. In ref. [10] the residual multiline signal was estimated to be 40% of that detected in untreated samples. If however this residual signal is compared to that observed in the Ca²⁺-reactivated sample, a value of 60% is obtained which is appreciably higher than the residual oxygen evolution detected under continuous illumination. This indicates that S-state advancement occurs in absence of Ca²⁺ even though oxygen evolution is inhibited. In ref. [11] the treatment was done in the dark with 5 mM-EGTA. Miyao and Murata [19] have clearly shown that even with 10 mM-EGTA such a dark treatment is not sufficient to decrease oxygen evolution appreciably and that the release of Ca²⁺ requires a treatment in the light. In this paper we have looked at the multiline signal formed in NaCl-washed material treated in the light. The multiline signal is formed by 200 K illumination to an extent comparable to that seen in the sample reconstituted with Ca²⁺. The signal seems to be slightly modified and a further modication seems to occur if reconstitution is done with Sr²⁺. It is important to note that the modification of the shape of the signal complicates the measurement of its amplitude when comparing different samples. In addition we have observed that treatment of NaCl-washed PS II particles with a high concentration of EGTA in the dark does lead to a decrease in multiline intensity but that the effect can be reversed by removing the EGTA. This observation may help to explain the decrease in multiline amplitude observed in ref.

[11]. The effect of EGTA on the multiline signal may be related to the binding of its acetate groups to, or close to, the Mn-cluster perhaps in the Cl^- site. From data reported earlier [2] it is clear that NaCl-washing in the light results in an inhibition of oxygen evolution which is reversed by Ca^{2+} addition. When oxygen evolution is inhibited in this way charge accumulation can occur up to the S_3Z^+ state ([8,9] and this work) and the S_2 multiline signal can still be formed.

Acknowledgement

A.B. and A.W.R are supported by the C.N.R.S.

References

1. Kok, B., Forbush, B. and McGloin, M., *Photochem. Photobiol.* **11**, 457 (1970).
2. Murata, N. and Miyao, M., in *Progress in Photosynthesis Research* (ed. J. Biggins) vol. 1, p. 453 (1987).
3. Miyao, M., Murata, N., Lavorel, J., Maison-Peteri, B., Boussac, A. and Etienne, A. L., *Biochim. Biophys. Acta* **890**, 151 (1987).
4. Ono, T., Zimmermann, J. L., Inoue, Y. and Rutherford, A. W., *Biochim. Biophys. Acta* **851**, 193 (1986).
5. Theg, S. M., Jursinic, P. A. and Homann, P. H., *Biochim. Biophys. Acta* **766**, 636 (1984).
6. Miyao, M. and Murata, N., *FEBS Lett.* **168**, 118 (1984).
7. Dekker, J. P., Ghanotakis, D. F., Plijter, J. J., van Gorkom, H. J. and Babcock, G. T., *Biochim. Biophys. Acta* **767**, 515 (1984).
8. Boussac, A., Maison-Peteri, B., Vernotte, C. and Etienne, A. L., *Biochim. Biophys. Acta.* **808**, 225 (1985).
9. Ono, T. A. and Inoue, Y., *Biochim. Biophys. Acta* **806**, 331 (1985).
10. Blough, N. V. and Sauer, K., *Biochim. Biophys. Acta* **767**, 377 (1984).
11. de Paula, J. C., Li, P. M., Miller, A.-F., Wu, B. W. and Brudvig, G. W., *Biochemistry* **25**, 6487 (1986).
12. Dismukes, G. C. and Siderer, Y., *FEBS Lett.* **121**, 78 (1980).
13. Cole, J. and Sauer, K., *Biochim. Biophys. Acta* **891**, 40 (1987).
14. Franzen, L. G., Hansson, O. and Andreasson, L. E., *Biochim. Biophys. Acta* **808**, 171 (1985).
15. Zimmermann, J. L. and Rutherford, A. W., *Biochim. Biophys. Acta* **767**, 160 (1984).
16. Boussac, A. and Etienne, A. L., *Biochim. Biophys. Acta* **682**, 281 (1982).
17. Ghanotakis, D. F., Babcock, G. T. and Yocum, C. F., *FEBS Lett.* **167**, 127 (1984).
18. Ghanotakis, D. F., Demetriou, D. M. and Yocum, C. F., *Biochim. Biophys. Acta* **891**, 15 (1987).
19. Miyao, M. and Murata, N., *Photosynth. Res.* **10**, 343 (1986).

Chemica Scripta 1988, **28A**, 127–131

Postscript

Helmut Beinert

Department of Biochemistry, Medical College of Wisconsin, Milwaukee, Wis. 53226, USA

and Tore Vänngård

Department of Biochemistry and Biophysics, University of Göteborg and Chalmers University of Technology, S-412 96 Göteborg, Sweden

Helmut Beinert...

The gulls are screeching and the sea is washing up and down the sturdy rocks of Sweden's West coast as ever. Barely noticed in this vast, somber and in its own way grandiose landscape, a small group of scientists from many countries, gathered there with Swedish colleagues in honor of one of their leading minds, is now all but dispersed, back to their laboratories, to other meetings or for a European vacation. Yet the memory remains strong from the two days of intensive discussions among congenial colleagues at Fiskebäckskil, amidst the impressive background given by Nature. If there were any regrets on the part of the participating guests, it was the hosts' decision to hide their own light under a bushel, by refraining from presenting any of their own work. As those assembled were well aware, it was Bo Malmström who has not only contributed experimentally for almost 30 years to the field of metal-enzymes and oxygen reduction, but has also continuously monitored progress in the field in critical writings and has provided direction way beyond the boundaries of his own laboratory.

The theme of the Fiskebäckskil meeting centered around the question: How does nature accomplish the feat, basic to all aerobic life, converting, at ambient temperature, gaseous oxygen into liquid water, as in respiration, and liquid water into gaseous oxygen as in green plant photosynthesis? The contributions in the first-mentioned area were mainly aimed in two directions. Some, and this was the majority, endeavoring to advance our knowledge by new attacks at the frontier, and others attempting to consolidate or mend the very foundations. The importance of the latter kind of work has become painfully evident in two recent pieces of research, from which definitive answers were expected on two vital aspects of cytochrome oxidase function; namely the oxidation-reduction potentials of individual redox couples and their pH and temperature [1] dependencies and the electronic structure and magnetic properties of the metal centers with particular attention to spin coupling between the components of the oxygen reduction site [2]. Both attempts, though excellent pieces of work, were flawed by the heterogeneity of the enzyme preparations used for this work, one set of mammalian and one of bacterial origin. The heterogeneity referred to here is not simply due to protein impurity. It is not revealed by the usual criteria of protein chemistry, but only by certain ligand-binding or function-related tests so that one may speak of 'active-site heterogeneity'. One presentation (Palmer) at the meeting was entirely devoted to solving this problem. It seems to be the first successful attempt to deliberately produce a homogeneous 'resting' (as isolated) enzyme. Another contribution (M. T. Wilson) pursuing a related goal, dealt with the vexing question as to whether the monomer ($aa_3\text{Cu}_A\text{Cu}_B$) of the oxidase or the dimer $(aa_3\text{Cu}_A\text{Cu}_B)_2$ is the active form in electron transfer and proton translocation. Wikström *et al.* [3] in their book of 1981 had already seriously grappled with this problem without coming to a definitive answer. Experiments to solve the problem are difficult to design. Thus, for example, in work with liposomes, even if the preparation used for incorporation clearly is in the monomeric form, it does not follow that the material incorporated does not dimerize. It is easily seen that many processes in reduction and re-oxidation of cytochrome oxidase can become immensely more complicated, once intradimer or inter-monomer oxidoreductions have to be considered. Wilson mapped out the course of re-oxidation of cytochrome oxidase under conditions, when electron transfer between individual functional units is favored. By graded poisoning with CO and thus setting the stage for excess of a and Cu_A over non-CO-bound a_3, inter-unit electron transfer could be convincingly demonstrated.

Although, as mentioned above, a unique set of magnetic parameters could not be obtained from the work done so far, there were nevertheless exciting news on the oxidase from *Thermus thermophilus* (Fee). This enzyme has two subunits, one of which contains $aa_3\text{Cu}_A\text{Cu}_B$ and the other a c cytochrome. Crystals of this enzyme have been obtained which might be suitable for diffraction work. In addition to this form of the enzyme, under different growth conditions, another cytochrome c oxidase has been encountered in this organism in which a is replaced by a b cytochrome, no c is present and $ba_3Cu_ACu_B$ are contained in a single subunit which contains only a single cysteine. This bears on ideas concerning the binding site for Cu_A, which generally has been thought to contain two histidine and two cysteine ligands. The properties of Cu_A of this enzyme form have so far not been found to differ from those of mammalian Cu_A. May we have to revise our picture of the Cu_A binding site? While Fee was barely able to accommodate all this interesting information in his lecture at the meeting, his written version contains an extensive discussion of observations on the reaction of H_2O_2 with the *Thermus* enzyme.

Wevers' verbal and written contributions differ, but they were variations on the central theme of electron-transfer reactions among electron carriers in individual functional units of cytochrome c oxidase. In elegant experiments, Wever and his colleagues had previously shown [4, 5] that substantial electron shifts can be observed, particularly well with the mixed valence form $(2e^-/aa_3)$ in the presence of CO. Dissociation of CO on illumination will then generate an electron flow from a_3^{2+} and Cu_B^+ toward a^{3+} and Cu_A^{2+} which will even spill over into cytochrome c, when this is also present. In the experiments reported at the meeting (written version) the pH dependence of this electron shift was now investigated. As the pH is increased, more a_3 becomes oxidized, while electrons flow toward a and Cu_A. These observations are of interest in view of the presently ongoing search for pH dependent phenomena with this enzyme, which might be implicated in the proton transfer catalyzed by it. In his oral presentation Wever examined the reaction of reduced oxidase with H_2O_2. With high H_2O_2 concentrations the oxidation of a_3 preceded the oxidation of a, allowing a determination of the a^{2+} to a_3^{3+} electron transfer rate, which turned out to be intermediate between those observed with the enzyme in the absence and in the presence of O_2, respectively. Thus the a_3^{3+} peroxide complex does enhance internal e^- transfer. Only slight effects of pH and D_2O were found, the significance of which is not clear.

Greenwood, who spoke about cytochrome c peroxidase, not oxidase, of *Pseudomonas aeruginosa* provided important clues to the MCD of iron in the ferryl, Fe(IV)=O, state, which could be useful in definitively identifying this state of iron in intermediates in the re-oxidation of cytochrome c oxidase. The paper is also an example of the intricate interactions and their expression in various types of spectroscopy which may occur in a hemeprotein with multiple prosthetic groups. Previous work on this enzyme that did not involve MCD is mentioned [6, 7] but there is no discussion of the disagreement in conclusions arrived at by the two research groups. The paper emphasizes the importance of bringing to bear as many of the decisive spectroscopic tools as possible to resolve such complicated cases.

Despite this statement one must admire the efforts and skill with which Orii has applied optical rapid-scanning techniques in the visible region to the analysis of the reoxidation of reduced cytochrome c oxidase. The quality of even the μsec kinetic traces is impressive and this very quality allowed the author to draw detailed conclusions as to discernible reaction steps and their rates. In this and related work at low as well as ambient temperature by other authors two observations and/or interpretations recur. The first step in the course of interaction of the enzyme with oxygen is not binding of oxygen to heme or Cu_B, but the admission or penetration of oxygen into a pocket or channel close to the oxygen-reduction site. The other recurrent interpretation of the observations is that a conformational change intervenes before or when electrons are delivered from aCu_A to the a_3Cu_B site. While just a few years ago the detailed resolution of the reaction steps at room temperature appeared to be a matter of the future, this future obviously has arrived and one can look forward to seeing more activity in this area in the next few years; particularly promising here are of course those spectroscopies that do not require low temperature, such as resonance Raman.

The events observable at low temperatures during the re-oxidation of cytochrome oxidase have been analyzed in much more detail. Chan presented his fascinating work in this area, particularly the analysis of the events occurring at the 'three electron' stage, that is when the bound oxygen molecule has taken up three electrons. Chan showed that at this stage the breaking of the oxygen–oxygen bound occurs. As in other instances with cytochrome oxidase, branched pathways exist. The ferryl state of iron also arises at this stage and so does the EPR signal of Cu_B. Chan emphasizes in his work that one may have to account for the presence of CO in the reaction medium, which enters into reactions [e.g. with Fe(IV)]! This is altogether a somewhat disquieting situation, because all the work on the re-oxidation of cytochrome oxidase is initiated by flashing-off CO from the reduced form. The consideration of interference by CO is therefore equally germane to the room-temperature work, e.g. Orii's as reported above. In a discussion remark Brunori emphasized this at the meeting. However, work which was reported since the meeting [8] indicates that experience with hemoglobin and myoglobin concerning geminal recombination of CO with heme is not directly transferable to the situation encountered with cytochrome oxidase. A relatively new development is Chan's finding that the oxy-ferryl heme $a_3Cu_B^{2+}$ intermediate can be formed by reaction of 'pulsed' oxidase with H_2O_2 or three-electron reduced oxidase with O_2, both at room temperature. This work provides a connection to work done by Wikström, who, according to the optical spectra observed in mitochondria or mitoplasts, arrives at the same states, by reversing electron flow at high phosphate potential in the presence of ferricyanide and cytochrome c.

As Chan emphasizes in his paper, the critical intermediate may be produced either by reducing O_2 by three electrons or by reversing the O_2 reduction reaction by a single electron transfer. Chan also notes that Nature obviously employs the same basic chemical capabilities of heme iron, e.g. formation of an oxyferryl species, in heme-containing enzymes as different in their overall reactions as peroxidases, cytochrome P-450 and cytochrome c oxidase. The convergence of Chan's and Wikström's observations and conclusions on the events occurring in the last steps of dioxygen reduction is one of the most exciting aspects of recent research in this field. The papers by both investigators are the result of mental synthesis out of many and often seemingly unrelated experimental observations and of broad knowledge in physical chemistry and bioenergetics, with Chan coming from the former and Wikström from the latter background. No doubt, there is plenty of risk in such expositions, but it is encouraging that some of our colleagues are willing to take them – and both have a history of success in doing so. Wikström has repeatedly shown his skill in such synthesis: in his classical paper proposing the 'neoclassical' model for the interpretation of spectrophotometrically monitored redox titrations [9], foremost in his book of 1981 [3] and again in the present paper and other related ones, partly in press at this stage. The bibliography of his contribution here in this volume is a good guide to the relevant Wikström library of things past and yet to come, all related to the theme of proton transfer coupled to dioxygen reduction by cytochrome c oxidase. A similarly daring part of Wikström's oral presentation at the meeting was an attempt of building a model of the active site of cytochrome c oxidase with fragments borrowed from various known structures, such as that of the reaction center of *Rhodopseudomonas viridis* or plastocyanin and from known segments and by hydropathy plots of cytochrome oxidase peptides. Those interested in this model can find it in [10].

M. Brunori, who is also a co-author on the paper by Wilson discussed above, described in his oral presentation a model for a voltage gated proton pump, which is based on the analysis of data derived from fast kinetic studies of cytochrome oxidase incorporated into phospholipid vesicles. The pump can operate in two states: maximally coupled (P) and slipping (S). The conversion from P to S is controlled by $\Delta\tilde{\mu}_{H^+}$ with $[P] \gg [S]$ at $\Delta\tilde{\mu}_{H^+} = 0$. An elaboration on the aspect of pump models, efficiency and control, not available at the time of the meeting, is found in a review: 'Hypothesis. The proton-pumping site of cytochrome oxidase, a model of its structure and mechanism', by Gelles, Blair and Chan [11].

No doubt, this volume is worthy of the occasion to which it is dedicated. The numerous new approaches, facts and ideas presented in it will serve as food for thought, a stimulus and as a beginning of more extensive and decisive research in the future.

Tore Vänngård...

The study of photosynthesis is as old as that of respiration, and Ken Sauer gave in his introductory talk a historical background, starting with Priestley's experiments more than 200 years ago. He described how our present understanding has developed from the work of Ingen-Housz, Van Niel, Ruben, Kamen, Hill and Kok, just to mention a few. However, less is now known about the structural features responsible for oxygen evolution in photosynthesis than about the corresponding features in respiration, but the crystallographic determination of a bacterial system has helped also in the case of the oxygen-evolving systems. Both Sauer and Dismukes pointed out that basic residues in a conserved C-terminal part of the so called D1 and D2 polypeptides in the Photosystem II reaction center – absent in the corresponding bacterial center – might provide ligands for manganese which is well known to participate in oxygen evolution.

How are the four manganese ions required for oxygen evolution arranged? Dismukes reported on X-ray absorption studies and analysis of EPR data that suggested a cluster of at least three manganese ions. This, of course, is what Brudvig earlier has proposed from his temperature studies in EPR. A symmetrical structure involving all four ions seems less likely, however, from the X-ray absorption data reported by Sauer in this volume. The oxidation states of manganese is another question under discussion. Optical absorption data presented by van Gorkom indicated an increase in manganese oxidation states on each of the four S_i-to-S_{i+1} transitions, but Renger did not want to speculate about the valence changes from his optical transients. At present, X-ray absorption methods may be better in defining the state of manganese in different S-states than optical techniques. Thus, Sauer observed that there is no change in the S_2-to-S_3 transition (contrasting the clear increase in oxidation states on the S_1-to-S_2 step) in agreement with Rutherford's measurements of the flash-induced changes of the interaction between the manganese center and the free radical D^+ (signal II_{slow}). This is somewhat surprising since it means that one single oxidation equivalent must be stored somewhere else, presumably not on a metal ion. Is this finding related somehow to the Renger's temperature studies, showing that the S_2-to-S_3 transition has a much larger activation energy than the other transitions? New experimental approaches to study these questions are welcome, and Dismukes reported on preliminary measurements of changes in the magnetic susceptibility after flashes.

Several speakers treated the effects of amines. Brudvig's EPR studies were only explicable with two binding sites with ammonia binding directly to manganese in one site. Junge pointed out that the high concentration needed for EPR and X-ray work possibly could give different results as compared with the low concentrations needed for optical investigations, and both he and Renger concluded that also with NH_2OH at least two molecules bind. Junge meant that several molecules bind cooperatively to manganese, in some contradiction to the conclusions from EPR studies. Amazingly, Rutherford showed a multiline EPR spectrum produced by the action of Sr^{2+} after Ca^{2+}-depletion, which was similar to that obtained after ammonia binding.

The so called $g = 4.1$ signal, derived from the S_2 state, is also affected by ammonia, and Brudvig had the interesting idea that it represents an inactive conformation in equilibrium with an active form, which has the 'usual' multiline EPR signal.

Research in photosynthesis can take advantage of the fact that the system can be advanced in precise steps through the action of flashes. Thus, proton liberation in water splitting is better defined than uptake in oxygen reduction. Junge and Renger both had studied the proton release, and their work strongly suggests that at least some protons are not released directly from the reacting water but probably from another protonizable group, such as an amino-acid residue. This, of course, has strong implications for any attempt to draw schemes of how water is split.

The structure and function of the Signal II radical, which is not directly involved in water splitting but receives electrons from the manganese cluster, was one of the highlights of the conference. Babcock, who earlier had been a strong proponent for the quinone nature of the radical, now unequivocally showed that this radical is derived from tyrosine. Such radicals had been identified many years ago in the enzyme ribonucleotide reductase by Ehrenberg, and at the conference he presented the latest on the magnetic interaction present in this system, in fact related to the interactions described above in photosystem II. This new identification opens up the possibility to localize the radical to specific amino acids in the polypeptide sequences. There are two tyrosine radicals in the reaction center, and Rutherford showed that one of them, the kinetically very sluggish species Signal II_{slow} seemingly has the function to oxidize the manganese center from S_0 to the chemically more stable S_1 state. S_0 may have Mn(II), which relatively easily could be lost.

The basis of both respiration and photosynthesis lies of course in the transport of electrons, and this was a theme of two of the introductory lectures. Sven Larsson gave an introduction to the theory of electron transport, mainly in the framework of the celebrated Marcus theory. He had made calculations on quite complicated systems, and although in general the rates decline with distance, the geometry of the intervening medium plays a role that presumably could be used in biological control. Harry Gray presented his ingenious experiments with derivatives of heme proteins, and his rates now decrease by a factor of exp (0.9) per Ångström. He also stressed the importance of keeping water out from the participating centers if rapid electron transfer is required. This, however, is what photosynthesis and respiration are about – water must be involved. Also, there is little structural information available on the oxygen reaction systems so that the more detailed theories of electron transfer cannot yet be applied.

Last, but not least, R. J. P. Williams gave a fascinating survey of oxygen chemistry in the opening lecture of the conference. He proposed that Cu came into use only subsequent to the start of oxygen evolution and gave arguments why Mn is used for water splitting, whereas Fe and Cu find their use in oxygen reduction. He also discussed how the problems with oxygen chemistry must be put into a wider perspective, 'Biology is chemistry in space'. Supposedly, this is what we all aim for in the end – an integration of the detailed understanding of electron transfer and bond making/breaking and of how the organism keeps this all under control.

References

1. Blair, D. F., Ellis, W. R., Jr., Wang, H., Gray, H. B. and Chan S. I., *J. Biol. Chem.* **261**, 11524 (1986).
2. Rusnak, F. M., Münck, E., Nitsche, C., Zimmermann, B. H. and Fee, J. A., *J. Biol. Chem.* (in the press).
3. Wikström, M., Krab, K. and Saraste, M., *Cytochrome Oxidase, A Synthesis*. Academic Press, London (1981).
4. Boelens, R. and Wever, R., *Biochim. Biophys. Acta* **547**, 296 (1979).
5. Boelens, R., Wever, R. and van Gelder, B. F., *Biochim. Biophys. Acta* **682**, 264 (1982).
6. Aasa, R., Ellfolk, N., Rönnberg, M. and Vänngård, T., *Biochim. Biophys. Acta* **670**, 170 (1981).
7. Ellfolk, N., Rönnberg, M., Aasa, R., Andréasson, L.-E. and Vänngård, T., *Biochim. Biophys. Acta* **743**, 23 (1983).
8. Findsen, E. W., Centeno, J., Babcock, G. T. and Ondrias, M. R., *J. Am. Chem. Soc.* **109**, 5367 (1987).
9. Wikström, M. K. F., Harmon, H. J., Ingledew, W. J. and Chance, B., *FEBS Lett.* **65**, 259 (1976).
10. Holm, L., Saraste, M. and Wikström, M., *EMBO J.* **6**, 2819 (1987).
11. Gelles, J., Blair, D. F. and Chan, S. I., *Biochim. Biophys. Acta* **853**, 205 (1987).